THE COSMOLOGICAL SINGULARITY

Written for researchers focusing on general relativity, supergravity, and cosmology, this is a self-contained exposition of the structure of the cosmological singularity in generic solutions of the Einstein equations, and an up-to-date mathematical derivation of the theory underlying the Belinski–Khalatnikov–Lifshitz (BKL) conjecture on this field.

Part I provides a comprehensive review of the theory underlying the BKL conjecture. The generic asymptotic behavior near the cosmological singularity of the gravitational field, and fields describing other kinds of matter, is explained in detail. Part II focuses on the billiard reformulation of the BKL behavior. Taking a general approach, this section does not assume any simplifying symmetry conditions and applies to theories involving a range of matter fields and space-time dimensions, including supergravities.

Overall, this book will equip theoretical and mathematical physicists with the theoretical fundamentals of the Big Bang, Big Crunch, Black Hole singularities, their billiard description, and emergent mathematical structures.

VLADIMIR BELINSKI holds a permanent professor position at the International Center for Relativistic Astrophysics Network, Italy. He is noted for his role in several key developments in theoretical physics, including the Belinski–Khalatnikov–Lifshitz conjecture on the behavior of generic solutions of Einstein equations near a cosmological singularity, the Belinski–Zakharov transform, and the "Inflationary Attractor." He is co-author of the book *Gravitational Solitons* (Cambridge University Press, 2001) and has received the Landau Prize of the Russian Academy of Sciences (1974) and Marcel Grossmann Award (2012).

MARC HENNEAUX is Full Professor at the Université Libre de Bruxelles and Director of the International Solvay Institutes for Physics and Chemistry. He has contributed significantly to understandings of gravity and black holes in three dimensions, and the geometrical and algebraic aspects of theories with a gauge freedom (in particular the Becchi–Rouet– Stora–Tyutin symmetry and its cohomology). He has received the Francqui Prize (2000), the Humboldt Prize (2009), the Bogoliubov Prize (2014), the FNRS Quinquennial Prize in Fundamental Exact Sciences (2015), and two ERC advanced grants (2011–2015 and 2016–2021).

CAMBRIDGE MONOGRAPHS ON MATHEMATICAL PHYSICS

General Editors: P. V. Landshoff, D. R. Nelson, S. Weinberg

S. J. Aarseth *Gravitational N-Body Simulations: Tools and Algorithms*†
J. Ambjørn, B. Durhuus and T. Jonsson *Quantum Geometry: A Statistical Field Theory Approach*†
A. M. Anile *Relativistic Fluids and Magneto-fluids: With Applications in Astrophysics and Plasma Physics*
J. A. de Azcárraga and J. M. Izquierdo *Lie Groups, Lie Algebras, Cohomology and Some Applications in Physics*†
O. Babelon, D. Bernard and M. Talon *Introduction to Classical Integrable Systems*†
F. Bastianelli and P. van Nieuwenhuizen *Path Integrals and Anomalies in Curved Space*†
D. Baumann and L. McAllister *Inflation and String Theory*
V. Belinski and M. Henneaux *The Cosmological Singularity*†
V. Belinski and E. Verdaguer *Gravitational Solitons*†
J. Bernstein *Kinetic Theory in the Expanding Universe*†
G. F. Bertsch and R. A. Broglia *Oscillations in Finite Quantum Systems*†
N. D. Birrell and P. C. W. Davies *Quantum Fields in Curved Space*†
K. Bolejko, A. Krasiński, C. Hellaby and M-N. Célérier *Structures in the Universe by Exact Methods: Formation, Evolution, Interactions*
D. M. Brink *Semi-Classical Methods for Nucleus-Nucleus Scattering*†
M. Burgess *Classical Covariant Fields*†
E. A. Calzetta and B.-L. B. Hu *Nonequilibrium Quantum Field Theory*
S. Carlip *Quantum Gravity in 2+1 Dimensions*†
P. Cartier and C. DeWitt-Morette *Functional Integration: Action and Symmetries*†
J. C. Collins *Renormalization: An Introduction to Renormalization, the Renormalization Group and the Operator-Product Expansion*†
P. D. B. Collins *An Introduction to Regge Theory and High Energy Physics*†
M. Creutz *Quarks, Gluons and Lattices*†
P. D. D'Eath *Supersymmetric Quantum Cosmology*†
J. Dereziński and C. Gérard *Mathematics of Quantization and Quantum Fields*
F. de Felice and D. Bini *Classical Measurements in Curved Space-Times*
F. de Felice and C. J. S Clarke *Relativity on Curved Manifolds*†
B. DeWitt *Supermanifolds, 2nd edition*†
P. G. O. Freund *Introduction to Supersymmetry*†
F. G. Friedlander *The Wave Equation on a Curved Space-Time*†
J. L. Friedman and N. Stergioulas *Rotating Relativistic Stars*
Y. Frishman and J. Sonnenschein *Non-Perturbative Field Theory: From Two Dimensional Conformal Field Theory to QCD in Four Dimensions*
J. A. Fuchs *Affine Lie Algebras and Quantum Groups: An Introduction, with Applications in Conformal Field Theory*†
J. Fuchs and C. Schweigert *Symmetries, Lie Algebras and Representations: A Graduate Course for Physicists*†
Y. Fujii and K. Maeda *The Scalar-Tensor Theory of Gravitation*†
J. A. H. Futterman, F. A. Handler, R. A. Matzner *Scattering from Black Holes*†
A. S. Galperin, E. A. Ivanov, V. I. Ogievetsky and E. S. Sokatchev *Harmonic Superspace*†
R. Gambini and J. Pullin *Loops, Knots, Gauge Theories and Quantum Gravity*†
T. Gannon *Moonshine beyond the Monster: The Bridge Connecting Algebra, Modular Forms and Physics*†
A. García-Díaz *Exact Solutions in Three-Dimensional Gravity*
M. Göckeler and T. Schücker *Differential Geometry, Gauge Theories, and Gravity*†
C. Gómez, M. Ruiz-Altaba and G. Sierra *Quantum Groups in Two-Dimensional Physics*†
M. B. Green, J. H. Schwarz and E. Witten *Superstring Theory Volume 1: Introduction*
M. B. Green, J. H. Schwarz and E. Witten *Superstring Theory Volume 2: Loop Amplitudes, Anomalies and Phenomenology*
V. N. Gribov *The Theory of Complex Angular Momenta: Gribov Lectures on Theoretical Physics*†
J. B. Griffiths and J. Podolský *Exact Space-Times in Einstein's General Relativity*†
S. W. Hawking and G. F. R. Ellis *The Large Scale Structure of Space-Time*†
F. Iachello and A. Arima *The Interacting Boson Model*†
F. Iachello and P. van Isacker *The Interacting Boson-Fermion Model*†
C. Itzykson and J. M. Drouffe *Statistical Field Theory Volume 1: From Brownian Motion to Renormalization and Lattice Gauge Theory*†
C. Itzykson and J. M. Drouffe *Statistical Field Theory Volume 2: Strong Coupling, Monte Carlo Methods, Conformal Field Theory and Random Systems*†

G. Jaroszkiewicz *Principles of Discrete Time Mechanics*
C. V. Johnson *D-Branes*†
P. S. Joshi *Gravitational Collapse and Spacetime Singularities*†
J. I. Kapusta and C. Gale *Finite-Temperature Field Theory: Principles and Applications, 2nd edition*†
V. E. Korepin, N. M. Bogoliubov and A. G. Izergin *Quantum Inverse Scattering Method and Correlation Functions*†
J. Kroon *Conformal Methods in General Relativity*
M. Le Bellac *Thermal Field Theory*†
Y. Makeenko *Methods of Contemporary Gauge Theory*†
S. Mallik and S. Sarkar *Hadrons at Finite Temperature*
N. Manton and P. Sutcliffe *Topological Solitons*†
N. H. March *Liquid Metals: Concepts and Theory*†
I. Montvay and G. Münster *Quantum Fields on a Lattice*†
P. Nath *Supersymmetry, Supergravity, and Unification*
L. O'Raifeartaigh *Group Structure of Gauge Theories*†
T. Ortín *Gravity and Strings, 2nd edition*
A. M. Ozorio de Almeida *Hamiltonian Systems: Chaos and Quantization*†
L. Parker and D. Toms *Quantum Field Theory in Curved Spacetime: Quantized Fields and Gravity*
R. Penrose and W. Rindler *Spinors and Space-Time Volume 1: Two-Spinor Calculus and Relativistic Fields*†
R. Penrose and W. Rindler *Spinors and Space-Time Volume 2: Spinor and Twistor Methods in Space-Time Geometry*†
S. Pokorski *Gauge Field Theories, 2ndedition*†
J. Polchinski *String Theory Volume 1: An Introduction to the Bosonic String*†
J. Polchinski *String Theory Volume 2: Superstring Theory and Beyond*†
J. C. Polkinghorne *Models of High Energy Processes*†
V. N. Popov *Functional Integrals and Collective Excitations*†
L. V. Prokhorov and S. V. Shabanov *Hamiltonian Mechanics of Gauge Systems*
S. Raychaudhuri and K. Sridhar *Particle Physics of Brane Worlds and Extra Dimensions*
A. Recknagel and V. Schiomerus *Boundary Conformal Field Theory and the Worldsheet Approach to D-Branes*
R. J. Rivers *Path Integral Methods in Quantum Field Theory*†
R. G. Roberts *The Structure of the Proton: Deep Inelastic Scattering*†
C. Rovelli *Quantum Gravity*†
W. C. Saslaw *Gravitational Physics of Stellar and Galactic Systems*†
R. N. Sen *Causality, Measurement Theory and the Differentiable Structure of Space-Time*
M. Shifman and A. Yung *Supersymmetric Solitons*
H. Stephani, D. Kramer, M. MacCallum, C. Hoenselaers and E. Herlt *Exact Solutions of Einstein's Field Equations, 2nd edition*†
J. Stewart *Advanced General Relativity*†
J. C. Taylor *Gauge Theories of Weak Interactions*†
T. Thiemann *Modern Canonical Quantum General Relativity*†
D. J. Toms *The Schwinger Action Principle and Effective Action*†
A. Vilenkin and E. P. S. Shellard *Cosmic Strings and Other Topological Defects*†
R. S. Ward and R. O. Wells, Jr *Twistor Geometry and Field Theory*†
E. J. Weinberg *Classical Solutions in Quantum Field Theory: Solitons and Instantons in High Energy Physics*
J. R. Wilson and G. J. Mathews *Relativistic Numerical Hydrodynamics*†

† Available in paperback

The Cosmological Singularity

VLADIMIR BELINSKI
International Center for Relativistic Astrophysics Network, Italy

MARC HENNEAUX
Université Libre de Bruxelles

CAMBRIDGE
UNIVERSITY PRESS

CAMBRIDGE
UNIVERSITY PRESS

University Printing House, Cambridge CB2 8BS, United Kingdom

One Liberty Plaza, 20th Floor, New York, NY 10006, USA

477 Williamstown Road, Port Melbourne, VIC 3207, Australia

314-321, 3rd Floor, Plot 3, Splendor Forum, Jasola District Centre, New Delhi - 110025, India

79 Anson Road, #06-04/06, Singapore 079906

Cambridge University Press is part of the University of Cambridge.

It furthers the University's mission by disseminating knowledge in the pursuit of education, learning and research at the highest international levels of excellence.

www.cambridge.org
Information on this title: www.cambridge.org/9781107047471
DOI: 10.1017/9781107239333

First published 2018

A catalogue record for this publication is available from the British Library

ISBN 978-1-107-04747-1 Hardback
ISBN 978-1-107-64430-4 Paperback

Contents

Preface

The first exactly solvable cosmological models of Einstein's theory revealed the presence of a very striking phenomenon: the Big Bang singularity. Since the time it was discovered in 1922 by Alexander Friedmann, a fundamental question has arisen as to whether this phenomenon is due to the special simplifying assumptions underlying the exactly solvable models or whether a singularity is a general property of the Einstein equations. This question was formulated for the first time by L. Landau in 1959.

The question was answered by V. Belinski, I. Khalatnikov and E. Lifshitz (BKL) in 1969. The BKL work showed that a singularity is a general property of a generic cosmological solution of the classical gravitational equations and not a consequence of the special symmetric structure of the exact models. Most importantly, BKL were able to find the analytical structure of this generic solution and show that its behavior is of an extremely complex oscillatory character, of chaotic type. Because it provides the description of a general solution of the Einstein equations (i.e., a solution depending on sufficiently many freely adjustable functions of space), the BKL analysis sheds light on intrinsic properties of Einstein gravity. Given the nonlinear character of the Einstein equations and the difficulty of finding exact solutions without symmetries, the BKL results are quite notable. They have a fundamental significance not only for cosmology but also for the evolution of collapsing matter forming a black hole. The last stage of collapsing matter will follow in general the BKL regime.

The chaotic oscillations discovered by BKL can be understood in terms of a "cosmological billiard" system, where the cosmological evolution is described at each spatial point as the relativistic motion of a fictitious billiard ball in the Lorentzian space of the logarithmic scale factors. This reformulation of the BKL behavior can be naturally extended to arbitrary matter couplings and dimensions of space-time, enabling one to show that the BKL regime is inherent not only to General Relativity but also to more general physical theories containing gravity, such as supergravity models. The dimension of the billiard table and the nature of the walls that bound it depend on the theory, but the billiard description remains universally valid.

The billiard point of view provides a remarkably simple description of the gravitational field in the vicinity of a spacelike singularity. In spite of the complexity of the Einstein-matter field equations, the asymptotic behavior of the fields near a cosmological singularity can be phrased in surprisingly elementary terms involving finite-dimensional dynamical systems. This description is valid generically, i.e., without making any symmetry assumption.

The billiard point of view has also unexpectedly led to the discovery of a remarkable connection with one of the most beautiful and active subjects of modern mathematics, namely hyperbolic Coxeter groups and the theory of indefinite Kac–Moody algebras. This connection emerges because the billiard region in which the cosmological billiard ball moves turns out to possess exceptional properties, which imply that the group of reflections in the billiard walls is a simplex crystallographic hyperbolic Coxeter group for the known theories containing gravity. This intriguing fact opens up the fascinating perspective that an underlying infinite-dimensional symmetry algebra might play a central role in the fundamental formulation of gravity. However, at the time of writing this book, a complete proof of the presence of such algebras has not been found, so that the origin of the observed emergence of the hyperbolic Coxeter groups in the BKL description remains something of a mystery.

The purpose of this book is to explain at length the BKL analysis, starting from the early work on the subject and going all the way to the most modern developments.

Acknowledgements

We are grateful to many colleagues for helpful discussions, including Claudio Bunster, Jacques Demaret, Bernard Julia, Axel Kleinschmidt, Daniel Persson, Philippe Spindel, and especially Thibault Damour, Isaak Khalatnikov, Evgeny Lifshitz and Hermann Nicolai for joint collaborations that led to the work presented in this book. We also thank Victor Lekeu for help with the figures.

MH wants to express his particular debt to Thibault Damour and Hermann Nicolai for the extremely fruitful scientific interactions that led to the generic billiard picture at the core of the second part of this monograph.

We also gratefully acknowledge the hospitality of various institutions which provided the environment and atmosphere propitious to the writing of this monograph, including the Institute for Advanced Study (Princeton), the Institut des Hautes Études Scientifiques (Bures-sur-Yvette) and the Max Planck Institute for Gravitational Physics (Potsdam).

Introduction

Our book is devoted to the structure of the general solution of the Einstein equations with a cosmological singularity. We cover Einstein-matter systems in four and higher space-time dimensions.

Under the terminology "cosmological singularity," we mean a singularity in time, i.e., a spacelike singularity on a "submanifold" that can be viewed as the limit of a family of regular spacelike hypersurfaces forming (locally) a Gaussian foliation, such that the curvature invariants together with invariant characteristics of matter fields diverge as one tends to this submanifold.

The nonlinearities of the Einstein equations are notably known to prevent the construction of an exact general solution. From this perspective, the BKL work which describes the asymptotic *general* behavior of the gravitational field in four space-time dimensions as one approaches a spacelike singularity, is quite unique and exceptional. The central attainment of the BKL theory is the analysis of the delicate relationship between the time derivatives and the spatial gradients in the gravitational field equations near the singularity. The main technical idea of the BKL approach consists in identifying among the huge number of spatial gradients, those terms that are of the same importance as the time derivatives. In the vicinity of the singularity, these terms are in no way negligible. They act during the whole course of evolution up to the singularity, and it is actually due to these spatial gradients that oscillations do arise.

A remarkable simplifying feature nevertheless emerges as one tends to the singularity. This is the fact that the spatial gradient terms that must be retained in the dynamical equations of motion can be asymptotically represented as the products of some functions of the undifferentiated (in space) "scale factors" (which represent how distances along independent spatial directions evolve with time) by some slowly varying coefficients containing spacelike derivatives. This nontrivial separation springing up in the vicinity of the singularity leads to gravitational equations of motion which effectively reduce to a system of ordinary differential equations in time for the scale factors – one such system at each point of 3-space – because in the leading approximation, all relevant coefficients containing spacelike derivatives enter these equations solely as external (albeit, dynamically crucial) time-independent parameters.

Our presentation of the structure of the general solution of the Einstein equations with a cosmological singularity starts with vacuum gravity in four dimensions, where we follow the original BKL derivation. All key ideas and ingredients are already present in this model. We derive the effective dynamics and exhibit its oscillatory, chaotic behavior in Chapter 1. The effect of the "rotation of the Kasner axes," which appears in addition to the never-ending changes of Kasner exponents, is in particular carefully discussed. Some of the more technical derivations regarding this chapter are relegated to Appendix A.

The effective description of the asymptotic evolution in terms of ordinary differential equations can be reformulated as the motion of a particle in some external potential. Furthermore, it is possible to mimic the essential features of the BKL system of ordinary differential equations at any given point by considering spatially homogeneous cosmological models that have the property that their (non-abelian) homogeneity group leads to a spatial curvature for which the aforementioned dominating spatial gradients are nonzero. Such is the case for the spatially homogeneous models of so-called Bianchi types VIII and IX. The spatial curvature terms become, in the particle picture, the reflecting sharp wall potentials responsible for the oscillatory regime.

For the case of diagonal homogeneous cosmological models of Bianchi IX type, the potential has been introduced and investigated by C. Misner and D. Chitre [130, 35]. A billiard picture grew out from the work of these authors, which was very inspirational for future developments.

Chapter 2 is devoted to homogeneous cosmological models and an explanation of these developments. We also exhibit the rotation of the Kasner axes for non-diagonal spatial metrics. Appendix B provides more information about spatial homogeneity and gives, in particular, the Bianchi classification of spatially homogeneous models.

Chapter 3 discusses then the nature of the chaotic behavior near the cosmological singularity. The interesting phenomenon of "gravitational turbulence" is exhibited. Due to this phenomenon, a systematic growth of the spatial gradients arises when one approaches the singularity. At the end of this chapter we indicate that such growth does not invalidate the BKL analysis.

The BKL analysis was originally carried out for pure gravity in four space-time dimensions. However, the BKL approach can readily be extended to include matter, or by going to higher dimensions. One finds then that the same analysis of the delicate relationship between the time derivatives and the spatial gradients in the field equations near the singularity goes through, ending up again with an effective description of the dynamics at each spatial point in terms of a system of ODEs with respect to time of the "scale factors" (which also now include some of the scalar fields, if any).

What is new is that for some systems, spatial gradients are subdominant and the assertion that "only time derivatives are relevant near the singularity" becomes literally correct. Such singularities are "velocity-dominated" in the terminology of [66]. In some of the velocity-dominated models, the general solution is of non-oscillatory character and has simple Kasner-like power-law asymptotics near the singularity (at each spatial point). Examples are given by gravity coupled to a scalar field [16, 18], or gravity coupled to matter with a stiff equations of state, or pure gravity in space-time dimensions greater than or equal to 11 [63, 62]. In those cases, one can actually write down an explicit power-law asymptotic form of the metric valid all the way to the singularity, which contains as many arbitrary functions of space as the general solution must do, and demonstrate directly that this explicit form of the metric asymptotes an exact solution as one approaches the singularity [18, 3, 52]. This provides an independent check of the validity of the BKL procedure for such systems, which is much more explicit than for oscillatory models, where there is only the original BKL argument, which is more indirect.

These developments are first given in Chapter 4 by following the BKL approach. Perfect and viscous fluids, Yang Mills and Electromagnetic fields and scalar fields in four space-time dimensions are treated, as well as pure gravity in higher dimensions. [The general solution involving a classical spinor field is also considered because it involves interesting algebra, but because of its somewhat unphysical nature, its discussion is given in Appendix C.] We find that while the oscillatory behavior can be suppressed for some Einstein-matter models, it is present for some others. Which case arises depends on the matter content and on the space-time dimension.

The second part of our book is devoted to the billiard reformulation of the BKL behavior. Although the billiard picture originally arose in the context of homogeneous cosmological models, it is important to realize, however – and this turns out to be crucial for extensions to more general models – that the billiard motion also captures the dynamical behavior in the inhomogeneous case and is by no means tied to spatial homogeneity. While a single billiard suffices for homogeneous models, one gets one such billiard at each spatial point in the generic case. In other words, the billiard description is quite general. This is explained in Chapter 5, where we construct the billiard for pure Einstein gravity in four dimensions without any simplifying symmetry assumption. We follow the modern billiard point of view [45, 46, 51], based on the Iwasawa decomposition of the spatial metric and on a radial projection of the motion of the scale factors on the relevant hyperbolic space. On the technical side, the modern derivation streamlines the original billiard analysis by using a description in which the potential and the reflecting walls remain stationary in the vicinity of the singularity, which makes the analysis of their influence and their geometrical structure much more transparent.

The billiard viewpoint also applies to gravitational theories involving other matter fields, and in different space-time dimensions. This is explained in Chapter 6. We extend there the billiard analysis to arbitrary space-time dimensions ≥ 4, and to general systems containing gravity consistently coupled to matter (scalar and p-form) fields through second-order partial differential equations. We also indicate how the billiard point of view enlightens some of the results obtained in Chapter 4 through the original BKL approach.

Somewhat unexpectedly, the generalization of the billiard description gave rise to a remarkable development: it led directly to the discovery of an intriguing connection between the BKL asymptotic regime and Coxeter groups of reflections in hyperbolic space. This connection holds for the majority of theories of interest from a fundamental physical point of view, which includes pure gravity in D dimensions as well as various supergravity models [47]. It emerges because the billiard table is described in those cases by a convex polyhedron in hyperbolic space bounded by hyperplanes that make acute angles that are integer submultiples of π. This is quite remarkable in view of the fact that the angles depend on various discrete or continuous parameters: the space-time dimension, the ranks of the p-forms (if any), as well as the dilaton couplings (if dilatons are present).

For all theories of physical interest, the relevant billiard region is thus a Coxeter polyhedron (i.e., a convex polyhedron with all diehedral angles equal to integer submultiples of π). This means that it is a fundamental domain for the group of reflections in the billiard walls. The motion is a succession of such reflections, and thus defines elements of that group. But there is more. (1) The billiard table is a simplex, and this is also remarkable. Indeed, the number of walls following from the Lagrangian grows much faster than the dimension of the billiard table, but only a small subset of these walls, yielding a simplex, are dominant and relevant for the billiard description. The Coxeter group is thus a "simplex Coxeter group." (2) Furthermore, the billiard walls come with a natural normalization and define through their normalized scalar products $2\frac{(\alpha_i|\alpha_j)}{(\alpha_i|\alpha_i)}$ a matrix wich turns out to be the Cartan matrix of a Lorentzian Kac–Moody algebra. The simplex Coxeter group is thus the Weyl group of a Lorentzian Kac–Moody algebra, and the billiard region can be identified with its fundamental Weyl chamber.

These demonstrated properties have led to a conjecture that goes beyond the BKL analysis and which asserts that the corresponding infinite-dimensional Kac–Moody algebra itself might be a symmetry of an appropriate completion of the theory, the BKL Coxeter group being the signal of this huge symmetry. If true, this so-called Hidden Symmetry Conjecture would create promising new perspectives for the development of gravitation theory. However, the Hidden Symmetry Conjecture has not been proven yet, and therefore falls outside the scope of this book, which concentrates only on well-established facts. We refer to the review [51] for more information.

These intriguing developments on the connection with Coxeter groups are treated in Chapter 7, with additional information of a mathematical nature given in Appendix D.

Part I

BKL Analysis

1

Basic Structure of Cosmological Singularity

In this chapter, we derive the BKL behavior for pure Einstein gravity in four space-time dimensions by following the original approach of BKL. This approach consists in a careful analysis of the relative importance of the various terms in the Einstein equations as one goes to the singularity.

The central result is that in the BKL limit, the dynamics of the gravitational field can be described, at each spatial point, in terms of ordinary differential equations with respect to time for the appropriately defined spatial scale factors.

The effective representation of the dynamics of the gravitational field in terms of ordinary differential equations with respect to time is sometimes rephrased in the literature as the statement that in the BKL approach only the time derivatives are important near the singularity. Such a statement, taken at face value, is wrong or at best misleading since, as we shall see, spacelike gradients of the metric tensor cannot be neglected for generic solutions of pure Einstein gravity in four space-time dimensions, and in fact play a crucial role in the appearance of the oscillatory regime.* It is true that one gets at each spatial point an effective description of the singularity in terms of a finite dimensional dynamical system described by ordinary differential equations with respect to time, but the spatial gradients do enter these equations nontrivially.

1.1 Synchronous Reference System

For cosmological applications, the most appropriate coordinate systems are the synchronous reference systems – also called normal Gaussian coordinate systems – defined by the following conditions:†

* However, there exist reformulations of Einstein theory in terms of new variables involving the relevant gradients, for example in Ashtekar-like variables, for which the statement about the dominant role of the time derivatives is correct; see [7].

† In Part I and Appendices A, B, C, we use units in which the Einstein gravitational constant and the velocity of light are equal to unity. The small Greek indices refer to the three-dimensional space and assume values 1, 2, 3. The small Latin indices refer to the

$$g_{00} = -1, \; g_{0\alpha} = 0. \tag{1.1}$$

We denote the time coordinate in such a system as $x^0 = t$. The interval takes the form:

$$-ds^2 = -dt^2 + g_{\alpha\beta}dx^\alpha dx^\beta. \tag{1.2}$$

The components of the tensor g^{ik} inverse to g_{ik} are: $g^{00} = -1$, $g^{0\alpha} = 0$ and $g^{\alpha\beta}$ is inverse to $g_{\alpha\beta}$.

The gauge conditions (1.1) do not fix the coordinate system completely. We still have the freedom of performing some coordinate transformations containing four arbitrary functions depending on the three spatial variables x^α, which are easily worked out in infinitesimal form:

$$x^i = \acute{x}^i + \xi^i(\acute{x}). \tag{1.3}$$

Here and in what follows, we denote the collections of the four old coordinates (t, x^α) and four new coordinates $(\acute{t}, \acute{x}^\alpha)$ by the symbols x and $\acute{x}$, respectively. The functions $\xi^i(\acute{x})$ together with their first derivatives are infinitesimally small quantities. After such a transformation, the four-dimensional interval takes the form:

$$-ds^2 = g_{ik}(x)dx^i dx^k = g_{ik}^{(new)}(\acute{x})d\acute{x}^i d\acute{x}^k, \tag{1.4}$$

where

$$g_{ik}^{(new)}(\acute{x}) = g_{ik}(\acute{x}) + g_{il}(\acute{x})\frac{\partial \xi^l(\acute{x})}{\partial \acute{x}^k} + g_{kl}(\acute{x})\frac{\partial \xi^l(\acute{x})}{\partial \acute{x}^i} + \frac{\partial g_{ik}(\acute{x})}{\partial \acute{x}^l}\xi^l(\acute{x}). \tag{1.5}$$

In the last formula, the $g_{ik}(\acute{x})$ are the same functions $g_{ik}(x)$ in which x should simply be replaced by $\acute{x}$. If we wish to preserve the gauge (1.1) also for the new metric tensor $g_{ik}^{(new)}(\acute{x})$ in the new coordinates $\acute{x}$, it is necessary to impose the following restrictions on the functions $\xi^i(\acute{x})$:

$$\frac{\partial \xi^0(\acute{x})}{\partial \acute{t}} = 0, \quad g_{\alpha\beta}(\acute{x})\frac{\partial \xi^\beta(\acute{x})}{\partial \acute{t}} - \frac{\partial \xi^0(\acute{x})}{\partial \acute{x}^\alpha} = 0. \tag{1.6}$$

The solutions of these equations are:

$$\xi^0 = f^0(\acute{x}^1, \acute{x}^2, \acute{x}^3), \qquad \xi^\alpha = \int g^{\alpha\beta}(\acute{x})\frac{\partial f^0(\acute{x}^1, \acute{x}^2, \acute{x}^3)}{\partial \acute{x}^\beta}d\acute{x}^0 + f^\alpha(\acute{x}^1, \acute{x}^2, \acute{x}^3), \tag{1.7}$$

where f^0 and f^α are four arbitrary functions depending only on the spatial coordinates $\acute{x}^\alpha$.

A geometrical way to understand the ambiguity in synchronous coordinate systems proceeds as follows (see, e.g., [120]). To construct such a reference frame

four-dimensional space-time and take values 0, 1, 2, 3. The interval we write in the old Landau–Lifschitz [120] fashion: $-ds^2 = g_{ik}dx^i dx^k$, where g_{ik} has signature $(-+++)$. Then any timelike vector has negative squared norm. Ordinary partial derivatives are denoted by a comma, while ordinary derivative with respect to the synchronous time t is also denoted by a dot. For covariant derivatives, we use the semicolon.

in some region of a given space-time, one needs to choose an initial spacelike hypersurface and build the congruence of the timelike geodesic curves normal to this hypersurface. These curves will serve as the coordinate lines of new synchronous time t. Introducing local coordinates X^i, the equation for the initial spacelike hypersurface will read $X^0 = F(X^1, X^2, X^3)$ for some function F. Such a construction clearly permits a free choice of the shape of such initial hypersurface [i.e., of the function $F(X^1, X^2, X^3)$] and arbitrary transformations of 3-space coordinates on it, that is $X^\alpha = X^\alpha(\acute{X}^1, \acute{X}^2, \acute{X}^3)$. This freedom corresponds precisely to the four nonphysical arbitrary functions of 3-space existing in a general synchronous system in agreement with the foregoing infinitesimal analysis.

When discussing general solutions $g_{\alpha\beta}$ of the field equations in synchronous gauges, it is necessary to keep in mind that the gravitational potentials $g_{\alpha\beta}$ contain, among all possible arbitrary functional parameters present in them, four arbitrary functions of 3-space just representing the gauge freedom and therefore of no direct physical significance.

1.2 The Gravitational Field Equations

In the synchronous system (1.1), it is convenient to introduce the special notation $\kappa_{\alpha\beta}$ for the time derivative of the three-dimensional metric tensor $g_{\alpha\beta}$ (in this system $\kappa_{\alpha\beta}$ is proportional to the so-called second fundamental form). The definitions and some evident relations are:

$$\kappa_{\alpha\beta} = \dot{g}_{\alpha\beta}\,, \quad \kappa_\alpha^\beta = g^{\beta\gamma}\kappa_{\gamma\alpha}\,, \quad \kappa = \kappa_\alpha^\alpha = \dot{g}/g\,, \; g = \det g_{\alpha\beta}\,. \tag{1.8}$$

Then for the complete set of Christoffel symbols Γ^i_{kl} we obtain:

$$\Gamma^0_{00} = \Gamma^\alpha_{00} = \Gamma^0_{0\alpha} = 0, \;\; \Gamma^0_{\alpha\beta} = \frac{1}{2}\kappa_{\alpha\beta}\,, \quad \Gamma^\alpha_{0\beta} = \frac{1}{2}\kappa^\alpha_\beta\,, \quad \Gamma^\gamma_{\alpha\beta} = \gamma^\gamma_{\alpha\beta}\,, \tag{1.9}$$

where $\gamma^\gamma_{\alpha\beta}$ are the three-dimensional gamma-symbols constructed from $g_{\alpha\beta}$:

$$\gamma^\gamma_{\alpha\beta} = \frac{1}{2}g^{\gamma\mu}(g_{\mu\alpha,\beta} + g_{\mu\beta,\alpha} - g_{\alpha\beta,\mu}). \tag{1.10}$$

With the Christoffel symbols (1.9), the components $R^i_k = g^{il}R_{lk}$ of the Ricci tensor can be written in the form:

$$R^0_0 = \frac{1}{2}\dot{\kappa} + \frac{1}{4}\kappa^\beta_\alpha\kappa^\alpha_\beta, \tag{1.11}$$

$$R^0_\alpha = \frac{1}{2}(\kappa_{,\alpha} - \kappa^\beta_{\alpha;\beta}), \tag{1.12}$$

$$R^\beta_\alpha = \frac{1}{2\sqrt{g}}(\sqrt{g}\kappa^\beta_\alpha)^\cdot + P^\beta_\alpha, \tag{1.13}$$

where covariant differentiation in (1.12) is performed with respect to the three-dimensional metric $g_{\alpha\beta}$ with three-dimensional Christoffel symbols $\gamma^\gamma_{\alpha\beta}$ and P^β_α in (1.13) is a three-dimensional Ricci tensor constructed from $g_{\alpha\beta}$:

$$P_\alpha^\beta = g^{\beta\gamma} P_{\gamma\alpha} \ , \quad P_{\alpha\beta} = \gamma^\gamma_{\alpha\beta,\gamma} - \gamma^\gamma_{\gamma\alpha,\beta} + \gamma^\gamma_{\alpha\beta}\gamma^\mu_{\gamma\mu} - \gamma^\gamma_{\alpha\mu}\gamma^\mu_{\beta\gamma} \ . \tag{1.14}$$

It follows from (1.11)–(1.13) that the Einstein equations $R_i^k = T_i^k - \frac{1}{2}\delta_i^k T_l^l$ (with the components of the energy–momentum tensor $T_0^0 = -T_{00}$, $T_\alpha^0 = -T_{0\alpha}$, $T_\alpha^\beta = g^{\beta\gamma}T_{\gamma\alpha}$) become in a synchronous coordinate system:

$$\frac{1}{2}\dot{\kappa} + \frac{1}{4}\kappa_\beta^\alpha \kappa_\alpha^\beta = T_0^0 - \frac{1}{2}(T_0^0 + T_\alpha^\alpha) \ , \tag{1.15}$$

$$\frac{1}{2}(\kappa_{,\alpha} - \kappa_{\alpha;\beta}^\beta) = T_\alpha^0 \ , \tag{1.16}$$

$$\frac{1}{2\sqrt{g}}(\sqrt{g}\kappa_\alpha^\beta)^{\cdot} + P_\alpha^\beta = T_\alpha^\beta - \frac{1}{2}\delta_\alpha^\beta(T_0^0 + T_\gamma^\gamma) \ . \tag{1.17}$$

1.3 General Solution

We define a *general solution of the gravitational equations (1.15)–(1.17)* to be *a solution containing as many arbitrary three-dimensional functional parameters (i.e., functions depending only on three space coordinates x^α) as are necessary to match arbitrary initial data.*

Let us determine the number of functional parameters that should enter a general solution by first considering the dynamical equations of motion for the gravitational potentials $g_{\alpha\beta}$ and the matter fields, namely, the equations (1.17) together with the T_i^k-conservation law,

$$T_{i;k}^k = 0. \tag{1.18}$$

Let us assume that we found the general solution of the equations (1.17) and (1.18). Because the equations (1.17) are six differential equations of second order in time for the six quantities $g_{\alpha\beta}$, their general solution contains 12 arbitrary three-dimensional functions. The total number of arbitrariness is therefore at this stage $12{+}M$, where M is the number of arbitrary three-dimensional functions appearing in the matter fields.

We now have to satisfy the additional equations (1.15) and (1.16). It is well known that the only role of these additional equations is to generate four restrictions (the so-called "initial data constraints") on the set of $12{+}M$ arbitrary functions obtained in the course of integration of the dynamical equations (1.17) and (1.18). This phenomenon is a consequence of the Bianchi identity $(R_i^k - \frac{1}{2}\delta_i^k R_l^l)_{;k} = 0$. To see this, let us denote $R_i^k - (T_i^k - \frac{1}{2}\delta_i^k T_l^l) = A_i^k$. It is easy to check that if $T_{i;k}^k = 0$, then from the Bianchi identity follows $A_{i;k}^k = \frac{1}{2}A_{k,i}^k$. When written in a synchronous system under the condition that the equations (1.17) are also satisfied (i.e., $A_\beta^\alpha = 0$), this last relation gives $\frac{1}{\sqrt{g}}(\sqrt{g}A_i^k)_{,k} = \frac{1}{2}A_{0,i}^0$. This is a system of four linear homogeneous differential equations of first order in time for the four quantities A_0^0, A_α^0. With zero initial conditions for A_0^0, A_α^0, the only solution of such a system is the solution which is everywhere zero. Consequently, in order to satisfy the gravitational equations (1.15) and (1.16) (i.e., $A_0^0 = 0$,

$A^0_\alpha = 0$) once the equations (1.17) and (1.18) have been solved, it is sufficient to put to zero the values A^0_0 and A^0_α on the initial spacelike hypersurface.

It is clear that these values will not contain any new arbitrariness because they will be constructed only from those 12+M three-dimensional arbitrary functions which already appeared in the course of integration of equations (1.17) and (1.18). In any regular part of space-time there is no difficulty in calculating A^0_0 and A^0_α on the initial hypersurface using the Taylor expansions of all quantities in its vicinity. Near a cosmological singularity, one has to kill only the first non-vanishing terms in the asymptotic expansions of A^0_0 and A^0_α (the form of which is different for the different types of singularity).

It follows that the constraints (1.15) and (1.16) on the initial data reduce the degree of arbitrariness of the generic solution of the basic dynamical equations (1.17) and (1.18) by four three-dimensional functions*.

Furthermore, as we have already shown, among all arbitrary three-dimensional functions in the general solution, four of them represent a nonphysical gauge freedom of the synchronous system. Consequently, the number of physically significant three-dimensional functional parameters in the general solution is $12 + M - 4 - 4 = 4 + M$. In vacuum, $M = 0$. Hence, a vacuum solution must contain four arbitrary three-dimensional functional parameters in order to be general. In the presence of matter, the number M depends on the type of energy–momentum tensor. Each type needs individual consideration, but the analysis proceeds along the same lines. For instance, for a scalar field the number M is equal to 2, while for the electromagnetic field, which carries also its own gauge invariance, $M = 4$. For a simple perfect liquid with definite equation of state, $M = 4$ corresponding to three functional parameters for the velocity and one for the energy density.

1.4 Definition of Cosmological Singularity

This book is devoted to cosmological singularities. How are these defined?

* In a more transparent way, and independently of the structure of the initial manifold (singular or regular), everything that has been stated in this section can be also derived by using the gravitational equations in the equivalent form $R^k_i - \frac{1}{2}\delta^k_i R^m_m = T^k_i$. Again, if the equation of motion (1.18) are satisfied, we have the identity $B^k_{i;k} = 0$, where $B^k_i = R^k_i - \frac{1}{2}\delta^k_i R^m_m - T^k_i$. If we solve the dynamical set of equations $B^\beta_\alpha = 0$ (which are again six differential equations of second order in time for the six potentials $g_{\alpha\beta}$), then we get from the identity $B^k_{i;k} = 0$ the following two relations in synchronous coordinates: $(\sqrt{g}B^0_\alpha)_{,0} = 0$ and $(\sqrt{g}B^0_0)_{,0} + (\sqrt{g}B^\beta_0)_{,\beta} = 0$. This means that the quantities B^0_α and B^0_0 have the structure $\sqrt{g}B^0_\alpha = F^0_\alpha(x^1, x^2, x^3)$ and $\sqrt{g}B^0_0 = \int(\sqrt{g}g^{\beta\alpha}B^0_\alpha)_{,\beta}dt + F^0_0(x^1, x^2, x^3)$ where F^0_α and F^0_0 are some three-dimensional functions. These four functions are constructed from those 12+M arbitrary functions that appeared already in the solution through the integration of the equations (1.18) and $B^\beta_\alpha = 0$. Consequently, to satisfy the gravitational equations $B^0_\alpha = 0$ and $B^0_0 = 0$, it is enough to demand $F^0_\alpha = 0$ and $F^0_0 = 0$, which only fix in some way four arbitrary three-dimensional functions among the $12 + M$ that are already present in the solution.

In General Relativity, one encounters singularities of different types. The general definition of a physical (i.e., not removable by coordinate transformations) singularity follows naturally from the corner-stone of the Einstein theory, that is from the equivalence principle. A space-time point is singular if at this point the equivalence principle is violated. In other words *a point is singular if it has no vicinity homeomorphic to Minkowski space-time.*

Each physical singularity is covered by this definition, but each one also has additional more detailed characteristics, and can be classified accordingly. In this book we are interested in the most violent singular phenomenon of General Relativity, namely in the singularities with divergent strengths of the gravitational field and, in particular, those which are relevant to the origin (or to the end) of the cosmological evolution.

We adopt then the following definition. *A cosmological singularity is a singularity that is both (i) a singularity in time, i.e., such that the singular three-dimensional manifold is spacelike; and (ii) a curvature singularity, i.e., such that the curvature invariants together with invariant characteristics of matter fields (like the energy density) if any, diverge on this manifold.* One could equivalently call such singularities "spacelike singularities" but for historical reasons we shall stick to the original BKL terminology.

We say that a singular 3-manifold is spacelike if it can be viewed as the limit of a family of regular spacelike hypersurfaces in a Gaussian coordinate system. In a general synchronous coordinate system, the singular manifold has an equation of the form $t = f(x^1, x^2, x^3)$, but it is always possible to choose the initial spacelike hypersurface in the geometrical construction of a synchronous system so as to make the points in the singular set "simultaneous," with equation $t = 0$. If the singularity is an initial singularity (i.e., occurs in the past), one has $t \geq 0$ and the hypersurfaces $t > 0$ are regular, with the spatial metric becoming singular as $t \to 0$.

The definition of a generic cosmological singularity implies that in its vicinity, nothing outstanding happens in the regions of the spacelike hypersurfaces $t = const > 0$ we are interested in. In other words, in the (sufficiently small) three-dimensional volumes under examination, one should not encounter any additional singularity at some special values of the space coordinates x^α. From this follows that in such volumes, the orders of magnitude of the space derivatives of the field variables can be evaluated as products of the absolute values of these variables with some universal characteristic number k representing the averaged value of the dominant wave numbers. Even if some singularities with respect to the space coordinates x^α were to exist on the spacelike hypersurfaces up to the limiting manifold $t = 0$, our analysis would remain correct in all smooth regions between them. This is because our analysis is essentially local and is insensitive to the global structure of the three-dimensional region under investigation. The only exception which could spoil such an approach would be the exotic case (if it exists at all) when the multitude of points singular with respect to the space coordinates x^α tends to form a *dense* set on the limiting manifold $t = 0$. However, such a wild structure will be discarded here since it does not appear to bear any relation to a typical cosmological situation in any reasonable physical sense.*

These considerations also mean that the "cosmological singularity" does not have to embrace the entirety of three-dimensional space. Our results can be applied equally well to a finite part of it, however small (but of finite measure). In this sense the terminology "singularity in time" would be more precise than "cosmological singularity." For example, all findings in this book about the structure of a cosmological singularity are literally applicable to the collapse of an isolated body in its comoving system, i.e., to the properties of the internal singularity inside a black hole.

Also to be noted is that we should make a clear distinction between a real physical singularity and fictitious ones. Fictitious singularities in time are unavoidably present in any synchronous system [120] because the normal geodesics refocus, but all invariant characteristics of the gravitational field and matter are regular at such points. Singularities of this kind can be removed simply by passing to another system of coordinates. We will work in this book in the vicinity of a physical singularity, far enough from any fictitious one.

It is important to realize that in a synchronous coordinate system adapted to the simultaneous singularity (located at $t = 0$), one of the four arbitrary three-dimensional functions characterizing the residual coordinate freedom of the synchronous gauge has been fixed. Therefore, the general solution with *simultaneous cosmological singularity* in a synchronous system contains only three arbitrary nonphysical gauge functional parameters. The residual freedom is completely captured by the time-independent changes of spatial coordinates $x^\alpha = x^\alpha(\acute{x}^1, \acute{x}^2, \acute{x}^3)$.

1.5 Kasner-Like Singularities of Power Law Asymptotics

If we are interested in singularities in time, it is natural to examine first what happens when the metric tensor $g_{\alpha\beta}$ effectively depends only on time, that is when we neglect the three-dimensional Ricci tensor P_α^β in the equations (1.17). It is reasonable to study such an approximation first in empty space, where $T_i^k = 0$, and to take into account later all the changes that may be observed in the presence of matter. This procedure is sensible and consistent since, in general,

* The aforesaid observation bears a direct relationship with the so-called "spike solutions" discovered in [26] and analyzed later in many articles (see, e.g., [23, 125]). In these solutions, the spatial gradients on the singular manifold $t = 0$ diverge at some isolated points. Then at such points no wave number k exists. However, the set of such points is not dense. Consequently, spikes represent some small set of zero measure which can be added to the general BKL picture. The physical application of spikes is unclear but it is worth keeping their existence in mind. It is interesting that spikes are pure classical phenomena that disappear under quantization [42].

the influence of matter upon the solution in the vicinity of the singularity appears to be either negligible or can be put under control (see Chapter 4).

So let us assume that the tensor T_i^k is equal to zero, and let us retain in (1.17) only the terms involving the time derivatives of the fields. In such an approximation, the dynamical gravitational equations are:

$$(\sqrt{g}\kappa_\alpha^\beta)^{\cdot} = 0 \ . \tag{1.19}$$

Taking the contraction of these equations over α and β yields $(\sqrt{g})^{\cdot\cdot} = 0$, that is $\sqrt{g} = \Lambda t + \Theta$, where Λ and Θ are arbitrary functions of 3-space.

However, as we observed in the previous section, we are free to assume that the singularity occurs simultaneously everywhere in the region under consideration. This choice does not mean any loss of generality. We take the singular hypersurface to be $t = 0$. Correspondingly, we put $\Theta = 0$. *Here and everywhere in the subsequent analysis, we follow the evolution towards the singularity, i.e., we study the dynamics as time decreases from certain positive value $t > 0$ down to $t = 0$.*

It is easy to check that for a simultaneous singularity, the most general solution of the equations (1.19) can be represented in the following triad form:

$$g_{\alpha\beta} = \eta_{\bar{\alpha}\bar{\beta}} l_\alpha^{\bar{\alpha}} \, l_\beta^{\bar{\beta}} \ , \quad \eta_{\bar{\alpha}\bar{\beta}} = diag(t^{2p_{\bar{1}}}, t^{2p_{\bar{2}}}, t^{2p_{\bar{3}}}). \tag{1.20}$$

In (1.20) and in the sequel, we use letters with a bar for the frame indices. In formula (1.20), the three time-independent vectors $l_\alpha^{\bar{\alpha}}(x^1, x^2, x^3)$ determine three independent directions in three-dimensional space and the three scale factors $t^{2p_{\bar{\alpha}}}$ describe expansion or contraction of space along these three principal directions. Since $g \sim t^2$ the exponents $p_{\bar{\alpha}}(x^1, x^2, x^3)$ should satisfy the restriction

$$p_{\bar{1}} + p_{\bar{2}} + p_{\bar{3}} = 1. \tag{1.21}$$

Once the dynamical equations (1.19) are satisfied, one must impose the constraints. The initial data constraint (1.15) gives only one requirement which represents yet another condition for the exponents $p_{\bar{\alpha}}$:

$$p_{\bar{1}}^2 + p_{\bar{2}}^2 + p_{\bar{3}}^2 = 1. \tag{1.22}$$

As for the 3-vectorial constraint (1.16), it turns out that its left-hand side takes the form $a_\alpha t^{-1} \ln t + b_\alpha t^{-1}$ where a_α and b_α are three-dimensional functions. The function a_α is proportional to the gradient $(p_{\bar{1}}^2 + p_{\bar{2}}^2 + p_{\bar{3}}^2)_{,\alpha}$ and hence, it is automatically zero by virtue of the relation (1.22). So, we need to impose only the conditions $b_\alpha = 0$. These conditions give three differential equations of the first order with respect to the coordinates x^α connecting the three-dimensional functions $l_\alpha^{\bar{\alpha}}$, $p_{\bar{\alpha}}$. The concrete form of these equations will not be reported here since these constraints are of no relevance for the dynamics of the *general* solution near the cosmological singularity (an interested reader can find them in [124]). It is important only to take into account their existence, not their specific form, when calculating the number of free three-dimensional arbitrary functional parameters appearing in the solution.

All these assertions can readily be checked taking into account that the metric $g^{\alpha\beta}$ inverse to $g_{\alpha\beta}$ is:

$$g^{\alpha\beta} = \eta^{\bar{\alpha}\bar{\beta}} l^{\alpha}_{\bar{\alpha}}\, l^{\beta}_{\bar{\beta}}\,, \quad \eta^{\bar{\alpha}\bar{\beta}} = diag(t^{-2p_{\bar{1}}}, t^{-2p_{\bar{2}}}, t^{-2p_{\bar{3}}}), \tag{1.23}$$

where $\eta^{\bar{\alpha}\bar{\beta}}$ is inverse to $\eta_{\bar{\alpha}\bar{\beta}}$ and where *the three vectors* $l^{\alpha}_{\bar{\alpha}}$ *form the triad inverse to the one constructed from* $l^{\bar{\alpha}}_{\alpha}$, *that is,* $l^{\alpha}_{\bar{\alpha}} l^{\bar{\alpha}}_{\beta} = \delta^{\alpha}_{\beta}$ *and* $l^{\bar{\alpha}}_{\alpha} l^{\alpha}_{\bar{\beta}} = \delta^{\bar{\alpha}}_{\bar{\beta}}$. For raising and lowering the (unbarred) vector indices, one uses the metric $g_{\alpha\beta}$. The frame metric $\eta_{\bar{\alpha}\bar{\beta}}$ is used for the same operations on the frame indices. The components of $\kappa_{\alpha\beta} = \dot{g}_{\alpha\beta}$ in the triad frame is $\kappa_{\bar{\alpha}\bar{\beta}} = l^{\alpha}_{\bar{\alpha}} l^{\beta}_{\bar{\beta}} \kappa_{\alpha\beta} = \dot{\eta}_{\bar{\alpha}\bar{\beta}}$ and

$$\kappa^{\bar{\beta}}_{\bar{\alpha}} = \eta^{\bar{\beta}\bar{\mu}} \kappa_{\bar{\mu}\bar{\alpha}} = diag\left(\frac{2p_{\bar{1}}}{t}, \frac{2p_{\bar{2}}}{t}, \frac{2p_{\bar{3}}}{t}\right), \quad \kappa = \kappa^{\mu}_{\mu} = \kappa^{\bar{\mu}}_{\bar{\mu}} = \frac{2}{t}. \tag{1.24}$$

For the determinant g, we have:

$$g = t^2 \Lambda^2, \quad \Lambda = \varepsilon^{\alpha\beta\gamma} l^{\bar{1}}_{\alpha} l^{\bar{2}}_{\beta} l^{\bar{3}}_{\gamma}, \tag{1.25}$$

where $\varepsilon^{\alpha\beta\gamma}$ is the totally antisymmetric three-dimensional Levi-Civita symbol ($\varepsilon^{123} = 1$). Owing to the independence of the triad vectors on time we can pass in the equations (1.19) directly to the frame quantities $\kappa^{\bar{\beta}}_{\bar{\alpha}}$ instead of κ^{β}_{α}. It is easy to see that due to the relations (1.24) and (1.25), the equations (1.19) are trivially satisfied.

It follows from the properties (1.21) and (1.22) of the exponents $p_{\bar{\alpha}}$ that at each 3-space point, one of the $p_{\bar{\alpha}}$ is negative while the other two are positive. Consequently, when $t \to 0$, the three-dimensional distances around each point expand in one direction and contract in two others. However, the value of any three-dimensional volume element decreases in time monotonically, since the determinant of $g_{\alpha\beta}$ decreases proportionally to t^2 at each space point. It is a simple task to show that for general values of the exponents $p_{\bar{1}}, p_{\bar{2}}, p_{\bar{3}}$ (excluding the special zero measure cases when two of the exponents vanish and the third one is equal to unity), this singularity is physical, that is all curvature invariants diverge in the limit $t \to 0$.

It is obvious that for the particular case when $l^{\bar{\alpha}}_{\alpha}$ and $p_{\bar{\alpha}}$ are constants, the metric (1.20)–(1.22) becomes an exact solution of the Einstein equations in vacuum. This solution is nothing else but the well-known Kasner solution [112]. In the considered inhomogeneous generalization, we also call for that reason the directions along $l^{\bar{\alpha}}_{\alpha}\,(x^1, x^2, x^3)$ "the Kasner axes" and the functions $p_{\bar{\alpha}}(x^1, x^2, x^3)$ "the Kasner exponents." The metric (1.20)–(1.22) describes "the Kasner asymptotics."

To deal in the sequel with the exponents $p_{\bar{\alpha}}$, it is convenient to have some definite order for them. Without loss of generality we can take $p_{\bar{1}} < p_{\bar{2}} < p_{\bar{3}}$ in the vicinity of some arbitrary chosen spatial point and represent $p_{\bar{\alpha}}$ in this vicinity in the following parametric way:

$$p_{\bar{1}} = \frac{-u}{1+u+u^2}, \; p_{\bar{2}} = \frac{1+u}{1+u+u^2}, \; p_{\bar{3}} = \frac{u(1+u)}{1+u+u^2}, \tag{1.26}$$

where the functional parameter $u(x^1, x^2, x^3)$ takes values in the region $u \geqslant 1$. This representation automatically meets the requirements (1.21) and (1.22). With the chosen ordering, the domain of variation of the Kasner exponents is

$$-\frac{1}{3} \leqslant p_{\bar{1}} \leqslant 0, \; 0 \leqslant p_{\bar{2}} \leqslant \frac{2}{3}, \; \frac{2}{3} \leqslant p_{\bar{3}} \leqslant 1. \tag{1.27}$$

The behavior of $p_{\bar{\alpha}}$ as functions of u is shown in Figure 1.1.

Let us now count the number of physically significant three-dimensional arbitrary functional parameters appearing in the metric (1.20). We have nine arbitrary functions in the vectors $l_{\alpha}^{\bar{\alpha}}$ plus one arbitrary function among the Kasner exponents $p_{\bar{\alpha}}$. Therefore, the total number is ten. Since the equation (1.15) is already satisfied, we have to impose on these ten functions only three additional constraints following from the "initial conditions" (1.16). Furthermore, three arbitrary functions in the solution represent nonphysical gauge freedom (we recall that the singularity is chosen to be simultaneous, which removes one gauge function). The final count thus gives $10 - 3 - 3 = 4$ arbitrary physical functional parameters, which is exactly the number of arbitrary functions that a general solution of the gravitational equations in vacuum should contain.

It seems therefore that we found the behavior of a general solution of the vacuum Einstein equations near a cosmological singularity. This is not the case,

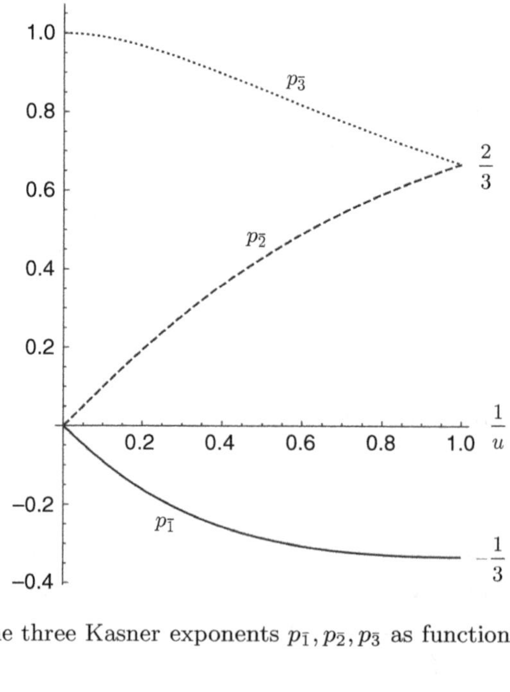

Figure 1.1: The three Kasner exponents $p_{\bar{1}}, p_{\bar{2}}, p_{\bar{3}}$ as functions of the parameter u.

however, because further analysis of the corrections to the Kasner asymptotics (1.20)–(1.22) shows that this solution actually ceases to be valid in the vicinity of the singularity.

1.6 Instability of Kasner Dynamics

The point is that in order to confirm the validity of the approximation made in the previous section, one has to show that the three-dimensional Ricci tensor P_{α}^{β} in equations (1.17) can indeed be neglected in first approximation. Only in that case is the procedure legitimate. To find whether this is so, we must compute the corrections to the metric (1.20)–(1.22) generated by P_{α}^{β} and show that these corrections are negligibly small in the vicinity of the singularity $t \to 0$. The peculiarity of our problem permits one to analyze this question in a simple way, without having to solve the equations explicitly for the perturbations. The conclusion is that the above procedure is not legitimate for arbitrary choices of initial data. We derive the exact solution for the corrections in Appendix A.1 and the result confirms the simplified procedure used below.

The corrections $\delta g_{\alpha\beta}$ generate also corrections $\delta\kappa_{\alpha\beta}$ to the second fundamental form. As usual, any quantity F is decomposed as $F = F_{(0)} + \delta F$ where $F_{(0)}$ represents the background value of F and δF the correction. We write the exact vacuum gravitational equations (1.17) as

$$(\sqrt{g}\kappa_{\alpha}^{\beta})^{\cdot} = -2\sqrt{g}P_{\alpha}^{\beta}\,. \tag{1.28}$$

In accordance with the standard procedure, we express that the corrections to the main part of these equations, that is to their left-hand side, are generated by the first non-vanishing terms of the right-hand side (which is the side that has been neglected in the first approximation) calculated with the metric $\overset{(0)}{g}_{\alpha\beta}$ (1.20) appearing in the first approximation. In this way we obtain the following equations for the variations:

$$\left[\delta(\sqrt{g}\kappa_{\alpha}^{\beta})\right]^{\cdot} = -2\left[\sqrt{g}P_{\alpha}^{\beta}\right]_{(0)}\,. \tag{1.29}$$

It is essential that we need to evaluate the corrections only in the vicinity of the singularity, that is in the limit $t \to 0$. This circumstance, together with the fact that all terms in the right-hand side of (1.29) are essentially of power law character with respect to time*, makes such an evaluation straightforward. From (1.29) we find, in the limit $t \to 0$:

$$\delta(\sqrt{g}\kappa_{\alpha}^{\beta}) \sim t\left[\sqrt{g}P_{\alpha}^{\beta}\right]_{(0)}\,. \tag{1.30}$$

* The terms containing the factor $\ln t$, which originate from differentiation of powers $t^{p_{\bar{\alpha}}}$ with arbitrary exponents $p_{\bar{\alpha}}(x^1, x^2, x^3)$, can be ignored since $\ln t$ in the limit $t \to 0$ effectively behaves like a constant in expressions of the type $t^{p_{\bar{\alpha}}} \ln t$. Taking into account these logarithmic terms does not change the qualitative results of our evaluation.

We omitted all possible time independent arbitrary additional terms in the right-hand side of this relation, that is arbitrary "constants of integration" of equations (1.29), because we need only that particular part of the solution of the equations (1.29) that is directly generated by the three-dimensional Ricci tensor. The aforementioned omitted additional terms yield nothing new since they just repeat the structure already exhibited by the main approximation $g^{(0)}_{\alpha\beta}$ (such additional terms correspond to arbitrary infinitesimal variations of the vectors $l^{\bar{\alpha}}_{\alpha}$ and of the exponents $p_{\bar{\alpha}}$ within the class of three-dimensional functions).

In the vicinity of the singularity, the conditions of smallness of the corrections require

$$\delta(\sqrt{g}\kappa^{\beta}_{\alpha}) \ll \left[\sqrt{g}\kappa^{\beta}_{\alpha}\right]_{(0)} \,. \tag{1.31}$$

Using the result (1.30), one can write these conditions in the triad basis of the $l^{\alpha}_{\bar{\alpha}}$'s and get

$$P^{(0)}_{\bar{\alpha}\bar{\beta}} \ll t^{-1}\kappa^{(0)}_{\bar{\alpha}\bar{\beta}}. \tag{1.32}$$

These inequalities (and their consequences of similar form derived below) should be treated only in the sense of comparing the absolute values of both sides without taking into account their signs.

Because

$$\kappa^{(0)}_{\bar{\alpha}\bar{\beta}} = diag(2p_{\bar{1}}t^{2p_{\bar{1}}-1}, 2p_{\bar{2}}t^{2p_{\bar{2}}-1}, 2p_{\bar{3}}t^{2p_{\bar{3}}-1}) \tag{1.33}$$

the diagonal components of the inequalities (1.32) give:

$$P^{(0)}_{\bar{1}\bar{1}} \ll t^{2p_{\bar{1}}-2},\ P^{(0)}_{\bar{2}\bar{2}} \ll t^{2p_{\bar{2}}-2},\ P^{(0)}_{\bar{3}\bar{3}} \ll t^{2p_{\bar{3}}-2}. \tag{1.34}$$

For the non-diagonal components, the conditions (1.32) have the meaning $P^{(0)}_{\bar{\alpha}\bar{\beta}} \ll \sqrt{t^{-1}\kappa^{(0)}_{\bar{\alpha}\bar{\alpha}} \cdot t^{-1}\kappa^{(0)}_{\bar{\beta}\bar{\beta}}}$ (here $\bar{\alpha} \neq \bar{\beta}$ and there is no sum over repeated indices). Consequently, the non-diagonal inequalities stand for

$$P^{(0)}_{\bar{1}\bar{2}} \ll t^{p_{\bar{1}}+p_{\bar{2}}-2},\ P^{(0)}_{\bar{1}\bar{3}} \ll t^{p_{\bar{1}}+p_{\bar{3}}-2},\ P^{(0)}_{\bar{2}\bar{3}} \ll t^{p_{\bar{2}}+p_{\bar{3}}-2}. \tag{1.35}$$

The result is then that all the corrections will be negligibly small in the limit $t \to 0$ if for any choice of values of the indices $\bar{\alpha}$ and $\bar{\beta}$ (equal to each other or not), the following conditions are satisfied:

$$P^{(0)}_{\bar{\alpha}\bar{\beta}} \ll t^{p_{\bar{\alpha}}+p_{\bar{\beta}}-2},\ t \to 0. \tag{1.36}$$

The task is therefore to calculate the three-dimensional Ricci tensor corresponding to $g^{(0)}_{\alpha\beta}$ (1.20) and to check whether its frame components are in agreement with the inequalities (1.36). To do this it is necessary to adopt some definite order for the exponents $p_{\bar{\alpha}}$. Which one one takes does not matter due to the freedom of re-numeration of these exponents and the corresponding frame vectors. We choose the order following from the representation (1.26)–(1.27) for $u > 1$ that is $p_{\bar{1}} < p_{\bar{2}} < p_{\bar{3}}$. Then $p_{\bar{1}}$ is negative and $t^{p_{\bar{1}}} \to \infty$ while $t^{p_{\bar{2}}} \to 0$,

$t^{p_{\bar{3}}} \to 0$ (and $t^{p_{\bar{2}}} \gg t^{p_{\bar{3}}}$) in the limit $t \to 0$. The calculations* show that in the limit $t \to 0$, and with the chosen order for $p_{\bar{\alpha}}$, we have the following asymptotical behavior for $P^{(0)}_{\bar{\alpha}\bar{\beta}}$:

$$\begin{aligned} P^{(0)}_{\bar{1}\bar{1}} &= \left[\frac{1}{2}t^2(\sigma^{\bar{1}}_{\bar{2}\bar{3}})^2 + O\right] t^{2p_{\bar{1}}-2}, \\ P^{(0)}_{\bar{2}\bar{2}} &= \left[-\frac{1}{2}t^2(\sigma^{\bar{1}}_{\bar{2}\bar{3}})^2 + O\right] t^{2p_{\bar{2}}-2}, \\ P^{(0)}_{\bar{3}\bar{3}} &= \left[-\frac{1}{2}t^2(\sigma^{\bar{1}}_{\bar{2}\bar{3}})^2 + O\right] t^{2p_{\bar{3}}-2}. \end{aligned} \tag{1.37}$$

$$\begin{aligned} P^{(0)}_{\bar{1}\bar{2}} &= \left\{-t^2\left[\sigma^{\bar{1}}_{\bar{2}\bar{3}}\sigma^{\bar{1}}_{\bar{1}\bar{3}} + \frac{1}{2}\left(\sigma^{\bar{1}}_{\bar{2}\bar{3}}\right)_{,\nu} l^{\nu}_{\bar{3}} t^{-p_{\bar{3}}}\right] + O\right\} t^{p_{\bar{1}}+p_{\bar{2}}-2}, \\ P^{(0)}_{\bar{1}\bar{3}} &= \left\{t^2\left[\sigma^{\bar{1}}_{\bar{2}\bar{3}}\sigma^{\bar{1}}_{\bar{1}\bar{2}} - \frac{1}{2}\left(\sigma^{\bar{1}}_{\bar{2}\bar{3}}\right)_{,\nu} l^{\nu}_{\bar{2}} t^{-p_{\bar{2}}}\right] + O\right\} t^{p_{\bar{1}}+p_{\bar{3}}-2}, \\ P^{(0)}_{\bar{2}\bar{3}} &= O \cdot t^{p_{\bar{2}}+p_{\bar{3}}-2}, \end{aligned} \tag{1.38}$$

where by the symbol O, we designate a collection of terms each of which tends to zero when $t \to 0$.

The σ-coefficients in these formulas are:

$$\begin{aligned} \sigma^{\bar{1}}_{\bar{2}\bar{3}} &= \left[(l^{\bar{1}}_{\mu}t^{p_{\bar{1}}})_{,\nu} - (l^{\bar{1}}_{\nu}t^{p_{\bar{1}}})_{,\mu}\right] l^{\mu}_{\bar{2}} l^{\nu}_{\bar{3}} t^{-p_{\bar{2}}-p_{\bar{3}}}, \\ \sigma^{\bar{1}}_{\bar{1}\bar{2}} &= \left[(l^{\bar{1}}_{\mu}t^{p_{\bar{1}}})_{,\nu} - (l^{\bar{1}}_{\nu}t^{p_{\bar{1}}})_{,\mu}\right] l^{\mu}_{\bar{1}} l^{\nu}_{\bar{2}} t^{-p_{\bar{1}}-p_{\bar{2}}}, \\ \sigma^{\bar{1}}_{\bar{1}\bar{3}} &= \left[(l^{\bar{1}}_{\mu}t^{p_{\bar{1}}})_{,\nu} - (l^{\bar{1}}_{\nu}t^{p_{\bar{1}}})_{,\mu}\right] l^{\mu}_{\bar{1}} l^{\nu}_{\bar{3}} t^{-p_{\bar{1}}-p_{\bar{3}}}. \end{aligned} \tag{1.39}$$

From the first formula (1.39), it follows that

$$\sigma^{\bar{1}}_{\bar{2}\bar{3}} = (l^{\bar{1}}_{\mu,\nu} - l^{\bar{1}}_{\nu,\mu}) l^{\mu}_{\bar{2}} l^{\nu}_{\bar{3}} t^{p_{\bar{1}}-p_{\bar{2}}-p_{\bar{3}}}. \tag{1.40}$$

The relations (1.21) show then that near the singularity, $t^2(\sigma^{\bar{1}}_{\bar{2}\bar{3}})^2$ behaves as

$$t^2(\sigma^{\bar{1}}_{\bar{2}\bar{3}})^2 \sim t^{4p_{\bar{1}}} \to \infty, \; t \to 0. \tag{1.41}$$

That is, all three factors in the rectangular brackets in the expressions (1.37) for the diagonal components of the Ricci tensor tend to infinity. This evidently

* To perform these calculations, it is convenient to work with the time-dependent orthonormal frame $L^{\bar{\alpha}}_{\alpha}$ defined as $L^{\bar{\alpha}}_{\alpha} = l^{\bar{\alpha}}_{\alpha} t^{p_{\bar{\alpha}}}$ (no sum over repeated indices). The preceding identities $l^{\alpha}_{\bar{\alpha}} l^{\bar{\alpha}}_{\beta} = \delta^{\alpha}_{\beta}$, $l^{\bar{\alpha}}_{\alpha} l^{\alpha}_{\bar{\beta}} = \delta^{\bar{\alpha}}_{\bar{\beta}}$ imply similar ones for the new triad: $L^{\alpha}_{\bar{\alpha}} L^{\bar{\alpha}}_{\beta} = \delta^{\alpha}_{\beta}$, $L^{\bar{\alpha}}_{\alpha} L^{\alpha}_{\bar{\beta}} = \delta^{\bar{\alpha}}_{\bar{\beta}}$. With this definition, the metric tensor (1.20) of the first approximation takes the form $g_{\alpha\beta} = \gamma_{\bar{\alpha}\bar{\beta}} L^{\bar{\alpha}}_{\alpha} L^{\bar{\beta}}_{\beta}$, $\gamma_{\bar{\alpha}\bar{\beta}} = diag(1,1,1)$. For the computation of the projections $L^{\alpha}_{\bar{\alpha}} L^{\beta}_{\bar{\beta}} P_{\alpha\beta}$ of the three-dimensional Ricci tensor in the triad $L^{\alpha}_{\bar{\alpha}}$, we can use directly the formula (A.19) from Appendix A.2. Doing this, and using the relations between the triads $L^{\bar{\alpha}}_{\alpha}$ and $l^{\bar{\alpha}}_{\alpha}$, it is easy to get the components $P_{\bar{\alpha}\bar{\beta}} = l^{\alpha}_{\bar{\alpha}} l^{\beta}_{\bar{\beta}} P_{\alpha\beta}$ in the original time-independent triad $l^{\bar{\alpha}}_{\alpha}$. These components are given by formulas (1.37)–(1.39).

violates the requirements (1.34) for the validity of the first approximation (1.20)–(1.22). This means that the Kasner behavior is unstable near the singularity due to the presence in the three-dimensional Ricci tensor of terms which sooner or later (with decreasing of time) destroy this regime.

The only way to save the Kasner asymptotics would be to put artificially the three-dimensional function $(l^{\bar{1}}_{\mu,\nu} - l^{\bar{1}}_{\nu,\mu})l^{\mu}_{\bar{2}}l^{\nu}_{\bar{3}}$ equal to zero. In this case $\sigma^{\bar{1}}_{\bar{2}\bar{3}} = 0$ and, as can be seen from (1.37)–(1.38), all components $P_{\bar{\alpha}\bar{\beta}}$ will satisfy the requirements (1.36) of the validity of the first approximation. However, in this case we lose one of the arbitrary functional parameters. We are thus left with a solution containing only three such parameters instead of the four that are necessary for the general solution.

The particular solution fulfilling the additional restriction

$$(l^{\bar{1}}_{\mu,\nu} - l^{\bar{1}}_{\nu,\mu})l^{\mu}_{\bar{2}}l^{\nu}_{\bar{3}} = 0$$

is called the *generalized Kasner solution*. It was constructed in [124].

1.7 Transition to the New Regime

In the previous section we found that in the course of approaching the cosmological singularity, some critical time t_c will unavoidably appear after which the Kasner behavior (1.20)–(1.27) will be destroyed by the terms in the three-dimensional Ricci tensor containing the coefficient $\sigma^{\bar{1}}_{\bar{2}\bar{3}}$. Now the question is: what form will the solution take after this critical time?

The natural way to reach an answer is to modify the first approximation by including in it also those terms from the three-dimensional Ricci tensor which are responsible for the instability of Kasner dynamics, namely all the terms containing the "dangerous" coefficient $\sigma^{\bar{1}}_{\bar{2}\bar{3}}$. These new equations can be formulated easily and solved exactly [14, 19]. The result is that after a short period centered around the critical time t_c, the solution follows again the Kasner-like behavior but with new Kasner exponents $\acute{p}_{\bar{\alpha}}$ and new directional vectors $\acute{l}^{\bar{\alpha}}_{\alpha}$ of the Kasner axes.

To find the evolution after the end of the first Kasner regime, let us formulate again the conditions of applicability of the generalized Kasner solution, but in more general form and in a way independent on any particular choice of the order of the exponents $p_{\bar{1}}, p_{\bar{2}}, p_{\bar{3}}$.

In the Kasner asymptotics the spatial metric near the singularity can be written in the form:

$$g_{\alpha\beta} = \eta_{\bar{\alpha}\bar{\beta}} l^{\bar{\alpha}}_{\alpha}\, l^{\bar{\beta}}_{\beta}\,, \quad \eta_{\bar{\alpha}\bar{\beta}} = diag(a^2, b^2, c^2), \tag{1.42}$$

where without loss of generality we take all three functions a, b, c to be positive. We also assume the components of the vectors $l^{\bar{\alpha}}_{\alpha}(x^1, x^2, x^3)$ to be of order of unity. In other words, the quantities which determine the order of magnitude of the components of the metric tensor $g_{\alpha\beta}$ are included in the functions:

$$(a^2, b^2, c^2) = (a_0^2 t^{2p_{\bar{1}}}, b_0^2 t^{2p_{\bar{2}}}, c_0^2 t^{2p_{\bar{3}}}) \tag{1.43}$$

with some factors a_0, b_0, c_0 which are three-dimensional parameters depending on the coordinates x^α. The "Euclidean" modules $\left[(l_1^{\bar{\alpha}})^2 + (l_2^{\bar{\alpha}})^2 + (l_3^{\bar{\alpha}})^2\right]^{1/2}$ are equal to unity and these three degrees of freedom of the frame are included into the parameters a_0, b_0, c_0.

The Kasner asymptotics has been obtained by neglecting the 3-space components of the Ricci tensor in the vacuum gravitational equations (1.17). The conditions for such a neglect to be valid follow from an analysis of the same type as in the previous section and lead to a slight modification of the inequalities (1.34)–(1.35) that takes into account that the vectors $l_\alpha^{\bar{\alpha}}$ are slightly different from those used in the preceding section. These requirements are now:

$$\begin{aligned} &P_{\bar{1}\bar{1}} \ll a^2 t^{-2}, \ P_{\bar{2}\bar{2}} \ll b^2 t^{-2}, \ P_{\bar{3}\bar{3}} \ll c^2 t^{-2}, \\ &P_{\bar{1}\bar{2}} \ll abt^{-2}, \ P_{\bar{1}\bar{3}} \ll act^{-2}, \ P_{\bar{2}\bar{3}} \ll bct^{-2}, \end{aligned} \tag{1.44}$$

where the components of the three-dimensional Ricci tensor are again defined with respect to the frame $l_\alpha^{\bar{\alpha}}$ of the metric (1.42), that is $P_{\bar{\alpha}\bar{\beta}} = P_{\alpha\beta} l_{\bar{\alpha}}^\alpha l_{\bar{\beta}}^\beta$. If the inequalities (1.44) are satisfied, one may disregard completely the three-dimensional Ricci tensor in equations (1.17) in the leading order in the limit $t \to 0$.

For the metric of the form (1.42), the frame components $P_{\bar{\alpha}\bar{\beta}}$ of the Ricci tensor $P_{\alpha\beta}$ in the triad $l_\alpha^{\bar{\alpha}}(x^1, x^2, x^3)$ can be calculated by the formulas (A.16) and (A.19) of Appendix A.2, where the frame $L_\alpha^{\bar{\alpha}}$ should be taken as $L_\alpha^{\bar{1}} = al_\alpha^{\bar{1}}$, $L_\alpha^{\bar{2}} = bl_\alpha^{\bar{2}}$, $L_\alpha^{\bar{3}} = cl_\alpha^{\bar{3}}$ (see previous footnote). This calculation shows that each quantity $a^{-2}P_{\bar{1}\bar{1}}$, $b^{-2}P_{\bar{2}\bar{2}}$, $c^{-2}P_{\bar{3}\bar{3}}$ contains (apart from many other terms of different magnitudes) a linear combination of the following three terms:

$$\begin{aligned} &\frac{a^4}{2a^2b^2c^2}\left[\left(l_{\alpha,\beta}^{\bar{1}} - l_{\beta,\alpha}^{\bar{1}}\right) l_{\bar{2}}^\alpha l_{\bar{3}}^\beta\right]^2 \sim \frac{k^2 a^4}{s^2 t^2}, \\ &\frac{b^4}{2a^2b^2c^2}\left[\left(l_{\alpha,\beta}^{\bar{2}} - l_{\beta,\alpha}^{\bar{2}}\right) l_{\bar{3}}^\alpha l_{\bar{1}}^\beta\right]^2 \sim \frac{k^2 b^4}{s^2 t^2}, \\ &\frac{c^4}{2a^2b^2c^2}\left[\left(l_{\alpha,\beta}^{\bar{3}} - l_{\beta,\alpha}^{\bar{3}}\right) l_{\bar{1}}^\alpha l_{\bar{2}}^\beta\right]^2 \sim \frac{k^2 c^4}{s^2 t^2}. \end{aligned} \tag{1.45}$$

On the right-hand side of formulas (1.45), we show the orders of magnitude of these terms in the limit $t \to 0$. The quantity k^{-1} denotes the order of magnitude of the spatial distances over which the metric changes substantially and $s = a_0 b_0 c_0$. The exact way in which the terms (1.45) enter the diagonal projections of the three-dimensional Ricci tensor is:

$$\begin{aligned} a^{-2}P_{\bar{1}\bar{1}} &= \frac{1}{2a^2b^2c^2}\left(a^4\lambda^2 - b^4\mu^2 - c^4\nu^2\right) + \ldots, \\ b^{-2}P_{\bar{2}\bar{2}} &= \frac{1}{2a^2b^2c^2}\left(-a^4\lambda^2 + b^4\mu^2 - c^4\nu^2\right) + \ldots, \\ c^{-2}P_{\bar{3}\bar{3}} &= \frac{1}{2a^2b^2c^2}\left(-a^4\lambda^2 - b^4\mu^2 + c^4\nu^2\right) + \ldots, \end{aligned} \tag{1.46}$$

where we introduced the notations:

$$\lambda = \left(l^{\bar{1}}_{\alpha,\beta} - l^{\bar{1}}_{\beta,\alpha}\right) l^{\alpha}_{\bar{2}} l^{\beta}_{\bar{3}} \ , \ \mu = \left(l^{\bar{2}}_{\alpha,\beta} - l^{\bar{2}}_{\beta,\alpha}\right) l^{\alpha}_{\bar{3}} l^{\beta}_{\bar{1}} \ , \ \nu = \left(l^{\bar{3}}_{\alpha,\beta} - l^{\bar{3}}_{\beta,\alpha}\right) l^{\alpha}_{\bar{1}} l^{\beta}_{\bar{2}} \ . \quad (1.47)$$

By the triple dots in (1.46), we denote all the other terms in the diagonal projections of the Ricci tensor.

Taking into account the evaluations (1.45), the requirements (1.44) lead to the following three inequalities:

$$a\sqrt{ks^{-1}} \ll 1 \ , b\sqrt{ks^{-1}} \ll 1 \ , c\sqrt{ks^{-1}} \ll 1. \quad (1.48)$$

It is remarkable that these inequalities are not only necessary, but also sufficient conditions for the existence of the solution (1.42). In other words, once the conditions (1.48) are satisfied, all other terms in $P_{\bar{1}\bar{1}}$, $P_{\bar{2}\bar{2}}$, $P_{\bar{3}\bar{3}}$ [those designated by the triple dots in (1.46)] as well as *all* terms in $P_{\bar{1}\bar{2}}$, $P_{\bar{1}\bar{3}}$, $P_{\bar{2}\bar{3}}$ automatically satisfy the conditions (1.44). Indeed, an estimate of these terms leads to the requirements:

$$k^2 s^{-2} \left(a^2b^2, a^2c^2, b^2c^2, a^3b, a^3c, b^3a, b^3c, c^3a, c^3b, a^2bc, b^2ac, c^2ab\right) \ll 1 \ , \quad (1.49)$$

(i.e., $k^2s^{-2}a^2b^2 \ll 1$, $k^2s^{-2}a^2c^2 \ll 1, \ldots,$ $k^2s^{-2}c^2ab \ll 1$). All these twelve inequalities contain on the left the products of powers of two or three of the quantities which enter the inequalities (1.48), and therefore are evidently true if the latter are satisfied.

As t decreases, there eventually occurs a critical instant t_c when one of the conditions (1.48) becomes violated. Thus, if during a given Kasner epoch the negative Kasner exponent refers to the function $a(t)$, then at the instant t_c we will have

$$a(t_c)\sqrt{ks^{-1}} \sim 1. \quad (1.50)$$

Since, during that epoch, the functions $b(t)$ and $c(t)$ decrease with decreasing t, the other two inequalities (1.48) remain valid and at $t \sim t_c$ we shall have

$$b(t_c) \ll a(t_c), \ c(t_c) \ll a(t_c). \quad (1.51)$$

It is essential that at the same time all the conditions (1.49) continue to hold. This means that *even at $t \sim t_c$ all off-diagonal projections of the equations (1.17) can be disregarded, as before.* In the critical region around $t \sim t_c$, the only terms that become important in the diagonal projections are those containing $a^4\lambda^2$. It follows that during the transition region around the critical time t_c the Kasner axes $l^{\bar{\alpha}}_{\alpha}(x^1, x^2, x^3)$ remain unchanged and the process of alternation of the scale factors a, b, c is governed by the equations:

$$\begin{aligned} \frac{(\dot{a}bc)^{\cdot}}{abc} &= -\frac{a^4\lambda^2}{2a^2b^2c^2}, \\ \frac{(a\dot{b}c)^{\cdot}}{abc} &= \frac{a^4\lambda^2}{2a^2b^2c^2}, \\ \frac{(ab\dot{c})^{\cdot}}{abc} &= \frac{a^4\lambda^2}{2a^2b^2c^2}. \end{aligned} \quad (1.52)$$

By adding these equations and combining the result with equation (1.15) for empty space, we obtain a relation which contains only the first derivatives of the functions a, b, c and represents a first integral of the system (1.52):

$$\left(\frac{\dot{a}}{a}\right)^2 + \left(\frac{\dot{b}}{b}\right)^2 + \left(\frac{\dot{c}}{c}\right)^2 - \left(\frac{\dot{a}}{a} + \frac{\dot{b}}{b} + \frac{\dot{c}}{c}\right)^2 + \frac{a^4\lambda^2}{2a^2b^2c^2} = 0. \tag{1.53}$$

We see that at the initial epoch, when the terms containing λ^2 in the equations (1.52)–(1.53) can be neglected, the solution coincides with a Kasner-like one (1.42)–(1.43). The condition of validity of the Kasner regime following from (1.44) and (1.46) is $a^4\lambda^2(abc)^{-2} \ll t^{-2}$. Consequently the critical time t_c at which the Kasner evolution will cease follows from the relation $\left[a^4\lambda^2(abc)^{-2}t^2\right]_{t=t_c} \sim 1$, which coincides with the evaluation (1.50).

The exact solution of equations (1.52)–(1.53) together with its asymptotics in the regions $t \gg t_c$ and $t \ll t_c$ are derived in Section A.3 of Appendix A and can be summarized as follows. During the initial epoch before the critical time t_c, we have the asymptotics of the solution given by formulas (1.42)–(1.43):

$$a^2 = a_0^2 t^{2p_{\bar{1}}},\ b^2 = b_0^2 t^{2p_{\bar{2}}},\ c^2 = c_0^2 t^{2p_{\bar{3}}}, \tag{1.54}$$

and during the final epoch after t_c the asymptotics is:

$$a^2 = \acute{a}_0^2 t^{2\acute{p}_{\bar{1}}},\ b^2 = \acute{b}_0^2 t^{2\acute{p}_{\bar{2}}},\ c^2 = \acute{c}_0^2 t^{2\acute{p}_{\bar{3}}}. \tag{1.55}$$

That is, it has again the power-type behavior but with new three-dimensional parameters. It is a crucial fact that the new exponents $\acute{p}_\alpha$ are again Kasner exponents, in the sense that they fulfill the Kasner relations. Explicitly, they are:

$$\acute{p}_{\bar{1}} = -\frac{p_{\bar{1}}}{1+2p_{\bar{1}}},\ \acute{p}_{\bar{2}} = \frac{p_{\bar{2}}+2p_{\bar{1}}}{1+2p_{\bar{1}}},\ \acute{p}_{\bar{3}} = \frac{p_{\bar{3}}+2p_{\bar{1}}}{1+2p_{\bar{1}}}, \tag{1.56}$$

from which one easily sees that the $\acute{p}_\alpha$ *satisfy indeed the same Kasner relations* $\acute{p}_{\bar{1}} + \acute{p}_{\bar{2}} + \acute{p}_{\bar{3}} = 1$, $\acute{p}_{\bar{1}}^2 + \acute{p}_{\bar{2}}^2 + \acute{p}_{\bar{3}}^2 = 1$. The new factors $\acute{a}_0^2, \acute{b}_0^2, \acute{c}_0^2$ are given in formulas (A.27) of Appendix A. It follows from these formulas that the new product $\acute{a}_0\acute{b}_0\acute{c}_0$, is connected to the old one $a_0b_0c_0$ by the relation:

$$\acute{a}_0\acute{b}_0\acute{c}_0 = a_0b_0c_0(1+2p_{\bar{1}}). \tag{1.57}$$

If at the first epoch we have the order $p_{\bar{1}} < p_{\bar{2}} < p_{\bar{3}}$ and a^2 was increasing ($p_{\bar{1}} < 0$), it is easy to check from (1.56) that at the second epoch we have $\acute{p}_{\bar{1}} > 0$ and $\acute{p}_{\bar{2}} < 0$, that is a^2 is now decreasing but the function b^2 increases instead. The negative Kasner exponent jumped from a^2 to b^2. The general law for this effect is that *the negative Kasner exponent always jumps to that scale factor which in the preceding epoch had the smallest positive Kasner exponent.*

It follows from the exact solution that the maximal value of the scale factor a^2, reached at the critical time $t = t_c$ (the transition instant to the second Kasner epoch), is equal to:

$$a^2_{\max} = \frac{2a_0b_0c_0\,|p_{\bar{1}}|}{|\lambda|}. \tag{1.58}$$

It is remarkable that the values of the exponents $\acute{p}_{\bar{1}}, \acute{p}_{\bar{2}}, \acute{p}_{\bar{3}}$ for the second Kasner epoch depend only on the values on the exponents $p_{\bar{1}}, p_{\bar{2}}, p_{\bar{3}}$ of the first epoch and not on the directional vectors $l^{\bar{\alpha}}_{\alpha}$ of the Kasner axes of the first epoch. In general the Kasner axes, in leading approximation, remain fixed only during the short transition period between two neighboring epochs, namely during that period when the Kasner exponents change their values accordingly the law (1.56). In the course of the subsequent evolution during the second epoch, the Kasner axes change their directions (see Section 1.9 on "Rotation of Kasner axes"). This effect, however, has no influence on the succeeding transformations of the Kasner exponents during the subsequent transitions to new epochs (if such transitions occur) because of the afore-stressed independence of the transition law (1.56) of the Kasner exponents on the vectors $l^{\bar{\alpha}}_{\alpha}$. Consequently the transformation law (1.56) represents a closed self-consistent map which can be continued as long as the process of alternation of the epochs goes on. In the next section we show that in fact this process has generically no end and will proceed up to the singularity. Then the map (1.56) reflects one of the basic dynamical features of the gravitational field as one approaches a cosmological singularity.

1.8 Oscillatory Nature of the Generic Singularity

We obtained in the previous section a good instrument to disclose the main features of the evolution of the gravitational field to the singularity, which is independent from the rotation of the Kasner axes. This instrument is the law (1.56) of change of the Kasner exponents. This law does not depend at all on the evolution of the triad vectors. We can therefore examine the evolution of the Kasner exponents as such, i.e., independently of the question of how they will be distributed among different directions of the three-dimensional space in each consecutive Kasner epoch.

We shall show in this section that the behavior of the Kasner exponents provides convincing evidence for the oscillatory character of the behavior of the gravitational field near a cosmological singularity. In the next section, we will also take into account the consequences of the rotation of the Kasner axes and will get the final, complete picture of the oscillatory regime. We stress again that the evolution of the Kasner exponents following from the already derived law (1.56) plays a basic role in this complete picture.

It is convenient to consider from now on that the three quantities $p_{\bar{1}}, p_{\bar{2}}, p_{\bar{3}}$ are defined once and for all by the relations (1.26)–(1.27) for $u > 1$, independently of how they are distributed among the scale factors a^2, b^2, c^2. Thus, for any value of the parameter u in this region, these quantities are ordered as $p_{\bar{1}} < p_{\bar{2}} < p_{\bar{3}}$, that is $p_{\bar{1}}$ is always negative. From the transformation law (1.56), it is easy to check that if during the first epoch we have the asymptotics (1.54):

$$(a^2, b^2, c^2) \sim (t^{2p_{\bar{1}}(u)}, t^{2p_{\bar{2}}(u)}, t^{2p_{\bar{3}}(u)}), \tag{1.59}$$

then, during the second epoch, the asymptotics with the new Kasner exponents (1.56) can be written as

$$(a^2, b^2, c^2) \sim (t^{2p_{\bar{2}}(u-1)}, t^{2p_{\bar{1}}(u-1)}, t^{2p_{\bar{3}}(u-1)}). \tag{1.60}$$

This result gives the first indication that the evolution to the singularity is of an endless oscillatory type, because during the second epoch we encounter the same instability problem caused by the new growing scale factor b^2, leading to another new epoch and so on. In this oscillatory evolution, the negative Kasner exponent changes its place among the factors a^2, b^2, c^2 during the successive Kasner-like epochs.

A support for such a picture comes from the symmetry of the gravitational equations with respect to the cyclic permutations of the scale factors a^2, b^2, c^2, together with accompanying permutations of the three-dimensional "dangerous" factors λ, μ, ν. Due to this symmetry no term in the dominant part (1.46) of the Ricci tensor can be neglected. The asymptotic form of the gravitational equations near the singularity should include all of them. That is, in the limit $t \to 0$, the relevant equations that replace (1.52) where only one term was kept must be:

$$\begin{aligned} \frac{(\dot{a}bc)^{\cdot}}{abc} &= -\frac{1}{2a^2b^2c^2}\left(a^4\lambda^2 - b^4\mu^2 - c^4\nu^2\right), \\ \frac{(a\dot{b}c)^{\cdot}}{abc} &= -\frac{1}{2a^2b^2c^2}\left(b^4\mu^2 - c^4\nu^2 - a^4\lambda^2\right), \\ \frac{(ab\dot{c})^{\cdot}}{abc} &= -\frac{1}{2a^2b^2c^2}\left(c^4\nu^2 - a^4\lambda^2 - b^4\mu^2\right). \end{aligned} \tag{1.61}$$

This system admits also a first integral, obtained as for (1.52) by adding the equations (1.61) and combining the result with equation (1.15). The resulting relation contains only first-order time derivatives and is:

$$\left(\frac{\dot{a}}{a}\right)^2 + \left(\frac{\dot{b}}{b}\right)^2 + \left(\frac{\dot{c}}{c}\right)^2 - \left(\frac{\dot{a}}{a} + \frac{\dot{b}}{b} + \frac{\dot{c}}{c}\right)^2 + \frac{1}{2a^2b^2c^2}\left(a^4\lambda^2 + b^4\mu^2 + c^4\nu^2\right) = 0. \tag{1.62}$$

It turns out that this system indeed describes, as anticipated, an endless oscillatory regime as one approaches the cosmological singularity. The evolution of the Kasner exponents goes in accordance with the evolution of the parameter u, starting from the initial value $u^{(1)} > 1$. It follows from (1.59) and (1.60) that this evolution yields the sequence $u^{(1)}, u^{(1)} - 1, u^{(1)} - 2, \ldots$ Because the range of the parameter u is fixed to be greater than unity, we can follow such a sequence up to the value $u^{(1)} - [u^{(1)}] + 1$, where $[u^{(1)}]$ stands for the integer value of $u^{(1)}$. During this sequence of epochs, which we call an *"era,"* the negative Kasner exponent and the smallest of the two positive exponents get permuted only between the scale factors a^2 and b^2 (i.e., these two factors oscillate), while the third factor c^2, which has the largest positive exponent during each epoch of the era, decreases monotonically. The last Kasner epoch of the first era corresponds to the parameter $u^{(1)} - [u^{(1)}] + 1 = 1 + x^{(1)}$ where $x^{(1)} < 1$.

After one has reached this value, a new era starts with new initial epoch corresponding to the parameter u equal to $x^{(1)}$ in accordance to the law $u \to u-1$. But this value is out of the range of u used here. However, the representation (1.26) for the Kasner exponents is invariant with respect to the transformation $u \to 1/u$ with simultaneous permutation of the positive exponents $p_{\bar{2}}$ and $p_{\bar{3}}$. Due to this freedom, we can arrange the initial parameter $u^{(2)}$ of the first epoch of the new era to be greater than unity by performing the transformation $u^{(2)} = 1/x^{(1)}$, with accompanying replacement of the two positive Kasner exponents by each other. Now the evolution of the Kasner exponents during the second era can be described by the same sequence $u^{(2)}, u^{(2)} - 1, u^{(2)} - 2, \ldots$ up to the last epoch in this second era when the parameter u is reduced to the value $u^{(2)} - \left[u^{(2)}\right] +1 = 1 + x^{(2)}$ with $x^{(2)} < 1$ and so on. It is important to realize that during the first epoch of the second era the largest positive exponent jumps to the scale factor a^2 (or b^2) while c^2 acquires the smallest of the two positive Kasner exponents. This means that, during the second era, the negative exponent and the smallest of the two positive exponents get permuted only between the scale factors b^2(or a^2) and c^2. In other words, it is now these two factors that oscillate, while the third factor a^2 (or b^2) keeps the largest positive exponent during the whole second era, i.e., decreases monotonically.

This process of interchange of epochs and eras never terminates up to the singular point $t = 0$ which is an accumulation point of oscillations. In synchronous time t, the frequencies of oscillations tend to infinity, the periods tend to zero and the number of epochs and eras between any instant $t > 0$ and the singularity $t = 0$ is infinite. It is also important to stress that as we approach the singularity, the values of *all three scale factors* a^2, b^2, c^2 *tend to zero* because their successive maxima tend to zero. This is evident from the transformation law (1.57) for the product $a_0b_0c_0$ and the formula (1.58) for the maximum of a^2 (and similar ones for b^2 and c^2). The factor $|p_{\bar{1}}/\lambda|$ in this expression can be ignored (it oscillates inside some restricted interval bounded from above), but $a_0b_0c_0$ is decreasing systematically since $1 + 2p_{\bar{1}} < 1$. This behavior is illustrated in Figure 1.2.

It thus also follows that the volume $\sqrt{g}$ goes to zero as one goes to the singularity, $\sqrt{g} \to 0$ at $t \to 0$ and that in the new time variable τ, defined through

$$d\tau = -(abc)^{-1}dt\,, \tag{1.63}$$

the singularity $t \to 0$ corresponds to $\tau \to \infty$.

For later purposes, it is useful to change the dynamical variables to new ones. Instead of the functions a, b, c, we introduce new dynamical variables $\beta^{\bar{\alpha}}$ by the relations:

$$|\lambda|\, a^2 = e^{-2\beta^{\bar{1}}}, \;\; |\mu|\, b^2 = e^{-2\beta^{\bar{2}}}, \;\; |\nu|\, c^2 = e^{-2\beta^{\bar{3}}}. \tag{1.64}$$

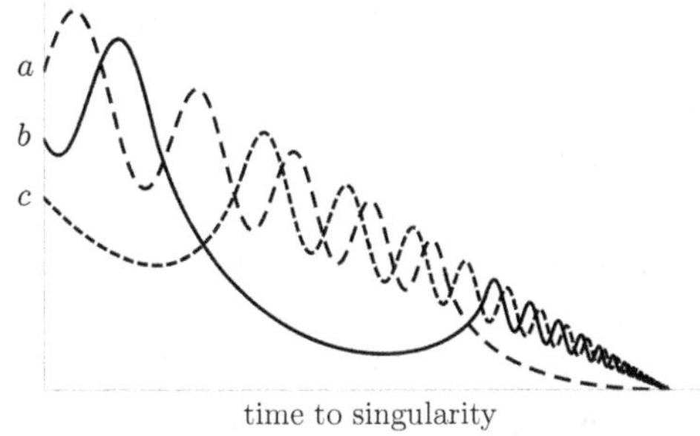

Figure 1.2: The character of the oscillations of the scale factors a^2, b^2, c^2 (symbolic representation).

In terms of these new variables, the equations (1.61)–(1.62) take the form:

$$2\frac{\partial^2 \beta^{\bar{1}}}{\partial \tau^2} = e^{-4\beta^{\bar{1}}} - e^{-4\beta^{\bar{2}}} - e^{-4\beta^{\bar{3}}}, \tag{1.65}$$
$$2\frac{\partial^2 \beta^{\bar{2}}}{\partial \tau^2} = e^{-4\beta^{\bar{2}}} - e^{-4\beta^{\bar{3}}} - e^{-4\beta^{\bar{1}}},$$
$$2\frac{\partial^2 \beta^{\bar{3}}}{\partial \tau^2} = e^{-4\beta^{\bar{3}}} - e^{-4\beta^{\bar{1}}} - e^{-4\beta^{\bar{2}}},$$

and

$$\left(\frac{\partial \beta^{\bar{1}}}{\partial \tau}\right)^2 + \left(\frac{\partial \beta^{\bar{2}}}{\partial \tau}\right)^2 + \left(\frac{\partial \beta^{\bar{3}}}{\partial \tau}\right)^2 - \left(\frac{\partial \beta^{\bar{1}}}{\partial \tau} + \frac{\partial \beta^{\bar{2}}}{\partial \tau} + \frac{\partial \beta^{\bar{3}}}{\partial \tau}\right)^2 = -\frac{1}{2}\left(e^{-4\beta^{\bar{1}}} + e^{-4\beta^{\bar{2}}} + e^{-4\beta^{\bar{3}}}\right). \tag{1.66}$$

1.9 Rotation of Kasner Axes

In spite of the discovery of the laws governing the change of the scale factors a^2, b^2, c^2 between two Kasner epochs made in the previous sections, we still have no complete picture of the transformation of the metric tensor $g_{\alpha\beta}$ during this process. What is missing is the phenomenon of "rotation of the Kasner axes," which is derived in this section.

Using the fact that all the non-diagonal projections of the three-dimensional Ricci tensor can be neglected in the region of transition between two Kasner epochs, we showed that the evolution in this very region can be described simply in terms of the time evolution of the diagonal scale factors. We used one and the same Kasner frame $l^{\bar{a}}_{\alpha}(x^1, x^2, x^3)$ for both epochs. More detailed analysis reveals, however, that while this is indeed valid for the appropriately short transition region between two epochs, the evolution during the whole course of the second epoch in fact does not follow the Kasner-like behavior with the same triad $l^{\bar{\alpha}}_{\alpha}(x^1, x^2, x^3)$. We now show that in addition to the variation of the functions a^2, b^2, c^2, there also occurs a rotation of the Kasner frame.

In the transition region around $t \sim t_c$, the functions b^2 and c^2 are much smaller then a^2, so that we have the relations $a^2 \gg b^2$ and $a^2 \gg c^2$ there. After the transition, b^2 starts to increase and a^2 decreases. This means that as time goes by during the second epoch, a period will unavoidably occur when b^2 gets much bigger than a^2 and c^2, i.e., $b^2 \gg a^2$ and $b^2 \gg c^2$ (recall that the function c^2 monotonically decreases during both epochs). This replacement of the order of magnitudes of the two Kasner scale factors a^2 and b^2 leads to the additional effect of rotation of the Kasner axes. The result, derived explicitly in Appendix A.4, is: if the first epoch has, in accordance with (1.42)–(1.43), the metric

$$g_{\alpha\beta} = \eta_{\bar{\alpha}\bar{\beta}} l^{\bar{\alpha}}_{\alpha} l^{\bar{\beta}}_{\beta} \ , \ \eta_{\bar{\alpha}\bar{\beta}} = diag(a_0^2 t^{2p_{\bar{1}}}, b_0^2 t^{2p_{\bar{2}}}, c_0^2 t^{2p_{\bar{3}}}), \tag{1.67}$$

then the second epoch acquires, in accordance with (1.55)–(1.56), the metric tensor

$$g_{\alpha\beta} = \eta_{\bar{\alpha}\bar{\beta}} \acute{l}^{\bar{\alpha}}_{\alpha} \acute{l}^{\bar{\beta}}_{\beta} \ , \ \eta_{\bar{\alpha}\bar{\beta}} = diag(\acute{a}_0^2 t^{2\acute{p}_{\bar{1}}}, \acute{b}_0^2 t^{2\acute{p}_{\bar{2}}}, \acute{c}_0^2 t^{2\acute{p}_{\bar{3}}}), \tag{1.68}$$

but with new frame vectors $\acute{l}^{\bar{\alpha}}_{\alpha}$ which are related to the old ones $l^{\bar{\alpha}}_{\alpha}$ by the following equations:

$$\acute{l}^{\bar{1}}_{\alpha} = l^{\bar{1}}_{\alpha} \ , \ \acute{l}^{\bar{2}}_{\alpha} = l^{\bar{2}}_{\alpha} + \sigma_{\bar{2}} l^{\bar{1}}_{\alpha} \ , \ \acute{l}^{\bar{3}}_{\alpha} = l^{\bar{3}}_{\alpha} + \sigma_{\bar{3}} l^{\bar{1}}_{\alpha} \ . \tag{1.69}$$

The coefficients $\sigma_{\bar{2}}$ and $\sigma_{\bar{3}}$ can be expressed in terms of the three-dimensional functions appearing in the first epoch and have been calculated exactly in [20] (see also the review paper [21]) for the case when the parameters a_0, b_0, c_0 can be considered as constants. In the general case when a_0, b_0, c_0 are also arbitrary functions of the 3-space coordinates x^{α}, the expressions for $\sigma_{\bar{2}}$ and $\sigma_{\bar{3}}$ are:

$$\begin{aligned} \sigma_{\bar{2}} &= \frac{2}{p_{\bar{2}} + 3p_{\bar{1}}} \left[\left(\frac{p_{\bar{1}}}{\lambda} \right)_{,\mu} l^{\mu}_{\bar{3}} + \frac{2p_{\bar{1}}}{\lambda} \lambda^{\bar{1}}_{\bar{1}\bar{3}} + \frac{p_{\bar{1}}}{\lambda} \left(\ln a_0 b_0 c_0 \right)_{,\mu} l^{\mu}_{\bar{3}} \right], \\ \sigma_{\bar{3}} &= -\frac{2}{p_{\bar{3}} + 3p_{\bar{1}}} \left[\left(\frac{p_{\bar{1}}}{\lambda} \right)_{,\mu} l^{\mu}_{\bar{2}} + \frac{2p_{\bar{1}}}{\lambda} \lambda^{\bar{1}}_{\bar{1}\bar{2}} + \frac{p_{\bar{1}}}{\lambda} \left(\ln a_0 b_0 c_0 \right)_{,\mu} l^{\mu}_{\bar{2}} \right], \end{aligned} \tag{1.70}$$

where the parameters $\lambda^{\bar{1}}_{\bar{1}\bar{3}}, \lambda^{\bar{1}}_{\bar{1}\bar{2}}$ are definite components of the set $\lambda^{\bar{\alpha}}_{\bar{\beta}\bar{\gamma}}$:

$$\lambda^{\bar{\alpha}}_{\bar{\beta}\bar{\gamma}} = \left(l^{\bar{\alpha}}_{\mu,\nu} - l^{\bar{\alpha}}_{\nu,\mu} \right) l^{\mu}_{\bar{\beta}} l^{\nu}_{\bar{\gamma}}. \tag{1.71}$$

For the quantities λ, μ, ν from (1.47), we have now*:

$$\lambda = \lambda^{\bar{1}}_{\bar{2}\bar{3}}, \ \mu = \lambda^{\bar{2}}_{\bar{3}\bar{1}}, \ \nu = \lambda^{\bar{3}}_{\bar{1}\bar{2}}. \tag{1.72}$$

This shows that after the short transition period, the Kasner-like metric (1.67) in the frame $l^{\bar{\alpha}}_{\alpha}$ ceases to be an acceptable approximation to the solution. Nevertheless, if one takes into account that the frame changes from $l^{\bar{\alpha}}_{\alpha}$ to $\acute{l}^{\bar{\alpha}}_{\alpha}$ in accordance with the transformation (1.69)–(1.70), the new metric (1.68) is a

* Note that in [20] the quantities λ, μ, ν have been defined with opposite signs.

valid approximation to the solution during the second epoch. This second epoch is again a standard Kasner-like period of evolution, but during this new period the space expands in the direction $l^{\acute{2}}_{\alpha}$ and contracts in the directions $l^{\acute{1}}_{\alpha}, l^{\acute{3}}_{\alpha}$.

In order to take into account a general transformation of the axes of a frame, one can represent its directional vectors as linear combinations (with time-dependent coefficients) of some fixed reference frame vectors (e.g., of the vectors directed along the Kasner axes of the initial epoch). Then instead of (1.42), we have to write

$$g_{\alpha\beta} = \Gamma_{\bar{\alpha}\bar{\beta}}(t,x)\left[A^{\bar{\alpha}}_{\bar{\mu}}(t,x)l^{\bar{\mu}}_{\alpha}(x)\right]\left[A^{\bar{\beta}}_{\bar{\nu}}(t,x)l^{\bar{\nu}}_{\beta}(x)\right], \quad \Gamma_{\bar{\alpha}\bar{\beta}} = diag(\Gamma_{\bar{1}}, \Gamma_{\bar{2}}, \Gamma_{\bar{3}}), \tag{1.73}$$

where the time-dependent matrix $A^{\bar{\alpha}}_{\bar{\beta}}(t,x)$ is responsible for the rotation effect we are interested in. We continue to call the components of the diagonal matrix $\Gamma_{\bar{\alpha}\bar{\beta}}$ the "scale factors" but now with respect to the rotating frame $A^{\bar{\alpha}}_{\bar{\mu}}(t,x)l^{\bar{\mu}}_{\alpha}(x)$. We denote these new scale factors by the new letters $\Gamma_{\bar{\alpha}}$ to stress that they are different from a^2, b^2, c^2 in (1.42). It is convenient to single out these new scale factors $\Gamma_{\bar{\alpha}}$ explicitly, that is, not to absorb them into the transformation matrix $A^{\bar{\alpha}}_{\bar{\beta}}$. With such a decomposition, we have only three physical degrees of freedom in this matrix, another three degrees of freedom of the metric tensor being included in the scale factors $\Gamma_{\bar{\alpha}}$.

The Kasner axes can be defined in an exact mathematical way for the entire evolution by the condition that both the metric tensor $g_{\alpha\beta}$ and the second fundamental form $\kappa_{\alpha\beta} = \dot{g}_{\alpha\beta}$ are diagonal when expressed in the corresponding frame. It is always possible to find such a global Kasner frame using the gauge freedom of orthogonal rotations of a triad $L^{\bar{\alpha}}_{\alpha}$ in the orthogonal frame representation of the metric $g_{\alpha\beta} = e_{\bar{\alpha}\bar{\beta}}L^{\bar{\alpha}}_{\alpha}L^{\bar{\beta}}_{\beta}$, $e_{\bar{\alpha}\bar{\beta}} = diag(1,1,1)$. Due to this freedom, the frame vectors can be chosen in such a way that $\kappa_{\alpha\beta}$ acquires the form $\kappa_{\alpha\beta} = \rho_{\bar{\alpha}\bar{\beta}}L^{\bar{\alpha}}_{\alpha}L^{\bar{\beta}}_{\beta}$ with $\rho_{\bar{\alpha}\bar{\beta}} = diag(\rho_{\bar{1}}, \rho_{\bar{2}}, \rho_{\bar{3}})$. The equation $\kappa_{\alpha\beta} = \dot{g}_{\alpha\beta}$ gives $\rho_{\bar{\alpha}\bar{\beta}} = e_{\bar{\alpha}\bar{\lambda}}\dot{L}^{\bar{\lambda}}_{\mu}L^{\mu}_{\bar{\beta}} + e_{\bar{\beta}\bar{\lambda}}\dot{L}^{\bar{\lambda}}_{\mu}L^{\mu}_{\bar{\alpha}}$. The three diagonal components of the last relation define the eigenvalues $\rho_{\bar{\alpha}}$ of the second fundamental form. The demand that the three non-diagonal components of $\rho_{\bar{\alpha}\bar{\beta}}$ must vanish gives restrictions on $L^{\bar{\alpha}}_{\alpha}$, which fix the choice of rotation gauge. To apply this procedure to the metric (1.73), we have to identify the orthogonal frame $L^{\bar{\alpha}}_{\alpha}$ as $L^{\bar{\alpha}}_{\alpha} = \gamma^{\bar{\alpha}}_{\bar{\lambda}}A^{\bar{\lambda}}_{\bar{\mu}}l^{\bar{\mu}}_{\alpha}$, where the matrix $\gamma^{\bar{\alpha}}_{\bar{\lambda}}$ is the square root of the matrix $\Gamma_{\bar{\alpha}\bar{\beta}}$,that is $\Gamma_{\bar{\alpha}\bar{\beta}} = e_{\bar{\lambda}\bar{\mu}}\gamma^{\bar{\lambda}}_{\bar{\alpha}}\gamma^{\bar{\mu}}_{\bar{\beta}}$ and $\gamma^{\bar{\alpha}}_{\bar{\beta}} = diag(\sqrt{\Gamma_{\bar{1}}}, \sqrt{\Gamma_{\bar{2}}}, \sqrt{\Gamma_{\bar{3}}})$.

The rotation of the Kasner frames makes their use inconvenient for the analytical description of the asymptotic regime near the singularity because this rotation never stops. Fortunately, it turns out that other frames exist, called "Iwasawa frames," the rotation of which asymptotically goes to zero in the limit $t \to 0$. These frames are also orthogonal so that the metric tensor in such "asymptotically frozen" triads is still a diagonal matrix. The explicit definition of the Iwasawa frames, and a discussion of the freezing effect and of the dynamics of the scale factors in these new frames will be discussed in Chapter 5 in the context of the billiard description of the dynamics. This freezing effect in well chosen non-Kasner frames will be highlighted in Chapter 2.

1.10 Final Comments

The physicist who inspired the beginning of a rigorous analysis of the cosmological singularity was L. D. Landau. In 1959, he formulated the crucial question as to whether the cosmological singularity is a general phenomenon of General Relativity, or whether it appears only in particular solutions constrained by some special symmetries. Landau regarded this question as one of the most important problems of theoretical physics, the other two being the problems of phase transitions and superconductivity.

This fundamental question of the existence of a general solution containing a cosmological singularity was the first and main goal of the BKL team. By exhibiting such a solution, they demonstrated that the singularity was not a consequence of peculiar symmetric initial conditions.

An intuitive feeling that there is no reason to doubt in the existence of a general solution with cosmological singularity was born in the Landau school already in 1964, but another five years had to elapse before the concrete structure was discovered. In 1965 appeared an important theorem of R. Penrose [138], stating that, under special conditions of global character precisely spelled out in the theorem, the appearance of incomplete geodesics in space-time is unavoidable. Geodesic incompleteness is also a form of singularity, but of a different type since in general, geodesic incompleteness does not mean that some curvature invariants diverge (although the opposite is true: the divergence of some curvature invariants imply incompleteness). Penrose theorem establishes geodesic incompleteness but does not provide the analytical structure of the fields near the points where the geodesics terminate. For that reason, the Penrose theorem was not of direct help in the search of the BKL team for a general solution near a cosmological singularity. Nevertheless, this result had some stimulating influence on the Russian research group. Today it is reasonable to consider the BKL analysis and the Penrose theorem as representing two complementary sides of the singularity phenomenon, but the links are still far from being understood. This is because the BKL approach deals with the asymptotics in the vicinity of the singularity while the Penrose theorem has to do with the global structure of space-time.

There are (at least) two misconceptions that the foregoing results are sometimes thought to imply, but which they by no means do.

First one finds sometimes in the literature the belief that the existence of a general solution of Einstein equations with a cosmological singularity means that such a singularity is unavoidable in General Relativity. In fact this conclusion does not follows neither from the BKL theory nor from Penrose theorem. BKL showed that a general solution containing such a singularity exists, but that this solution is general in the sense that initial data under which the cosmological singularity is bound to appear represent a set of nonzero measure in the space of all possible data. However, we do not know "how big" this measure is and we have no idea on how to evaluate that part of the totality of initial conditions which can be covered by this set. In nonlinear systems, many general solutions (i.e., solutions containing the maximal number of arbitrary functional parameters) of different types can exist, and it is not excluded that a general solution without singularity also exists. In fact, there is a proof [36] of the global stability of Minkowski space-time, which means that, at least in some small neighborhood around it, a general solution without any singularity at any time exists. The same is true in relation to all versions of Penrose theorem: for these versions to be applicable, some nontrivial, strictly essential conditions must be satisfied [147, 148], but an infinity of solutions exist which do not meet such requirements. The main reason why we are so interested in the class of general solutions with singularities goes in the end to experiments, since observations show that Nature has chosen a cosmological evolution with an origin of Big Bang type.

Another misconception is that the general solution with a cosmological singularity can be equally applied both to a singularity in the future (Big Crunch) and to a singularity in the past (Big Bang), ignoring the fact that these two situations are quite different physically. To describe the conditions that are prevailing near a cosmological Big Crunch (as well as near the final stage of gravitational collapse of an isolated object in its co-moving system), one really needs a general solution since in the course of evolution, arbitrary perturbations will inevitably arise and these will reorganize any regime into the general one. The Big Bang is not the same phenomenon. We do not know in principle the initial conditions at the singularity, and there is no reason to expect that they should be generic. We cannot rule out the possibility that the Universe in its classical phase started in some very special way (e.g., just the Friedmann solution) and it may be true that this does not imply any fine tuning from the point of view of the still unknown physics near such an exotic state. Of course, the arbitrary perturbations familiar from the present day physics will appear after the Big Bang but this is another story. The conclusion is that if somebody found the general cosmological solution in the framework of Einstein gravity, this does not mean that he knows how the Universe really started. However, he has grounds to think that he knows at least something about its end.

2

Homogeneous Cosmological Models

2.1 Homogeneous Models of Bianchi Types IX and VIII

Homogeneous cosmological models are by definition such that the three-dimensional hypersurfaces $t = const.$ in an appropriate synchronous system are three-dimensional homogeneous spaces of Euclidean signature. The advantage of such models over the general case is that they are mathematically much simpler. But at the same time some of them, namely the models of Bianchi types IX and VIII, exhibit all the key features of the dynamics of the general inhomogeneous case near a cosmological singularity. Historically, the discovery of the existence of a general solution of the gravitational equations with a cosmological singularity and its analytical behavior near a singular point was actually started in the framework of homogeneous models of Bianchi type IX [14]. It is only after these initial developments that the generalization to inhomogeneous space-times was constructed.

In a synchronous system, the four-dimensional interval of homogeneous cosmological models is:

$$- ds^2 = -dt^2 + \eta_{\bar{\alpha}\bar{\beta}}(t) l^{\bar{\alpha}}_{\alpha}(x^\mu) l^{\bar{\beta}}_{\beta}(x^\mu) dx^\alpha dx^\beta, \tag{2.1}$$

where the three-dimensional part of the interval corresponds to a three-dimensional homogeneous space of a definite type. The frame metric $\eta_{\bar{\alpha}\bar{\beta}}$ depends only on time and the frame vectors $l^{\bar{\alpha}}_{\alpha}$ are functions only of the 3-space coordinates x^α. The general theory of three-dimensional homogeneous spaces, their classification due to Bianchi and the Einstein equations for the general homogeneous cosmological models are given in Appendix B. Here we only use those basic points of the theory that are necessary for a self-consistent description of the oscillatory dynamics of the most general homogeneous space-time near a cosmological singularity.

For any homogeneous space, the frame vectors $l^{\bar{\alpha}}_{\alpha}(x^\mu)$ in (2.1) satisfy the conditions:

$$\left(l^{\bar{\alpha}}_{\mu,\nu} - l^{\bar{\alpha}}_{\nu,\mu} \right) l^{\mu}_{\bar{\beta}} l^{\nu}_{\bar{\gamma}} = C^{\bar{\alpha}}_{\bar{\beta}\bar{\gamma}} = const., \tag{2.2}$$

where the structure constants $C^{\bar{\alpha}}_{\bar{\beta}\bar{\gamma}}$ of the corresponding group of isometry of space (group of motions) can be represented in the following way:

$$C^{\bar{\gamma}}_{\bar{\alpha}\bar{\beta}} = \varepsilon_{\bar{\alpha}\bar{\beta}\bar{\lambda}} C^{\bar{\lambda}\bar{\gamma}}. \tag{2.3}$$

Here, $\varepsilon_{\bar{\alpha}\bar{\beta}\bar{\lambda}}$ is the totally antisymmetric three-dimensional Levi-Civita symbol ($\varepsilon_{\bar{1}\bar{2}\bar{3}} = 1$) and $C^{\bar{\alpha}\bar{\beta}}$ are nine arbitrary constants that can equivalently be used instead of the nine independent structure constants $C^{\bar{\gamma}}_{\bar{\alpha}\bar{\beta}}$. In terms of $C^{\bar{\alpha}\bar{\beta}}$ the Jacobi identity for the structure constants takes the form:

$$\varepsilon_{\bar{\alpha}\bar{\beta}\bar{\lambda}} C^{\bar{\beta}\bar{\lambda}} C^{\bar{\alpha}\bar{\gamma}} = 0. \tag{2.4}$$

The case of the three-dimensional homogeneous spaces of Bianchi types IX and VIII corresponds to the case where the matrix $C^{\bar{\alpha}\bar{\beta}}$ can be inverted. Then it follows from the last formula that $C^{\bar{\alpha}\bar{\beta}}$ must be symmetric:

$$C^{\bar{\alpha}\bar{\beta}} = C^{\bar{\beta}\bar{\alpha}}, \;\; \det\left(C^{\bar{\alpha}\bar{\beta}}\right) \neq 0. \tag{2.5}$$

For all other types of homogeneous spaces, the matrix $C^{\bar{\alpha}\bar{\beta}}$ has a more specific structure, namely, its determinant vanishes. In that sense, the Bianchi types VIII and IX are the most general. The difference between types IX and VIII is that for type IX the matrix $C^{\bar{\alpha}\bar{\beta}}$ is positive definite and $\det\left(C^{\bar{\alpha}\bar{\beta}}\right) > 0$, while for type VIII we have $\det\left(C^{\bar{\alpha}\bar{\beta}}\right) < 0$.

2.2 Equations of Motion for Homogeneous Models

The frame components $P^{\bar{\beta}}_{\bar{\alpha}} = P^{\beta}_{\alpha} l^{\alpha}_{\bar{\alpha}} l^{\bar{\beta}}_{\beta}$ of the three-dimensional Ricci tensor P^{β}_{α} corresponding to the homogeneous three-dimensional metric $g_{\alpha\beta} = \eta_{\bar{\alpha}\bar{\beta}} l^{\bar{\alpha}}_{\alpha} l^{\bar{\beta}}_{\beta}$ appearing in (2.1), take the form:

$$\begin{aligned} P^{\bar{\beta}}_{\bar{\alpha}} =& \frac{1}{2\eta}\{4C^{\bar{\beta}\bar{\mu}}C^{\bar{\gamma}\bar{\nu}}\eta_{\bar{\mu}\bar{\gamma}}\eta_{\bar{\nu}\bar{\alpha}} - 2C^{\bar{\beta}\bar{\mu}}C^{\bar{\gamma}\bar{\nu}}\eta_{\bar{\mu}\bar{\alpha}}\eta_{\bar{\nu}\bar{\gamma}} \\ &+ \delta^{\bar{\beta}}_{\bar{\alpha}}\left[\left(C^{\bar{\gamma}\bar{\nu}}\eta_{\bar{\nu}\bar{\gamma}}\right)^2 - 2C^{\bar{\lambda}\bar{\mu}}C^{\bar{\gamma}\bar{\nu}}\eta_{\bar{\mu}\bar{\gamma}}\eta_{\bar{\nu}\bar{\lambda}}\right]\}, \end{aligned} \tag{2.6}$$

where $\eta = \det\left(\eta_{\bar{\alpha}\bar{\beta}}\right)$. The three-dimensional Bianchi identities are:

$$P^{\bar{\beta}}_{\bar{\alpha}} C^{\bar{\alpha}}_{\bar{\beta}\bar{\gamma}} \equiv 0. \tag{2.7}$$

It is a crucial fact that the spatial homogeneity ansatz (2.1) is compatible with the Einstein equations. Because the frame $l^{\bar{\alpha}}_{\alpha}$ does not depend on time, it is easy to project the gravitational equations (1.15)–(1.17) into this frame, which yields the following system of ordinary differential equations in time:

$$\frac{1}{2}\dot{\kappa} + \frac{1}{4}\kappa^{\bar{\alpha}}_{\bar{\beta}}\kappa^{\bar{\beta}}_{\bar{\alpha}} = T^0_0 - \frac{1}{2}(T^0_0 + T^{\bar{\mu}}_{\bar{\mu}}) \,, \tag{2.8}$$

$$\frac{1}{2}\kappa^{\bar{\beta}}_{\bar{\gamma}} C^{\bar{\gamma}}_{\bar{\beta}\bar{\alpha}} = T^0_{\bar{\alpha}} \,, \tag{2.9}$$

$$\frac{1}{2\sqrt{\eta}}(\sqrt{\eta}\kappa^{\bar{\beta}}_{\bar{\alpha}})^{\cdot} + P^{\bar{\beta}}_{\bar{\alpha}} = T^{\bar{\beta}}_{\bar{\alpha}} - \frac{1}{2}\delta^{\bar{\beta}}_{\bar{\alpha}}(T^0_0 + T^{\bar{\mu}}_{\bar{\mu}}) \,. \tag{2.10}$$

Here, $P^{\bar\beta}_{\bar\alpha}$ is given by (2.6), $\kappa_{\bar\alpha\bar\beta} = \dot\eta_{\bar\alpha\bar\beta}$, $\kappa^{\bar\beta}_{\bar\alpha} = \eta^{\bar\beta\bar\gamma}\kappa_{\bar\gamma\bar\alpha}$ and the energy–momentum tensor should have that special structure guaranteeing that all the right-hand sides of equations (2.8)–(2.10) depend also only on time.

We stress that for the description of the dynamics of homogeneous models, one needs to know only the set of structure constants but not the explicit form of the frame vectors $l^{\bar\alpha}_{\alpha}(x^\mu)$. These vectors can be found from the system of partial differential equations (2.2) if the constants $C^{\bar\alpha}_{\bar\beta\bar\gamma}$ are given. An interested reader can find the explicit expressions for the frame vectors $l^{\bar\alpha}_{\alpha}(x^\mu)$ for canonical sets of Bianchi IX and VIII structure constants in Appendix B, Section B.3 on "Frame vectors."

We now turn to the vacuum case and therefore drop the stress–energy tensor in equations (2.8)–(2.10). We first consider the $\binom{\bar\beta}{\bar\alpha}$ and $\binom{0}{0}$ components of the Einstein equations. The $\binom{0}{\bar\alpha}$ components establish connections among the arbitrary constants of integration which appear in the solution of the $\binom{\bar\beta}{\bar\alpha}$ and $\binom{0}{0}$ equations and we postpone their discussion. Discarding, therefore, the right-hand side in equations (2.8) and (2.10) and passing to a new time τ defined by the relation

$$dt = -\sqrt{\eta}d\tau, \tag{2.11}$$

(see (1.63)), we obtain the following system of ordinary differential equations in time:

$$\begin{aligned} 2\eta R^{\bar\beta}_{\bar\alpha} = {} & \frac{d}{d\tau}\left(\eta^{\bar\beta\bar\gamma}\frac{d\eta_{\bar\gamma\bar\alpha}}{d\tau}\right) + 4C^{\bar\beta\bar\mu}C^{\bar\gamma\bar\nu}\eta_{\bar\mu\bar\gamma}\eta_{\bar\nu\bar\alpha} - 2C^{\bar\beta\bar\mu}C^{\bar\gamma\bar\nu}\eta_{\bar\mu\bar\alpha}\eta_{\bar\nu\bar\gamma} \\ & + \delta^{\bar\beta}_{\bar\alpha}\left[\left(C^{\bar\gamma\bar\nu}\eta_{\bar\nu\bar\gamma}\right)^2 - 2C^{\bar\lambda\bar\mu}C^{\bar\gamma\bar\nu}\eta_{\bar\mu\bar\gamma}\eta_{\bar\nu\bar\lambda}\right] = 0\,, \end{aligned} \tag{2.12}$$

$$\begin{aligned} 2\eta\left(R^0_0 - R^{\bar\alpha}_{\bar\alpha}\right) = {} & \frac{1}{2}\eta^{\bar\beta\bar\gamma}\eta^{\bar\alpha\bar\lambda}\frac{d\eta_{\bar\gamma\bar\alpha}}{d\tau}\frac{d\eta_{\bar\lambda\bar\beta}}{d\tau} - \frac{1}{2}\left(\eta^{\bar\alpha\bar\beta}\frac{d\eta_{\bar\alpha\bar\beta}}{d\tau}\right)^2 \\ & + 2C^{\bar\lambda\bar\mu}C^{\bar\gamma\bar\nu}\eta_{\bar\mu\bar\gamma}\eta_{\bar\nu\bar\lambda} - \left(C^{\bar\gamma\bar\nu}\eta_{\bar\nu\bar\gamma}\right)^2 = 0\,. \end{aligned} \tag{2.13}$$

Here we replaced the equation $R^0_0 = 0$ by the equivalent relation $R^0_0 - R^{\bar\alpha}_{\bar\alpha} = 0$, which does not contain the second time derivatives. It is easy to show that the left-hand side of the first-order equation (2.13) represents an exact first integral of the second-order dynamical equations (2.12).

To work with the system (2.12)–(2.13) it is convenient to use matrix notations. The matrices are written in bold face and we follow the rules: *upper indices enumerate the rows and lower indices correspond to the columns. If both indices are arranged on the same line (up or down), then the first index enumerates the rows and the second corresponds to the columns. Tilde over a matrix means transposition.*

We will develop the formalism simultaneously for types VIII and IX in order to show that the differences between the two models get erased as we go to the singularity. To that end, we introduce the constant symmetric matrix $\mathbf{B}$ which is the square root of the matrix $\mathbf{C}$ (with components $C^{\bar{\alpha}\bar{\beta}}$):

$$\mathbf{B}^2 = \mathbf{C}\ ,\ \mathbf{B} = \tilde{\mathbf{B}}\,. \tag{2.14}$$

For the homogeneous spaces of type IX, the matrix $\mathbf{B}$ is real because in this case $\mathbf{C}$ is positive definite, but for the type VIII some of the components of the matrix $\mathbf{B}$ must be complex since for the type VIII the determinant of the matrix $\mathbf{C}$ is negative.

Now, from the frame metric matrix η (with components $\eta_{\bar{\alpha}\bar{\beta}}$) we construct the symmetric matrix $\mathbf{U}$:

$$\mathbf{U} = \mathbf{B}\eta\mathbf{B} \tag{2.15}$$

and we rewrite equations (2.12)–(2.13) in terms of this matrix $\mathbf{U}$. The result is:

$$\frac{d}{d\tau}\left(\mathbf{U}^{-1}\frac{d}{d\tau}\mathbf{U}\right) + 4\mathbf{U}^2 - 2\mathbf{U}Sp\mathbf{U} + \mathbf{I}\left(Sp\mathbf{U}\right)^2 - 2\mathbf{I}Sp\left(\mathbf{U}^2\right) = 0, \tag{2.16}$$

$$\frac{1}{2}Sp\left[\left(\mathbf{U}^{-1}\frac{d}{d\tau}\mathbf{U}\right)^2\right] - \frac{1}{2}\left[Sp\left(\mathbf{U}^{-1}\frac{d}{d\tau}\mathbf{U}\right)\right]^2 + 2Sp\left(\mathbf{U}^2\right) - \left(Sp\mathbf{U}\right)^2 = 0, \tag{2.17}$$

where $\mathbf{I}$ is the unit matrix. It is remarkable that these equations contain only the matrix $\mathbf{U}$. The structure constants do not appear explicitly.

The only trace of the structure of the models of types IX and VIII is the fact that $\mathbf{U}$ is symmetric and regular (i.e., it has a nonzero determinant). It is known that such a matrix can be diagonalized by an *orthogonal* matrix $\mathbf{O}$, that is

$$\mathbf{U} = \tilde{\mathbf{O}}\mathbf{\Gamma}\mathbf{O}\ ,\ \tilde{\mathbf{O}}\mathbf{O} = \mathbf{I}\ , \tag{2.18}$$

where $\mathbf{\Gamma}$ is a diagonal matrix representing the scale factors relative to the rotating frame:

$$\mathbf{\Gamma} = diag(\Gamma_{\bar{1}}, \Gamma_{\bar{2}}, \Gamma_{\bar{3}}). \tag{2.19}$$

The orthogonal matrix $\mathbf{O}$ can be parametrized in the standard way by the three Euler angles ψ, θ, φ, so:

$$\begin{aligned}\mathbf{O} &= \begin{pmatrix} \cos\psi & \sin\psi & 0 \\ -\sin\psi & \cos\psi & 0 \\ 0 & 0 & 1 \end{pmatrix}\begin{pmatrix} 1 & 0 & 0 \\ 0 & \cos\theta & \sin\theta \\ 0 & -\sin\theta & \cos\theta \end{pmatrix}\begin{pmatrix} \cos\varphi & \sin\varphi & 0 \\ -\sin\varphi & \cos\varphi & 0 \\ 0 & 0 & 1 \end{pmatrix} \\ &= \begin{pmatrix} \cos\psi\cos\varphi - \sin\psi\cos\theta\sin\varphi & \cos\psi\sin\varphi + \sin\psi\cos\theta\cos\varphi & \sin\psi\sin\theta \\ -\sin\psi\cos\varphi - \cos\psi\cos\theta\sin\varphi & -\sin\psi\sin\varphi + \cos\psi\cos\theta\cos\varphi & \cos\psi\sin\theta \\ \sin\theta\sin\varphi & -\sin\theta\cos\varphi & \cos\theta \end{pmatrix}\end{aligned} \tag{2.20}$$

We define the matrix $\mathbf{\Omega}$ of "angular velocities" as:

$$\mathbf{\Omega} = \left(\frac{d}{d\tau}\mathbf{O}\right)\mathbf{O}^{-1} = \begin{pmatrix} 0 & \Omega_{\bar{3}} & -\Omega_{\bar{2}} \\ -\Omega_{\bar{3}} & 0 & \Omega_{\bar{1}} \\ \Omega_{\bar{2}} & -\Omega_{\bar{1}} & 0 \end{pmatrix}, \tag{2.21}$$

and from this definition and (2.20) we get:

$$\Omega_{\bar{1}} = \frac{d\varphi}{d\tau}\sin\psi\sin\theta + \frac{d\theta}{d\tau}\cos\psi,$$
$$\Omega_{\bar{2}} = \frac{d\varphi}{d\tau}\cos\psi\sin\theta - \frac{d\theta}{d\tau}\sin\psi,$$
$$\Omega_{\bar{3}} = \frac{d\varphi}{d\tau}\cos\theta + \frac{d\psi}{d\tau}. \tag{2.22}$$

We now substitute the form (2.18) for the matrix $\mathbf{U}$ into the matrix equation (2.16) and multiply the result from the left by the matrix $\mathbf{O}$ and from the right by the matrix $\tilde{\mathbf{O}}$. After that, we split the resulting equation in its antisymmetric and symmetric parts. In this way we derive the following system of two matrix equations which is equivalent to the original matrix equation (2.16):

$$\frac{d}{d\tau}\left(\mathbf{K} - \tilde{\mathbf{K}}\right) = 0, \tag{2.23}$$

$$\mathbf{O}\left[\frac{d}{d\tau}\left(\mathbf{K} + \tilde{\mathbf{K}}\right)\right]\tilde{\mathbf{O}} + 8\mathbf{\Gamma}^2 - 4\mathbf{\Gamma}Sp\mathbf{\Gamma} + 2\mathbf{I}\left(Sp\mathbf{\Gamma}\right)^2 - 4\mathbf{I}Sp\left(\mathbf{\Gamma}^2\right) = 0. \tag{2.24}$$

Here, we have introduced the matrix $\mathbf{K}$ by the relation:

$$\mathbf{K} = \mathbf{U}^{-1}\frac{d}{d\tau}\mathbf{U}\,. \tag{2.25}$$

With this notation the constraint equation (2.17) takes the form:

$$\frac{1}{4}Sp\left(\mathbf{K}^2\right) - \frac{1}{4}\left(Sp\mathbf{K}\right)^2 + Sp\left(\mathbf{\Gamma}^2\right) - \frac{1}{2}\left(Sp\mathbf{\Gamma}\right)^2 = 0\,. \tag{2.26}$$

It can be shown that once the antisymmetric matrix equation (2.23) is satisfied, it is sufficient to fulfill only the three diagonal components of the symmetric matrix equation (2.24). All the non-diagonal components of the symmetric equation (2.24) are automatically solved.

The antisymmetric part (2.23) gives the Euler equations for a freely rotating asymmetric top:

$$\frac{d}{d\tau}\left(I_{\bar{1}}\Omega_{\bar{1}}\right) + \left(I_{\bar{3}} - I_{\bar{2}}\right)\Omega_{\bar{2}}\Omega_{\bar{3}} = 0,$$
$$\frac{d}{d\tau}\left(I_{\bar{2}}\Omega_{\bar{2}}\right) + \left(I_{\bar{1}} - I_{\bar{3}}\right)\Omega_{\bar{1}}\Omega_{\bar{3}} = 0,$$
$$\frac{d}{d\tau}\left(I_{\bar{3}}\Omega_{\bar{3}}\right) + \left(I_{\bar{2}} - I_{\bar{1}}\right)\Omega_{\bar{1}}\Omega_{\bar{2}} = 0, \tag{2.27}$$

where the "moments of inertia" $I_{\bar{\alpha}}$ are:

$$I_{\bar{1}} = \frac{\left(\Gamma_{\bar{2}} - \Gamma_{\bar{3}}\right)^2}{\Gamma_{\bar{2}}\Gamma_{\bar{3}}},\quad I_{\bar{2}} = \frac{\left(\Gamma_{\bar{1}} - \Gamma_{\bar{3}}\right)^2}{\Gamma_{\bar{1}}\Gamma_{\bar{3}}},\quad I_{\bar{3}} = \frac{\left(\Gamma_{\bar{1}} - \Gamma_{\bar{2}}\right)^2}{\Gamma_{\bar{1}}\Gamma_{\bar{2}}}\,. \tag{2.28}$$

Of course, this top is not rigid since the scale factors $\Gamma_{\bar{\alpha}}$ evolve in time in accordance with the three diagonal components of the symmetric matrix equation (2.24). These components are:

$$\frac{d^2 \ln \Gamma_{\bar{1}}}{d\tau^2} + (\Gamma_{\bar{1}})^2 - (\Gamma_{\bar{2}} - \Gamma_{\bar{3}})^2 \tag{2.29}$$
$$-\frac{\Gamma_{\bar{1}}\Gamma_{\bar{2}}(\Gamma_{\bar{1}} + \Gamma_{\bar{2}})}{(\Gamma_{\bar{1}} - \Gamma_{\bar{2}})^3}(I_{\bar{3}}\Omega_{\bar{3}})^2 - \frac{\Gamma_{\bar{1}}\Gamma_{\bar{3}}(\Gamma_{\bar{1}} + \Gamma_{\bar{3}})}{(\Gamma_{\bar{1}} - \Gamma_{\bar{3}})^3}(I_{\bar{2}}\Omega_{\bar{2}})^2 = 0,$$

$$\frac{d^2 \ln \Gamma_{\bar{2}}}{d\tau^2} + (\Gamma_{\bar{2}})^2 - (\Gamma_{\bar{1}} - \Gamma_{\bar{3}})^2 \tag{2.30}$$
$$+\frac{\Gamma_{\bar{1}}\Gamma_{\bar{2}}(\Gamma_{\bar{1}} + \Gamma_{\bar{2}})}{(\Gamma_{\bar{1}} - \Gamma_{\bar{2}})^3}(I_{\bar{3}}\Omega_{\bar{3}})^2 - \frac{\Gamma_{\bar{2}}\Gamma_{\bar{3}}(\Gamma_{\bar{2}} + \Gamma_{\bar{3}})}{(\Gamma_{\bar{2}} - \Gamma_{\bar{3}})^3}(I_{\bar{1}}\Omega_{\bar{1}})^2 = 0,$$

$$\frac{d^2 \ln \Gamma_{\bar{3}}}{d\tau^2} + (\Gamma_{\bar{3}})^2 - (\Gamma_{\bar{1}} - \Gamma_{\bar{2}})^2 \tag{2.31}$$
$$+\frac{\Gamma_{\bar{1}}\Gamma_{\bar{3}}(\Gamma_{\bar{1}} + \Gamma_{\bar{3}})}{(\Gamma_{\bar{1}} - \Gamma_{\bar{3}})^3}(I_{\bar{2}}\Omega_{\bar{2}})^2 + \frac{\Gamma_{\bar{2}}\Gamma_{\bar{3}}(\Gamma_{\bar{2}} + \Gamma_{\bar{3}})}{(\Gamma_{\bar{2}} - \Gamma_{\bar{3}})^3}(I_{\bar{1}}\Omega_{\bar{1}})^2 = 0.$$

Finally, the constraint (2.26) can be written in terms of the scale factors and the "angular velocities" as:

$$\frac{1}{4}\left[\left(\frac{d\ln\Gamma_{\bar{1}}}{d\tau}\right)^2 + \left(\frac{d\ln\Gamma_{\bar{2}}}{d\tau}\right)^2 + \left(\frac{d\ln\Gamma_{\bar{3}}}{d\tau}\right)^2\right]$$
$$-\frac{1}{4}\left[\frac{d\ln\Gamma_{\bar{1}}}{d\tau} + \frac{d\ln\Gamma_{\bar{2}}}{d\tau} + \frac{d\ln\Gamma_{\bar{3}}}{d\tau}\right]^2 \tag{2.32}$$
$$+\frac{1}{2}I_{\bar{1}}(\Omega_{\bar{1}})^2 + \frac{1}{2}I_{\bar{2}}(\Omega_{\bar{2}})^2 + \frac{1}{2}I_{\bar{3}}(\Omega_{\bar{3}})^2$$
$$+\frac{1}{2}\left[(\Gamma_{\bar{1}})^2 + (\Gamma_{\bar{2}})^2 + (\Gamma_{\bar{3}})^2 - 2\Gamma_{\bar{1}}\Gamma_{\bar{2}} - 2\Gamma_{\bar{1}}\Gamma_{\bar{3}} - 2\Gamma_{\bar{2}}\Gamma_{\bar{3}}\right] = 0.$$

The constraint equation (2.32) represents the most important instrument for the qualitative analysis of the dynamics of the models under consideration because it provides the Lagrangian for the equation of motion (2.27)–(2.31). In a sense, the constraint (2.32) (as usual in General Relativity) can be treated as the zero "energy" condition if the sum of all terms in the left-hand side of (2.32) containing the time derivatives are identified with the "kinetic energy" and the sum of all the other terms with the "potential energy." In this way we conclude that the Lagrangian, that is the difference between the "kinetic energy" and the "potential energy," should be:

$$\mathcal{L} = \frac{1}{4}\left[\left(\frac{d\ln\Gamma_{\bar{1}}}{d\tau}\right)^2 + \left(\frac{d\ln\Gamma_{\bar{2}}}{d\tau}\right)^2 + \left(\frac{d\ln\Gamma_{\bar{3}}}{d\tau}\right)^2\right] \tag{2.33}$$
$$-\frac{1}{4}\left[\frac{d\ln\Gamma_{\bar{1}}}{d\tau} + \frac{d\ln\Gamma_{\bar{2}}}{d\tau} + \frac{d\ln\Gamma_{\bar{3}}}{d\tau}\right]^2$$

$$+\frac{1}{2}I_{\bar{1}}\left(\Omega_{\bar{1}}\right)^2+\frac{1}{2}I_{\bar{2}}\left(\Omega_{\bar{2}}\right)^2+\frac{1}{2}I_{\bar{3}}\left(\Omega_{\bar{3}}\right)^2$$
$$-\frac{1}{2}\left[\left(\Gamma_{\bar{1}}\right)^2+\left(\Gamma_{\bar{2}}\right)^2+\left(\Gamma_{\bar{3}}\right)^2-2\Gamma_{\bar{1}}\Gamma_{\bar{2}}-2\Gamma_{\bar{1}}\Gamma_{\bar{3}}-2\Gamma_{\bar{2}}\Gamma_{\bar{3}}\right].$$

The variables $\ln\Gamma_{\bar{1}}$, $\ln\Gamma_{\bar{2}}$, $\ln\Gamma_{\bar{3}}$, ψ, θ, φ are the generalized coordinates and their time derivatives are the generalized velocities.

This Lagrangian (2.33) actually coincides, in the case of the homogeneous models of types IX and VIII under consideration, with the Einstein–Hilbert Lagrangian $\eta\left(R_0^0+R_{\bar{\alpha}}^{\bar{\alpha}}\right)$, after dropping an inessential total derivative and absorbing an unimportant constant proportional to the coordinate volume of the three-dimensional space. It is thus the correct Lagrangian. A direct verification shows that the equations of motion (2.27)–(2.31) indeed follow from (2.33).

More information on the Lagrangian and Hamiltonian formulations, and on the constraints, are given in Chapter 5. The Hamiltonian formulation is indeed the starting point for the general billiard analysis.

2.3 Models of Types IX and VIII with Fixed Kasner Axes

The equations (2.27)–(2.32) for the general Bianchi IX and VIII models can be simplified due to the fact that the Euler equations (2.27) admit three exact integrals of the motion. This is a consequence of the three-dimensional Bianchi identities (2.7) and of the equations (2.10), in which we neglected the energy–momentum tensor. From these relations, it follow that the conservation law $\frac{d}{d\tau}\left[\eta^{\beta\lambda}\left(\frac{d}{d\tau}\eta_{\alpha\lambda}\right)C_{\bar{\beta}\bar{\gamma}}^{\bar{\alpha}}\right]=0$ holds in the time τ. In our matrix notations, this is nothing other than the Euler equations (2.23).

Thus we have $\mathbf{K}-\tilde{\mathbf{K}}=\mathbf{J}$, where $\mathbf{J}$ is some arbitrary constant antisymmetric matrix. Without loss of generality, this arbitrary matrix can be reduced to a form containing only two nonzero entries. Namely, we can represent the conserved integral in the form:

$$\mathbf{K}-\tilde{\mathbf{K}}=\mathbf{U}^{-1}\left(\frac{d}{d\tau}\mathbf{U}\right)-\left(\frac{d}{d\tau}\mathbf{U}\right)\mathbf{U}^{-1}=\begin{pmatrix}0&-J&0\\J&0&0\\0&0&0\end{pmatrix}, \tag{2.34}$$

where J is an arbitrary constant. Such a reduction of the matrix $\mathbf{J}$ is always possible due to the presence of arbitrary non-physical gauge transformation $l_\alpha^{\bar{\alpha}}=A_{\bar{\beta}}^{\bar{\alpha}}\acute{l}_\alpha^{\bar{\beta}}$ (with constant coefficients $A_{\bar{\beta}}^{\bar{\alpha}}$) of the frame vectors in (2.1). If by $\mathbf{A}$ we denote the transformation matrix with components $A_{\bar{\beta}}^{\bar{\alpha}}$, then it is easy to show that a transformation of the form $\mathbf{A}=\mathbf{B}\rho\mathbf{B}^{-1}$, where $\mathbf{B}$ has been introduced in (2.14) and ρ is an arbitrary orthogonal matrix, does not change the structure constants; that is, the new matrix $\acute{C}^{\bar{\alpha}\bar{\beta}}$ calculated from the new vectors $\acute{l}_\alpha^{\bar{\beta}}$ in accordance with the formulas (2.2)–(2.3) coincides with the old one $C^{\bar{\alpha}\bar{\beta}}$. However, this transformation does change the matrix $\mathbf{J}$, which transforms according to the rule $\acute{\mathbf{J}}=\tilde{\rho}\mathbf{J}\rho$.

By an appropriate choice of parameters of the arbitrary orthogonal matrix ρ, we can reduce $\mathbf{J}$ to the particular form shown in (2.34).

Now, in the vacuum case, the $\binom{0}{\alpha}$ components of the Einstein equations imply $\mathbf{K}-\tilde{\mathbf{K}}=\mathbf{0}$, i.e., $\mathbf{J}=\mathbf{0}$. This forces the matrix $\mathbf{\Omega}$ to vanish in the case of unequal $\Gamma_{\bar{\alpha}}$, which one can assume (generic case). Hence, the matrix $\mathbf{O}$ is a constant, which we can set equal to unity ($\psi = 0,\ \theta = 0,\ \varphi = 0$) by using the above ρ-freedom. The rotations of the Kasner axes are absent and their directional vectors coincide with the fixed frame vectors $l^{\bar{\alpha}}_{\alpha}(x)$ from the outset up to the singularity.

In the absence of rotations of the Kasner axes, the equations of motion simplify. With the notations

$$\Gamma_{\bar{1}} = e^{-2\beta^{\bar{1}}},\ \Gamma_{\bar{2}} = e^{-2\beta^{\bar{2}}},\ \Gamma_{\bar{3}} = e^{-2\beta^{\bar{3}}} \tag{2.35}$$

one gets

$$d\tau = -e^{\beta^{\bar{1}}+\beta^{\bar{2}}+\beta^{\bar{3}}}dt. \tag{2.36}$$

(up to a constant that can be absorbed by rescaling of the axes if necessary) and equations (2.29)–(2.32) take the much simpler form:

$$\begin{aligned} 2\frac{d^2\beta^{\bar{1}}}{d\tau^2} &= e^{-4\beta^{\bar{1}}} - \left(e^{-2\beta^{\bar{2}}} - e^{-2\beta^{\bar{3}}}\right)^2, \\ 2\frac{d^2\beta^{\bar{2}}}{d\tau^2} &= e^{-4\beta^{\bar{2}}} - \left(e^{-2\beta^{\bar{3}}} - e^{-2\beta^{\bar{1}}}\right)^2, \\ 2\frac{d^2\beta^{\bar{3}}}{d\tau^2} &= e^{-4\beta^{\bar{3}}} - \left(e^{-2\beta^{\bar{1}}} - e^{-2\beta^{\bar{2}}}\right)^2, \end{aligned} \tag{2.37}$$

and

$$\begin{aligned} &\left(\frac{d\beta^{\bar{1}}}{d\tau}\right)^2 + \left(\frac{d\beta^{\bar{2}}}{d\tau}\right)^2 + \left(\frac{d\beta^{\bar{3}}}{d\tau}\right)^2 - \left(\frac{d\beta^{\bar{1}}}{d\tau} + \frac{d\beta^{\bar{2}}}{d\tau} + \frac{d\beta^{\bar{3}}}{d\tau}\right)^2 \\ &= -\frac{1}{2}\left(e^{-4\beta^{\bar{1}}} + e^{-4\beta^{\bar{2}}} + e^{-4\beta^{\bar{3}}} - 2e^{-2\beta^{\bar{1}}-2\beta^{\bar{2}}} - 2e^{-2\beta^{\bar{1}}-2\beta^{\bar{3}}} - 2e^{-2\beta^{\bar{2}}-2\beta^{\bar{3}}}\right). \end{aligned} \tag{2.38}$$

These are the exact Einstein equations for the so-called "diagonal" models of types IX and VIII in empty space. The term "diagonal" here means that for the canonical sets of Bianchi IX and VIII structure constants (i.e., when the matrix $C^{\bar{\alpha}\bar{\beta}}$ is diagonal, see below), the frame metric $\eta_{\bar{\alpha}\bar{\beta}}$ is diagonal. Furthermore, the Lagrangian for the dynamical equations (2.37) becomes:

$$\begin{aligned} \mathcal{L} = &\left(\frac{d\beta^{\bar{1}}}{d\tau}\right)^2 + \left(\frac{d\beta^{\bar{2}}}{d\tau}\right)^2 + \left(\frac{d\beta^{\bar{3}}}{d\tau}\right)^2 - \left(\frac{d\beta^{\bar{1}}}{d\tau} + \frac{d\beta^{\bar{2}}}{d\tau} + \frac{d\beta^{\bar{3}}}{d\tau}\right)^2 \\ &-\frac{1}{2}\left(e^{-4\beta^{\bar{1}}} + e^{-4\beta^{\bar{2}}} + e^{-4\beta^{\bar{3}}} - 2e^{-2\beta^{\bar{1}}-2\beta^{\bar{2}}} - 2e^{-2\beta^{\bar{1}}-2\beta^{\bar{3}}}\right. \\ &\qquad \left. -2e^{-2\beta^{\bar{2}}-2\beta^{\bar{3}}}\right). \end{aligned} \tag{2.39}$$

The frame metric $\eta_{\bar{\alpha}\bar{\beta}}(t)$ in the line element (2.1) depends on the choice of the set of constants $C^{\bar{\alpha}\bar{\beta}}$. One of the possible choices is the canonical one. For the Bianchi IX model, this canonical set is:

$$C^{\bar{\alpha}\bar{\beta}} = diag\,(1,1,1)\,, \tag{2.40}$$

and the matrix **B** defined by the relation (2.14) is then just the unit matrix. Then from (2.15) and (2.18), it follows that:

$$\eta = diag\,(\Gamma_{\bar{1}}, \Gamma_{\bar{2}}, \Gamma_{\bar{3}})\,, \tag{2.41}$$

where $\Gamma_{\bar{\alpha}}$ are solutions of equations (2.35)–(2.38) and all three $\Gamma_{\bar{\alpha}}$ should be chosen to be positive.

In the case of the Bianchi VIII models, the canonical set for the constants $C^{\bar{\alpha}\bar{\beta}}$ is:

$$C^{\bar{\alpha}\bar{\beta}} = diag\,(1,1,-1)\,. \tag{2.42}$$

For the matrix **B**, we have now **B** $=diag\,(1,1,i)$ and from (2.15) and (2.18), it follows that:

$$\eta = diag\,(\Gamma_{\bar{1}}, \Gamma_{\bar{2}}, -\Gamma_{\bar{3}})\,. \tag{2.43}$$

In this case the solutions $\Gamma_{\bar{\alpha}}$ of equations (2.35)–(2.38) should be chosen in such a way that $\Gamma_{\bar{1}}$ and $\Gamma_{\bar{2}}$ are positive and $\Gamma_{\bar{3}}$ is negative, i.e., the exponent $\beta^{\bar{3}}$ should be taken to have an imaginary part $i\pi/2$. Thus, the only difference between type IX and type VIII lies in the potential terms $e^{-2\beta^{\bar{1}}-2\beta^{\bar{3}}} + e^{-2\beta^{\bar{2}}-2\beta^{\bar{3}}}$, which appear with opposite signs. The imaginary constant in $\beta^{\bar{3}}$ disappears when taking derivatives as well as in $e^{-4\beta^{\bar{3}}}$.

The "mixed" exponentials such as $e^{-2\beta^{\bar{1}}-2\beta^{\bar{2}}}$ are dominated by the "pure" exponentials such as $e^{-4\beta^{\bar{1}}}$ as one goes to the singularity. This is because, for instance, $e^{-2\beta^{\bar{1}}-2\beta^{\bar{2}}} << e^{-4\beta^{\bar{1}}}$ if $\beta^{\bar{2}} >> \beta^{\bar{1}}$ and $e^{-2\beta^{\bar{1}}-2\beta^{\bar{2}}} << e^{-4\beta^{\bar{2}}}$ if $\beta^{\bar{1}} >> \beta^{\bar{2}}$. It is only when $\beta^{\bar{1}}$ and $\beta^{\bar{2}}$ are comparable that the mixed and the pure exponentials are comparable. Otherwise, one of the pure exponentials "wins." In fact, by repeating the argument of the previous section, one can see that in the asymptotic vicinity of the singularity, a pure exponential is always dominant so that all three products $\Gamma_{\bar{1}}\Gamma_{\bar{2}} = e^{-2\beta^{\bar{1}}-2\beta^{\bar{2}}}$, $\Gamma_{\bar{1}}\Gamma_{\bar{3}} = e^{-2\beta^{\bar{1}}-2\beta^{\bar{3}}}$, $\Gamma_{\bar{2}}\Gamma_{\bar{3}} = e^{-2\beta^{\bar{2}}-2\beta^{\bar{3}}}$ in the right-hand side of (2.37)–(2.38) can be neglected.

We thus see that the difference between the curvature terms of types VIII and IX gets erased as we go to the singularity: the leading terms that survive are identical for both. Furthermore, these terms all come with a positive sign, making the potential positive. This is a general phenomenon that we shall encounter again below in more general contexts.

The two models are described in the limit by the same Lagrangian

$$\mathcal{L} = \left(\frac{d\beta^{\bar{1}}}{d\tau}\right)^2 + \left(\frac{d\beta^{\bar{2}}}{d\tau}\right)^2 + \left(\frac{d\beta^{\bar{3}}}{d\tau}\right)^2 - \left(\frac{d\beta^{\bar{1}}}{d\tau} + \frac{d\beta^{\bar{2}}}{d\tau} + \frac{d\beta^{\bar{3}}}{d\tau}\right)^2 - V_0(\beta), \tag{2.44}$$

$$V_0(\beta) = \frac{1}{2}\left(e^{-4\beta^{\bar{1}}} + e^{-4\beta^{\bar{2}}} + e^{-4\beta^{\bar{3}}}\right), \tag{2.45}$$

and the same equations of motion

$$2\frac{d^2\beta^{\bar{1}}}{d\tau^2} = e^{-4\beta^{\bar{1}}} - \left(e^{-4\beta^{\bar{2}}} + e^{-4\beta^{\bar{3}}}\right), \tag{2.46}$$
$$2\frac{d^2\beta^{\bar{2}}}{d\tau^2} = e^{-4\beta^{\bar{2}}} - \left(e^{-4\beta^{\bar{3}}} + e^{-4\beta^{\bar{1}}}\right),$$
$$2\frac{d^2\beta^{\bar{3}}}{d\tau^2} = e^{-4\beta^{\bar{3}}} - \left(e^{-4\beta^{\bar{1}}} + e^{-4\beta^{\bar{2}}}\right),$$

which coincide exactly with the equations (1.65)–(1.66) describing the oscillatory regime near a cosmological singularity which were derived in the previous chapter.

Therefore, near the singularity, the equations (2.37)–(2.38) for the homogeneous models of type IX and VIII describe the same oscillatory asymptotics as in the general inhomogeneous case (but without taking into account the effect of rotations of the Kasner axes). This is one of the reasons why the diagonal homogeneous models of types IX and VIII have received much attention in the literature over recent decades.

We repeat, however, that the "diagonal" models considered in this section, although very useful for illustrating many basic properties of the oscillatory asymptotics near the cosmological singularity, still do not contain all the qualitative features of the general case, since the rotation of the axes are absent. To include this effect as well, we must return to the study of the general models of types IX and VIII that was started in the previous section.

2.4 Models of Types IX and VIII with Rotating Axes

General homogeneous Bianchi type IX or VIII models considered here can exist only in the presence of matter, since otherwise the equations $R_{0\alpha} = 0$ would lead, as we have seen, to the elimination of the rotational angles ψ, θ, ϕ. However, in the asymptotic vicinity of the singularity, which is the region of interest to us, the presence of the standard forms of matter has no decisive importance in the 00 and $\alpha\beta$ components of the Einstein equations (we will discuss some exceptional cases in Chapter 4). In these equations we can neglect in the leading approximation the energy–momentum tensor. In the 0α components, the right-hand side of the gravitational equations is of the same order as the left-hand side, even in the leading approximation, but these equations only put some restrictions on the parameters of the solution without direct influence on the character of its dynamics. In our particular case, including matter (or, in fact, a small amount of inhomogeneity) permits a nonzero integration constant J, and this is how we shall take into account matter – leaving all other equations unchanged. This enables one to mimic in the homogeneous case the general phenomenon of rotations of the Kasner axes and understand its central features.

Substituting into equation (2.34) the matrix **U** in accordance with (2.18) and using expressions (2.21)–(2.22) for the "angular velocities" we obtain the following form of the three integrals of motion:

$$I_{\bar{1}}\Omega_{\bar{1}} = J\sin\theta\sin\psi \ , \ I_{\bar{2}}\Omega_{\bar{2}} = J\sin\theta\cos\psi \ , \ I_{\bar{3}}\Omega_{\bar{3}} = J\cos\theta. \tag{2.47}$$

Equations (2.47) and (2.29)–(2.32), together with definitions (2.28) and (2.22) of the "moments of inertia" and the "angular velocities," constitute a self-consistent and closed system of equations for the six dynamical variables θ, ψ, φ, $\Gamma_{\bar{1}}$, $\Gamma_{\bar{2}}$, $\Gamma_{\bar{3}}$ governing the evolution of the general homogeneous models of types IX and VIII. This system, in spite of the assumption of homogeneity, is still too complicated to be solved analytically. But it turns out that in the asymptotic vicinity of the cosmological singularity, it simplifies drastically. We will from now on restrict our attention on Bianchi IX models.

The model of type VIII needs a more sophisticated analysis because in this case the matrix **B** is complex and some of the Euler angles in the orthogonal matrix **O** should also be complex in order to ensure the reality and physical signature of the frame metric tensor $\eta_{\bar{\alpha}\bar{\beta}}$. For example, for the canonical choice of the Bianchi VIII structure constants (2.42) we have $\mathbf{B} = diag\,(1,1,i)$. It is then easy to see that the Euler angles ψ and φ should remain real but θ must be pure imaginary. To keep the correct metric signature, the third scale factor $\Gamma_{\bar{3}}$ must also be chosen to be negative, as in (2.43). However, a separate study of the asymptotic behavior of the models of type VIII is of no great interest since qualitatively it is the same as for the models of type IX. This is evident if one adopts an approach that avoids the use of an orthogonal matrix for the description of the effect of rotation of the axes. Such an approach, the so-called "triangular" or Iwasawa decomposition of the metric tensor, will be developed later in the billiard context (Chapter 5). It is especially useful for the inhomogeneous generalization of the oscillatory regime near a cosmological singularity.

The three-dimensional metric tensor in (2.1) has the structure:

$$g_{\alpha\beta} = \eta_{\bar{\alpha}\bar{\beta}} l^{\bar{\alpha}}_{\alpha} l^{\bar{\beta}}_{\beta} = \Gamma_{\bar{\alpha}\bar{\beta}} \mathcal{L}^{\bar{\alpha}}_{\alpha} \mathcal{L}^{\bar{\beta}}_{\beta} \ , \ \mathcal{L}^{\bar{\alpha}}_{\alpha} = O^{\bar{\alpha}\bar{\mu}} \left(\mathbf{B}^{-1}\right)_{\bar{\mu}\bar{\nu}} l^{\bar{\nu}}_{\alpha} \ , \tag{2.48}$$

and for Bianchi IX models, the matrix **B** is real, symmetric and positive definite. All three Euler angles of the orthogonal matrix **O** are also real and all three scale factors $\Gamma_{\bar{\alpha}}$ are positive.

The crucial point is that near the singularity the principal axes $\mathcal{L}^{\bar{\alpha}}_{\alpha}$ become fixed with respect to the directions $l^{\bar{\alpha}}_{\alpha}$; that is, the matrix **O** ceases to depend on time. The principal values $\Gamma_{\bar{\alpha}}$ oscillate but they never cross each other: if at some instant they are ranked, for example, in the order $\Gamma_{\bar{1}} > \Gamma_{\bar{2}} > \Gamma_{\bar{3}}$ then this order is maintained from then on, and near the singularity the inequalities grow until

$$\Gamma_{\bar{1}} \gg \Gamma_{\bar{2}} \gg \Gamma_{\bar{3}}, \tag{2.49}$$

that is, the anisotropy of space grows without bound. This inequality means that the ratios $\Gamma_{\bar{2}}/\Gamma_{\bar{1}}$, $\Gamma_{\bar{3}}/\Gamma_{\bar{1}}$ and $\Gamma_{\bar{3}}/\Gamma_{\bar{2}}$ go to zero:

$$\Gamma_{\bar{2}}/\Gamma_{\bar{1}} \to 0,\ \Gamma_{\bar{3}}/\Gamma_{\bar{1}} \to 0,\ \Gamma_{\bar{3}}/\Gamma_{\bar{2}} \to 0. \tag{2.50}$$

From this it follows that all three "moments of inertia" (2.28) tend to infinity. The equations (2.47) show then that all three angular velocities $\Omega_{\bar{a}}$ go to zero near the singularity. This leads to the result that the Euler angles approach three arbitrary constants:

$$(\theta, \varphi, \psi) \to (\theta_0, \varphi_0, \psi_0)\ . \tag{2.51}$$

Indeed, suppose that near the singularity $\tau \to \infty$ the Euler angles tend to the limits (2.51). Let us introduce the notations:

$$\Gamma_{\bar{1}} \equiv A^2,\ \Gamma_{\bar{2}} J^2 \cos^2\theta_0 \equiv B^2,\ \Gamma_{\bar{3}} J^4 \sin^2\theta_0 \cos^2\theta_0 \sin^2\psi_0 \equiv C^2. \tag{2.52}$$

From equations (2.29)–(2.32) and under the asymptotic conditions (2.50), we obtain the following equations, where we have kept only the leading terms:

$$2\frac{\partial^2 \ln A}{\partial\tau^2} = \frac{B^2}{A^2} - A^4, \tag{2.53}$$

$$2\frac{\partial^2 \ln B}{\partial\tau^2} = A^4 - \frac{B^2}{A^2} + \frac{C^2}{B^2}, \tag{2.54}$$

$$2\frac{\partial^2 \ln C}{\partial\tau^2} = A^4 - \frac{C^2}{B^2}, \tag{2.55}$$

$$\left(\frac{\partial \ln A}{\partial\tau}\right)^2 + \left(\frac{\partial \ln B}{\partial\tau}\right)^2 + \left(\frac{\partial \ln C}{\partial\tau}\right)^2 - \left(\frac{\partial \ln A}{\partial\tau} + \frac{\partial \ln B}{\partial\tau} + \frac{\partial \ln C}{\partial\tau}\right)^2 = -\frac{1}{2}\left(A^4 + \frac{B^2}{A^2} + \frac{C^2}{B^2}\right). \tag{2.56}$$

Note that also in this approximation equation (2.56) can be shown to be an exact integral of equations (2.53)–(2.55).

It is now easy to verify (*post factum*) the asymptotics (2.49)–(2.50) by means of an analysis of equations (2.53)–(2.56) of the same type as that used in the previous sections. The solution of this system can be studied by dividing the evolution into epochs during each of which the right-hand sides of equations (2.53)–(2.56) can be neglected. For each epoch, we obtain again the Kasner solution for the functions A^2, B^2, C^2. In the right-hand side of equations (2.53)–(2.56) will again appear "dangerous" terms which will grow in the course of approaching the singularity and which sooner or later will destroy this Kasner regime. Each time there is only one dominating growing term among A^4, B^2/A^2, C^2/B^2, and each time the transition to the new regime can be studied separately under the influence of this term. The solution of equations (2.53)–(2.56), in the right-hand side of which we keep only one of these "dangerous" terms, shows that each of them begins to grow, reaches a maximum, and, after a relatively short time, dies away so that its influence once again becomes small.

If at the initial epoch it is the growing function A^4 that dominates, then the corresponding transition is exactly the same as the one already described in Section 1.7 by equations (1.52) and (1.53) (in which, in order to see the coincidence, we should take $\lambda = 1$ and pass to the time τ). We call this already known transition a *gravitational transformation* or, in the billiard context developed in Part II of this book, a *reflection in the gravitational walls* (see Chapter 5).

If, on the other hand, the growing dominant term is B^2/A^2 (or C^2/B^2) we will have a new type of transition which is called a *centrifugal transformation*, or a *reflection in the centrifugal walls* (see Chapter 5). Again, the term B^2/A^2 (or C^2/B^2) begins to grow, reaches a maximum, and after some time dies away. Indeed, consider equations (2.53)–(2.56) with only B^2/A^2 on the right-hand side (the action of the term C^2/B^2 has exactly the same character as that of B^2/A^2). The exact general solution of the system in this case can be written in the form:

$$\frac{B^2}{A^2} = \frac{Q^2(p_A - p_B)^2}{\cosh^2\left[Q(p_A - p_B)\left(\tau - \tau_0\right) - \frac{1}{2}\ln\frac{4Q^2(p_A - p_B)^2 A_0^2}{B_0^2}\right]}, \tag{2.57}$$

$$A^2B^2 = A_0^2B_0^2 e^{-2Q(p_A+p_B)(\tau-\tau_0)}, \quad C^2 = C_0^2 e^{-2Qp_C(\tau-\tau_0)},$$

where p_A, p_B, p_C are the standard Kasner exponents and A_0, B_0, C_0, τ_0 are arbitrary constants. Consequently, the total number of the arbitrary parameters in the solution is five, which is exactly the number that is necessary for the solution of equations (2.53)–(2.56) to be general. The constant Q is chosen in the following special way:

$$Q = kA_0B_0C_0, \; k = J^3(\det \mathbf{C})^{-1/2}\sin\theta_0\cos^2\theta_0\sin\psi_0\,. \tag{2.58}$$

This choice for Q and the way we distribute the constants in the solution (2.57) in no way violates its generality, and is dictated only by the desire to get the standard expressions for the Kasner asymptotics of the solution in synchronous time t before and after the transition.

It is obvious that under the condition $p_A - p_B > 0$ with τ increasing B^2/A^2 at first grows, reaching a maximum

$$\left(\frac{B^2}{A^2}\right)_{\max} = Q^2(p_A - p_B)^2 \tag{2.59}$$

at the critical point where the hyperbolic cosine in the denominator of (2.57) becomes equal to unity, and after that begins to decay. It is easy to show (using the connection between τ and t following from formulas (2.11), (2.14), (2.15), (2.18) and (2.52)) that before the transition ($\tau \to -\infty$, $t \to \infty$) the asymptotic form of the solution, transformed to the synchronous time t, is:

$$\left(A^2, B^2, C^2\right) = \left(A_0^2 t^{2p_A}, B_0^2 t^{2p_B}, C_0^2 t^{2p_C}\right), \tag{2.60}$$

while after the transition ($\tau \to \infty$, $t \to 0$), we have

$$\left(A^2, B^2, C^2\right) = \left(\acute{A}_0^2 t^{2\acute{p}_A}, \acute{B}_0^2 t^{2\acute{p}_B}, \acute{C}_0^2 t^{2\acute{p}_C}\right), \tag{2.61}$$

where

$$\acute{A}_0^2 = \frac{B_0^2}{4Q^2 (p_A - p_B)^2}, \ \acute{B}_0^2 = 4A_0^2 Q^2 (p_A - p_B)^2, \ \acute{C}_0^2 = C_0^2 \tag{2.62}$$

and

$$\acute{p}_A = p_B, \ \acute{p}_B = p_A, \ \acute{p}_C = p_C. \tag{2.63}$$

We see that this centrifugal transition conserves the product of the pre-exponential coefficients of the Kasner regimes, that is $\acute{A}_0^2 \acute{B}_0^2 \acute{C}_0^2 = A_0^2 B_0^2 C_0^2$, and the numerical values of the exponents do not change. Two of the three Kasner exponents simply get exchanged. A description of the transformation under the influence of C^2/B^2 is obtained from the above formulas by the following permutation of the components $(A^2, B^2, C^2) \to (B^2, C^2, A^2)$.

We find an endless succession of Kasner regimes between each of which there is one of two types of transformations, one due to the growth of the perturbation A^4 and another due to the perturbation B^2/A^2 or C^2/B^2. Now it is easy to see that as we approach the singularity ($\tau \to \infty$, $t \to 0$) all three quantities $A^4, B^2/A^2, C^2/B^2$ go to zero because their maxima go to zero. Actually, the longer a given stage of evolution, the smaller the value of $A_0 B_0 C_0$ in the last epoch of the stage (each gravitational transformation makes the product $A_0 B_0 C_0$ smaller by an amount given by (1.57)). From this it follows that the maxima (2.59) and (1.58) go to zero. We can ignore the quantity $(p_A - p_B)^2$ and $|p_{\bar{1}}|$ in these maxima because, during the entire evolution, each Kasner exponent is bounded from above. The foregoing arguments proves the asymptotic inequalities (2.49)–(2.50).

As for the limit (2.51) for the Euler angles, we remark that equations (2.47) can be written as

$$\frac{\partial \varphi}{\partial \tau} \sin\theta \sin\psi + \frac{\partial \theta}{\partial \tau} \cos\psi = \frac{J\Gamma_{\bar{2}}\Gamma_{\bar{3}} \sin\theta \sin\psi}{(\Gamma_{\bar{2}} - \Gamma_{\bar{3}})^2}, \tag{2.64}$$

$$\frac{\partial \varphi}{\partial \tau} \sin\theta \cos\psi - \frac{\partial \theta}{\partial \tau} \sin\psi = \frac{J\Gamma_{\bar{1}}\Gamma_{\bar{3}} \sin\theta \cos\psi}{(\Gamma_{\bar{1}} - \Gamma_{\bar{3}})^2}, \tag{2.65}$$

$$\frac{\partial \varphi}{\partial \tau} \cos\theta + \frac{\partial \psi}{\partial \tau} = \frac{J\Gamma_{\bar{1}}\Gamma_{\bar{2}} \cos\theta}{(\Gamma_{\bar{1}} - \Gamma_{\bar{2}})^2}, \tag{2.66}$$

and can be used to find the corrections $\delta\theta, \delta\varphi, \delta\psi$ to the first approximation $(\theta, \varphi, \psi) = (\theta_0, \varphi_0, \psi_0)$. Writing

$$\theta = \theta_0 + \delta\theta, \quad \varphi = \varphi_0 + \delta\varphi, \quad \psi = \psi_0 + \delta\psi, \tag{2.67}$$

it follows from (2.64)–(2.66) that in the first approximation:

$$\frac{\partial}{\partial \tau}\delta\theta = \frac{C^2}{B^2} \frac{\cos\psi_0}{J \sin\theta_0 \sin\psi_0}, \tag{2.68}$$

$$\frac{\partial}{\partial \tau}\delta\varphi = \frac{C^2}{B^2} \frac{1}{J \sin^2\theta_0}, \tag{2.69}$$

$$\frac{\partial}{\partial \tau}\delta\psi = \left(\frac{B^2}{A^2} \frac{1}{\cos^2\theta_0} - \frac{C^2}{B^2} \frac{1}{\sin^2\theta_0} \right) \frac{\cos\theta_0}{J}. \tag{2.70}$$

It can now be confirmed that the fact that all three derivatives $\frac{\partial\psi}{\partial\tau}, \frac{\partial\theta}{\partial\tau}, \frac{\partial\varphi}{\partial\tau}$ go to zero in the limit $\tau \to \infty$ (due to the inequalities (2.49)–(2.50)) actually leads to the existence of the limits given in (2.51) for the angles ψ, θ, φ. The corresponding detailed analysis appears in Appendix B, Section B.4. The phenomenon of cessation of rotation in the asymptotic vicinity to the singularity is called *freezing*.

The oscillatory asymptotics following from equations (2.53)–(2.56) can be described as endless oscillations of a particle against the walls of some potential in the three-dimensional β-space. These equations, after introducing the variables analogous to (1.64):

$$A = \exp(-2\beta^{\bar{1}}),\ B = \exp(-2\beta^{\bar{2}}),\ C = \exp(-2\beta^{\bar{3}}), \tag{2.71}$$

become

$$2\frac{\partial^2\beta^{\bar{1}}}{\partial\tau^2} = e^{-4\beta^{\bar{1}}} - e^{-2(\beta^{\bar{2}}-\beta^{\bar{1}})}, \tag{2.72}$$
$$2\frac{\partial^2\beta^{\bar{2}}}{\partial\tau^2} = -e^{-4\beta^{\bar{1}}} + e^{-2(\beta^{\bar{2}}-\beta^{\bar{1}})} - e^{-2(\beta^{\bar{3}}-\beta^{\bar{2}})},$$
$$2\frac{\partial^2\beta^{\bar{3}}}{\partial\tau^2} = -e^{-4\beta^{\bar{1}}} + e^{-2(\beta^{\bar{3}}-\beta^{\bar{2}})},$$

$$\left(\frac{\partial\beta^{\bar{1}}}{\partial\tau}\right)^2 + \left(\frac{\partial\beta^{\bar{2}}}{\partial\tau}\right)^2 + \left(\frac{\partial\beta^{\bar{3}}}{\partial\tau}\right)^2 - \left(\frac{\partial\beta^{\bar{1}}}{\partial\tau} + \frac{\partial\beta^{\bar{2}}}{\partial\tau} + \frac{\partial\beta^{\bar{3}}}{\partial\tau}\right)^2 \tag{2.73}$$
$$= -\frac{1}{2}\left[e^{-4\beta^{\bar{1}}} + e^{-2(\beta^{\bar{2}}-\beta^{\bar{1}})} + e^{-2(\beta^{\bar{3}}-\beta^{\bar{2}})}\right].$$

It is remarkable that these equations are exactly the same as the equations emerging at each spatial point in the inhomogeneous case, confirming the invaluable guiding role that the Bianchi IX model plays.

Let us mention that equations (2.53), (2.54), (2.55) and (2.56) were investigated in [43, 141] using the theory of dynamical systems.

From the above equations a billiard description can be derived. This was actually done first in the diagonal Bianchi IX case by D. Chitre and C. Misner [35, 130]. The effect of rotation of the axes and collisions due to the centrifugal potentials, which is triggered by any small amount of inhomogeneity or generic matter, was however not taken into account in the billiard model constructed in these pioneering works. The asymptotic behavior near the singularity for the general non-diagonal homogeneous Bianchi IX model exhibiting also these phenomena was obtained by V. Belinski, I. Khalatnikov and M. Ryan [22].

2.5 On the Extension to the Inhomogeneous Case

The extension of the analysis of homogeneous cosmological models to the general inhomogeneous case has been done mainly through the billiard description, which will be the subject of Part II. It is however worth mentioning that different approaches to the inhomogeneous case have been developed in [153, 83, 152] and in [6, 7].

In the first three papers, the so-called "Hubble-normalized variables" were introduced. The physical quantities used by BKL are singular in the limit $t = 0$. This needs a careful analytical approach and, moreover, creates a real problem for numerical simulations. The Hubble-normalized variables introduced by Uggla et al. in [153, 83, 152] remain finite at the singularity. This permits more efficient numerical computations. Furthermore, these variables make it possible to effectively apply the qualitative methods of dynamical systems theory.

In the two papers [6, 7], Ashtekar-like variables are introduced in the Hamiltonian formalism, which are better behaved as one approaches the singularity. These variables enable one to formulate the asymptotic equations for the general inhomogeneous case in a simpler dynamical system form.

All three of these approaches show that, also in the general inhomogeneous case, the qualitative character of the oscillatory cosmological singularity in the leading approximation remains the same as in homogeneous models.

3

On the Cosmological Chaos

3.1 Stochasticity of the Oscillatory Regime

In the asymptotic vicinity of the singularity, the oscillatory regime described above acquires a stochastic (chaotic) character.

Of course, any system of differential equations, no matter how complicated it is, has one and only one definite solution for each set of initial data. From this point of view we have no place for stochasticity. When we are speaking about stochastic behavior of the solution of differential equations we mean a quite different situation, namely the case when the initial data are not known exactly but are distributed in accordance to some statistical law. Together with the initial data, all characteristics of the solution at any subsequent time will also be distributed in some way.

In the framework of such a statistical approach there are two essentially different types of behavior. The first corresponds to the non-chaotic systems for which the initial distribution being picked around some points in the phase space generates an evolution which remains picked around the corresponding trajectories (i.e., usual classical solutions) starting from these points. The systems of the second type are chaotic in the sense that, for them, any initial distribution is going to spread out over the whole phase space in the course of the evolution, independently of how sharply localized the distribution was at the initial instant (some authors say that such systems have a source of stochasticity). Moreover, in the chaotic case, the final asymptotic regime which arises at a late time has a universal character completely independent of the form of the initial distribution. We stress that it is this last property (known as *mixing*) that represents the basic feature of chaos. For complicated nonlinear systems of differential equations, the chaotic behavior (if it is present) should be considered as a rather useful property because it gives the qualitative character of the general solution. Indeed for such a system only approximate forms of the general solution can be obtained and all parameters in such forms have no exact sense; they are unavoidably uncertain and can be interpreted as statistically distributed. However, the chaotic property makes the uncertainty of the final asymptotics universal, that is, dictated by the system itself and not by artificial inaccuracy we bring into the system through our approximations. This permits one to obtain some understanding of the proper asymptotic behavior of the system.

In our case, the chaotic character of the oscillatory regime near the cosmological singularity can already be seen from the analysis of the evolution of Kasner exponents or, equivalently, of the evolution of parameter u [123].

Consider an infinite sequence of positive numbers u which is divided into ordered series (corresponding to eras) with the following properties: every s-th series starts with $u^{(s)} = n^{(s)} + x^{(s)}$ (where $n^{(s)}$ are integer numbers and $x^{(s)} < 1$) and represents the chain $n^{(s)}+x^{(s)}$, $n^{(s)}+x^{(s)}-1$, $n^{(s)}+x^{(s)}-2, \ldots$ The transition from the series s to the next, $s+1$, is governed by the rule

$$u^{(s+1)} = 1/x^{(s)} = n^{(s+1)} + x^{(s+1)}, \tag{3.1}$$

where $n^{(s+1)}$ is the integer part of the number $1/x^{(s)}$ and $x^{(s+1)}$ is its fractional part.

Let us consider, instead of a definite initial value $u^{(1)} = n^{(1)} + x^{(1)}$, a distribution of $x^{(1)}$s over the interval $0 < x^{(1)} < 1$ governed by a given probability distribution $w_1(x^{(1)})$. Then $w_1(x^{(1)})dx^{(1)}$ is the probability that $x^{(1)}$ lies in the interval $(x^{(1)}, x^{(1)} + dx^{(1)})$. All numbers $x^{(s)}$ which define the ends of eras will also be distributed over the intervals $0 < x^{(s)} < 1$ with some probabilities $w_s(x^{(s)})$. The probability that $x^{(s+1)}$ lies in the interval $(x^{(s+1)}, x^{(s+1)}+dx^{(s+1)})$ is $w_{s+1}(x^{(s+1)})dx^{(s+1)}$. This probability should be equal to $w_s(x^{(s)})dx^{(s)}$, i.e., to the probability that in the preceding era s we find in the interval $(x^{(s)}, x^{(s)} + dx^{(s)})$ that value $x^{(s)}$ which generates the value $x^{(s+1)}$ in the interval $(x^{(s+1)}, x^{(s+1)} + dx^{(s+1)})$ at the end of the era $s+1$. Because the numbers $x^{(s)}$ and $x^{(s+1)}$ are connected by the relation (3.1) we have the following recurrence formula

$$w_{s+1}(x^{(s+1)})dx^{(s+1)} = \sum_{n=1}^{\infty} w_s\left(\frac{1}{n + x^{(s+1)}}\right)\left|d\left[\frac{1}{n + x^{(s+1)}}\right]\right|. \tag{3.2}$$

The sum over n appear due to the fact that we do not know $x^{(s)}$ of the preceding era exactly and we should sum all possibilities which produce any integer number $n^{(s+1)}$ in the era $s+1$ but with the same value of $x^{(s+1)}$. From the formula (3.2) we have the recurrence functional equation for the probability distributions w_s:

$$w_{s+1}(x) = \sum_{n=1}^{\infty} \frac{1}{(n+x)^2} w_s\left(\frac{1}{n+x}\right). \tag{3.3}$$

The important fact is that in the limit $s \to \infty$, the distributions $w_s(x)$ tend to the stationary distribution $w(x)$ independent of s and satisfying the equation:

$$w(x) = \sum_{n=1}^{\infty} \frac{1}{(n+x)^2}\, w\left(\frac{1}{n+x}\right). \tag{3.4}$$

The solution of this equation (normalized to unity) is:

$$w(x) = \frac{1}{(1+x)\ln 2}. \tag{3.5}$$

We see that in the vicinity of the singularity (i.e., in the limit $s \to \infty$) the process of interchange of the eras indeed acquires a stochastic character; that is, any detail of the initial distribution of the Kasner exponents is completely forgotten.

The map (3.1) was introduced and analyzed by Carl F. Gauss around 200 years ago in relation to his study of continued fractions and number theory. He obtained also the distribution (3.5) but never published its derivation [114]. The rigorous derivation was constructed only in 1928 by the Russian mathematician R. O. Kuzmin [119], and in the contemporary literature, this result is often quoted as the Gauss–Kuzmin theorem.

From this result it is easy to get the corresponding stationary distribution $W(n)$ for the lengths of the Kasner eras measured by the number n of Kasner epochs contained in them. In order that the length of the $s+1$ series be $n^{(s+1)}$, the last value $x^{(s)}$ of the parameter u in the preceding series must lie in the interval between $1/\left[n^{(s+1)}+1\right]$ and $1/n^{(s+1)}$. The probability W_{s+1} of such an event is

$$W_{s+1}(n^{(s+1)}) = \int_{1/\left[n^{(s+1)}+1\right]}^{1/n^{(s+1)}} w_s(x^{(s)})dx^{(s)}. \tag{3.6}$$

In the stationary limit $s \to \infty$, we obtain, upon substitution of the limiting distribution (3.5):

$$W(n) = \frac{1}{\ln 2} \ln \frac{(n+1)^2}{n(n+2)}. \tag{3.7}$$

There is a subtlety with this result due to the rather slow decrease of $W(n)$ at large values of n ($W \sim n^{-2}$). For example, the average value of n, calculated in accordance with the distribution (3.7), diverges. To construct stable statistical distributions and sensible mean values for the physical quantities (such as amplitudes of the scale factors a^2, b^2, c^2), it is necessary to pass to their logarithms and to the logarithmic time [123] (see also [19]).

The fact that it is the Gauss map (3.1) itself that is the source of stochasticity in the cosmological oscillatory regime near the singularity was indicated first in [123]. In this paper the statistical distributions and mean values for other characteristics of the regime were also found and the way how one should interpret them was proposed [the question of physical interpretation is important because, as we already mentioned, the distributions that follow from the Gauss–Kuzmin formula (3.5) are unstable (the fluctuations around the mean values are not small)]. The results reported in the paper [123] were obtained with the help of some approximations, the validity of which was not evident at first glance. The confirmation of their validity and the construction of the corresponding exact approach was made later in the work [113]. An analogous analysis of the statistical properties of the oscillatory cosmological singularity was presented also in the papers ([34, 8]). A further and more mathematically-oriented study of the stochasticity of the oscillatory regime can be found in the articles [37, 38].

3.2 Historical Remarks

After the discovery of the stochastic character of the oscillatory regime made in [123], a vast literature appeared dedicated to further developments of the theory of chaos near the cosmological singularity. The theory of cosmological chaos today represents a wide branch of mathematical physics, and its complete detailed exposition is beyond the scope of our book. We stress below only the main steps of these developments.

In 1972 D. Chitre [35] established that the Bianchi IX system is equivalent to a billiard on the Lobachevsky plane. It is well known that this kind of system has stochastic properties (see our Chapter 5, Section 5.9).

In 1973 some important results appeared from O. Bogoiavlenskii and S. Novikov [28], who, using the qualitative theory of differential equations, i.e., the theory of dynamical systems, discovered the existence in the Bianchi IX phase space of an attractor consisting of a dense set of periodic trajectories. Today we know that this is one of the characteristic features of "strange attractors," meaning (but not always, however) the presence of chaos.

In 1982 J. Barrow [8] showed that to the stochastic evolution of the parameter u, which was described in the preceding section, one can attribute a positive Kolmogorov–Sinai entropy. He calculated this entropy and proved that it is not zero. Since the Kolmogorov–Sinai entropy can be defined as the sum of the positive Lyapunov exponents, his result means that at least one of these exponents is positive, which is enough for the appearance of stochasticity.

One year later A. Zardecki [156] confirmed Barrow's result. By numerical calculations, A. Zardecki showed that the main Lyapunov exponent is indeed positive.

In 1983 D. Chernoff and J. Barrow [34] studied the stochastic aspects of Bianchi IX models in general and not just for the parameter u. They considered the complete set of Bianchi IX phase space variables (scale factors a, b, c and their first derivatives $\dot{a}, \dot{b}, \dot{c}$) and separated the non-stochastic monotonous degrees of freedom (such as the determinant of the metric tensor) from those two that are stochastic. These stochastic variables are the parameter u and the ratio $\ln b/\ln c$. For these two variables, they obtained a two-dimensional stochastic map that could be reduced to the known Baker's map, which had been much studied previously and which possesses Kolmogorov–Sinai and topological entropy, ergodicity and mixing. For this two-dimensional map, they also found the asymptotic stationary distribution.

In 1985 I. M. Khalatnikov, E. M. Lifshitz, Ya. G. Sinai and their collaborators K. M. Khanin and L. N. Shchur [113] completed the construction of the statistical theory of the Bianchi IX oscillatory regime, the basic aspects of which had been worked out in [123]. One of the main results of the paper [123] (the stable distributions for the scale factors a^2, b^2, c^2 and their stable mean values) was obtained only in an approximate way. In [113] the exact solution of this problem was given.

Then a long period (1988–1995) of debates occurred, during which different authors expressed doubts in the statement that the Bianchi IX dynamics is really chaotic (for a review, see the book [92]). At that time, some authors even expressed the opinion that Bianchi IX models can in fact be of the type of an integrable system. Such conjecture was erroneous. The discussion had its origin in the fact that the Liapunov exponents are not invariant under time reparameterizations. Some authors (see, e.g., [37, 38, 96]) indicated a way to bypass this difficulty or how to define these exponents in a general covariant way (on these matters, see the review paper by S. E. Rugh [145]). However, it should be emphasized that this problem turns out to be not so significant for the essence of the phenomenon. In fact, Liapunov exponents do not constitute an appropriate instrument in General Relativity because of time reparametrizations. But this does not mean that one cannot identify a chaotic regime because many others of its symptoms can be found. The best demonstration of the validity of this statement is given in the works of N. Cornish and J. Levin, to which we now turn.

In 1997 the final persuasive confirmation of the chaotic behavior of the Bianchi IX dynamics was made by N. Cornish and J. Levin [37, 38], who used the two-dimensional Chernoff–Barrow map and showed the existence of an attractor of fractal dimension, i.e., a strange attractor. This is the same attractor that had been identified earlier by O. Bogoiavlenski and S. Novikov, but these authors did not qualify it at that time as a fractal set. N. Cornish and J. Levin also calculated the topological entropy, constructed a symbolic dynamics (codification of trajectories by the words and phrases of an alphabet) and showed that the resulting discrete map represented a valid approximation for the real continuous evolution.

The conclusion that the oscillatory cosmological evolution exhibits stochastic behavior can also be reached from its billiard version described in Chapter 5 and from the wide literature dedicated to the different types of hyperbolic billiards. Geodesic motion in a billiard in hyperbolic space has indeed been much studied. It is known that this motion is chaotic or non-chaotic depending on whether the billiard has finite or infinite volume. In the finite volume case, the generic evolution exhibits an infinite number of collisions with the walls with strong chaotic features ("oscillating behavior"). By contrast, if the billiard has infinite volume, the evolution is non-chaotic. In this case there are only finitely many collisions with the walls for a generic evolution. The system generically settles after a finite time in a Kasner-like motion that lasts all the way to the singularity [3, 52, 63, 62].

The facts that we will establish in Chapter 5 for the billiard associated with vacuum gravity in four space-time dimensions are: (i) the billiard table is a region of the two-dimensional space of constant negative curvature bounded by straight lines, (ii) the volume of the billiard region is finite, (iii) the two-dimensional trajectories of the billiard ball between reflections against the cushions are geodesics, (iv) the reflections are of the mirror type, that is the angles of incidence and reflections are the same. It is well known that the motion under these four conditions is strongly chaotic (including mixing). This conclusion follows from the works [129, 94, 149, 158, 72, 150] but the pioneering studies of this field belong to E. Hopf [93] and D. V. Anosov [4]. All these results are in agreement with the chaotic character of the oscillatory cosmological regime derived solely on the basis of the evolution of the Kasner exponents as we explained in the first section of this chapter.

3.3 Gravitational Turbulence

In general, the typical wave numbers k of the three-dimensional Fourier harmonics of the fields approaching the singularity grow in time due to the endless production of perturbations of smaller and smaller characteristic lengths. It is known that, in a nonlinear system with an infinite number of degrees of freedom, an oscillatory regime is unstable with respect to partial decay into oscillations of smaller scales: among small perturbations with an arbitrary spectrum, there are always, in general, some whose amplitudes grow, drawing energy from the large scale processes. As a result, one finds a complex picture of multiscale motions with a certain energy distribution and with an exchange of energy among the oscillations of various scales. The only case in which excitations with arbitrary small wavelengths do not arise is the one in which the physical conditions of the problem rule out an unlimited growth of small-scale excitations. For this to happen, there must be a natural physical length which determines the minimum scale at which energy is extracted from the system of the dynamic degrees of freedom. Such is the case, for example, for a sufficiently viscous fluid. For a system such as the (classical) gravitational field in vacuum, however, there is no internal physical scale, so there is nothing to prevent the growth of oscillations of arbitrarily small wavelength.

This conclusion, which follows from general considerations, also finds support in a direct quantitative analysis of the cosmological evolution, again with the help of its representation as an alternation of Kasner epochs and an infinite number of changes of the Kasner exponents. It can be shown that the spatial gradients of the exponents increase rapidly and without bound as the singularity is approached. The gradients of all the other components of the metric also increase, because they are coupled with the exponents through the field equations. This observation was made by A. Kirillov and A. Kochnev [116] and independently by G. Montani [134]. This phenomenon was discussed later in [11]. G. Montani called it "fragmentation" (alternatively, it is a "formation of cells" in the terminology of A. Kirillov and A. Kochnev). If we do not trace the correspondence between the Kasner exponents and the Kasner axes (this correspondence is unimportant at the moment), and if we focus exclusively on the evolution of the three numerical quantities that give a triplet of exponents in each epoch in some order, then this effect can be easily seen by examining the evolution of the parameter u, in terms of which the Kasner exponents are expressed by the formulas (1.26). As u varies from 1 to ∞, the numerical values of the exponents take all possible values in the intervals indicated in (1.27).

We already know that values of u less than unity contribute nothing new, since the quantities p_α are invariant (within an interchange of no interest here) under the transformation $u \to 1/u$. The evolution of the parameter u is described by the infinite sequence $u_1, u_1 - 1, \ldots, x_1 \to 1/x_1 \equiv u_2, u_2 - 1, \ldots, x_2 \to 1/x_2 \equiv u_3, u_3 - 1, \ldots$. In homogeneous models, this evolution is the same for the entire space; that is, the Kasner exponents do not depend on the spatial coordinates. In general, the initial value of the parameter u differs from point to point. This circumstance leads to an unbounded growth of the spatial derivatives of the metric in the course of the evolution, no matter how close the initial distribution $u_1(x, y, z)$ of the parameter u is to a uniform distribution.

To make this point clear, let us consider the case in which the function $u_1(x, y, z)$ is initially given as a continuous function throughout space, taking once and only once each value between 1 and ∞ (we assume that the latter value is reached at spatial infinity). In the initial epoch, the entire three-dimensional space then constitutes a single natural "cell," in which the range of the exponents p_α spans the intervals (1.27) completely and only once (let this be the definition of a "cell"). In the next epoch a new spatial region arises. In this region the parameter u varies between 0 and 1, and (after the substitution $u \to 1/u$) it varies between 1 and ∞. Outside this region, throughout the rest of the space, the parameter u varies between 1 and ∞, as before. Thus we have now two cells, instead of only one as in the preceding epoch. In the course of the transition to the third epoch, the same mechanism leads to a further splitting of each of the two cells into two new ones, etc. As the result of the infinite number of transitions, the number of cells goes off to infinity, and the size of the cells approaches zero. Since the exponents within each cell take all possible values, their spatial gradients tend to infinity. This process demonstrates the tendency of a gravitational field to go toward an unbounded self-excitation of progressively smaller-scale oscillations as the cosmological singularity is approached.

The evolution of a single excitation with an arbitrarily large wave number does not by itself pose any problem for the BKL approach. The description of the key element of the oscillatory regime – the alternation of Kasner epochs – remains valid for any length scale. The problem is that this approach might seem incapable of dealing with the consequences of the interaction of a large number of excitations, with a broad spectrum of possible wave numbers. By virtue of the nonlinearity of the theory, this interaction plays a dominant role, and it could happen that the process cannot be thought of as a superposition of oscillations with different length scales, each described by the BKL mechanism.

To clarify these questions, we must consider the nature of the migration of energy through the spectrum. The situation in this regard is simplified by the circumstance that a collapsing gravitational field has a specific effect: a continuous pumping of energy into a system of oscillatory degrees of freedom. The pumping rate itself increases without bound as the singularity is approached. To see that this is the case, one needs to single out the determinant of the metric as a common multiplier in the metric tensor matrix in the Einstein equations in a synchronous system, while the other components (namely, a matrix with unit determinant) should be assigned to the oscillatory degrees of freedom. We then find that the equations describe the evolution of the oscillatory components under the influence of a continuously increasing flux of energy from the exterior as the singularity is approached. This increase is a consequence of the Landau–Raichoudhury theorem regarding the monotonic tendency of the determinant to go toward zero. From the physical standpoint, this effect is of the same nature as the conversion of the potential energy of a system of collapsing particles into the kinetic energy of their relative motion. Since we know that there is no special length scale, this influx of energy solves the problem of the energy flow through the spectrum in a first approximation: the continuous pumping of energy toward infinitely small scales should be dominant in this energy flow. Under these conditions, a restoration is improbable, and it may be a continuous excitation of progressively new small-scale modes which determines the behavior of the system. As a result, the asymptotic state of the field near the singularity might be called an "infinitely developing gravitational turbulence."

Consequently, one might a priori worry that the effective representation of the spatial gradients by terms that are ultralocal in space can be self-contradictory. In fact this is not the case, since the diminution of the characteristic lengths goes slower than the decrease of the horizon, so that in the limit $t \to 0$ the product tk also goes to zero. This means that near the singularity the gravitational turbulence develops slower than the aspiration of the system to its singular state.

A closer examination of this question was performed in [51], where it was shown that, in spite of the increase of the spatial gradients, the conclusions stand on a firm basis. This is an important point since the increase of the spatial gradients, if it were too violent, would certainly work against the validity of an effective ultralocal description involving only the time derivatives of the scale factors. To address the question of the consistency of the BKL derivation, one must verify that the asymptotic growth of the spatial gradients of the quantities that enter the prefactors of the dominant powers of t in the differential equations of Sections 1.6 and 1.7, does not violate the estimates that were made there.

This was achieved in [51] using the billiard picture which is explained in Chapter 5. We refer to that work for the details. The conclusion is that the potentially dangerous prefactors of the dominant powers of t can be self-consistently considered to evolve very little near the singularity so that the unlimited growth of some of the spatial gradients, even though quite intricate, does not affect the consistency of the BKL analysis.

4

On the Influence of Matter and Space-Time Dimension

4.1 Introduction

In the preceding sections we have established the existence of a general cosmological solution of the gravitational equations with a singularity with respect to time, and have studied the asymptotic properties of this solution near the singularity. We saw that in this limit the solution may be described by an infinite alternation of Kasner epochs. The notion of a Kasner epoch and the transition between two of these are the key elements in the dynamics of the oscillatory regime.

We have derived and studied these properties in the example of empty space in four space-time dimensions. In this chapter, we investigate how the analysis can accommodate matter sources, as well as a change in the space-time dimension.

We will successively consider the following systems. First, we consider perfect fluids in four space-time dimensions. Next, we consider gauge fields of the Yang-Mills and electromagnetic types and scalar fields, also in four space-time dimensions. Then, we discuss pure gravity in higher dimensions. Finally, we outline some results on the character of the cosmological singularity in presence of a viscous matter source. In Appendix C, we solve the case of a commuting spinor field. This case is mathematically interesting but its physical significance is not clear (because the spinors are treated as classical c-numbers).

We shall see that in certain important cases the influence of matter upon the solution in the vicinity of the singularity appears to be qualitatively negligible. This is, for instance, true for the case of standard perfect fluid. For other (less simple) types of the energy–momentum tensor, the influence of matter proves to be more important, but it can still be analyzed analytically with sufficient clarity and along the same general lines. We will also find the surprising result that in space-time dimensions equal to or higher than 11, $D \geq 11$, gravity near a cosmological singularity ceases to be chaotic [63, 62].

We follow in this chapter the original BKL approach. The same analysis can be performed using the billiard description. It is given in Chapter 6; see in particular Section 6.6.

In the BKL approach, one treats the problem of the influence of matter as follows: one first determines the behavior of matter during one Kasner epoch; one then finds out the way in which the succession of Kasner epochs takes place. With this method, one automatically determines whether the Kasner epochs and their alternation continue to exist in the presence of matter.

For this purpose, let us again formulate the conditions of applicability of the approximation leading to the so-called Kasner epoch. It is evident from (1.15) and (1.17) that in the limit $t \to 0$ and in the presence of matter it is sufficient to add the following requirements to the applicability conditions (1.44) of the solution (1.42)–(1.43):

$$T_0^0 - \frac{1}{2}\left(T_0^0 + T_{\bar{\mu}}^{\bar{\mu}}\right) \ll t^{-2}, \tag{4.1}$$

$$T_{\bar{\beta}}^{\bar{\alpha}} - \frac{1}{2}\left(T_0^0 + T_{\bar{\mu}}^{\bar{\mu}}\right) \ll t^{-2}, \; for \; \bar{\alpha} = \bar{\beta}, \tag{4.2}$$

$$T_{\bar{1}\bar{2}} \ll abt^{-2}, \; T_{\bar{1}\bar{3}} \ll act^{-2}, \; T_{\bar{2}\bar{3}} \ll bct^{-2}, \tag{4.3}$$

where $T_{\bar{\beta}}^{\bar{\alpha}} = l_{\alpha}^{\bar{\alpha}} l_{\bar{\beta}}^{\beta} T_{\beta}^{\alpha}$ and $T_{\bar{\alpha}\bar{\beta}} = l_{\bar{\alpha}}^{\alpha} l_{\bar{\beta}}^{\beta} T_{\alpha\beta}$. These requirements mean that, in the region where they are fulfilled, matter may be "inserted" into the vacuum solution (1.42)–(1.43) with all the three-dimensional arbitrary functions necessary for matching arbitrary initial data. In other words, in this region, the distribution and motion of matter are determined only by its equations of motion in the given gravitational field (1.42)–(1.43) and the back-reaction of the energy–momentum tensor on the gravitational field is negligible.

Depending on the form of the energy–momentum tensor, we may a priori encounter three different possibilities: (1) the oscillatory regime remains as it is in vacuum, i.e., the influence of matter may be completely ignored; (2) the presence of matter eliminates the oscillatory behavior because there exists now a region in parameter space in which the monotonic power-law behavior is not destabilized by "dangerous terms"; (3) alternating Kasner epochs in which one spatial direction expands while the other two contract exist as before, but matter strongly affects the process of their formation and alternation.

Actually, each of these possibilities can be realized. Before entering their detailed analysis, let us mention in turn one example of (1), (2) and (3). The underlying computations are provided in the subsequent sections.

(1) The first case, where matter can be "inserted" into the vacuum solution without changing the oscillatory regime to leading approximation, is realized in a space filled with a perfect fluid with the equation of state $p = q\varepsilon$ for $0 \leqslant q < 1$. For the ultrarelativistic equation of state ($q = 1/3$), this fact was proved in [124] in the particular case of the generalized Kasner solution (i.e., the solution which has Kasner asymptotics up to the singularity) and then confirmed in [19, 21] for the process of alternation of Kasner epochs in the course of the oscillatory evolution of a general solution. In [18] it was shown that the same

result remains valid for all values of the constant q within the above-indicated interval.

Nevertheless, it is worth stressing that some delicate details appear for the hydrodynamic flows around the value $q = 2/3$. Then it turns out that the character of the motion of matter for $0 \leqslant q < 2/3$ remains the same as in the case of the ultrarelativistic equation of state $q = 1/3$. But for $2/3 < q < 1$, there is a difference between the periods of evolution during which the higher positive Kasner exponent $p_{\bar{3}}$ is greater than q and those periods where $p_{\bar{3}}$ is smaller than q. For $p_{\bar{3}} > q$, the time component of the 4-velocity of the fluid and the square of its spatial components (i.e., the quantities u^0 and $u^\alpha u_\alpha$) increase as the singularity is approached, and the motion of matter again tends to become the same as for the ultrarelativistic equation of state. This motion may be called relativistic since it corresponds to velocities close to the velocity of light (when $u^\alpha u_\alpha \sim \left(u^0\right)^2 \gg 1$). If the period of evolution under study is sufficiently long-lasting or if, at its start, the quantities u^0 and $u^\alpha u_\alpha$ are not small, then the motion of matter does actually achieve this relativistic stage. In those periods during which $p_{\bar{3}} < q$, the component u^0 and the square of $u^\alpha u_\alpha$ decrease as the singularity is approached. Depending on the duration and the relative number of the periods when the motion of matter slows down ($u^\alpha u_\alpha \ll 1$ and $u^0 \approx 1$), we shall encounter stages of evolution during which matter must be "inserted" into the solution (1.42)–(1.43) in another manner (a nonrelativistic manner).

(2) If $q = 1$ we have the "stiff matter" equation of state $p = \varepsilon$ proposed by Ya. B. Zeldovich [157]. This is one of the above-mentioned possibilities when both standard Kasner epochs with one expanding direction and the oscillatory regime disappear in the vicinity of a singular point. This case has been investigated in [16, 18], where it was shown that the condition (4.1) cannot be satisfied, since in the neighborhood of a singular point the right-hand side $T_0^0 - \frac{1}{2}\left(T_0^0 + T_{\bar{\mu}}^{\bar{\mu}}\right)$ of equation (1.15) behaves as t^{-2}, i.e., as its left-hand side (the remaining conditions (4.2) and (4.3) are still fulfilled). The influence of the "stiff matter" (equivalent, as was pointed out in [16], to a massless scalar field) results in the violation of the Kasner relation (1.22) for the asymptotic exponents. Then, thanks to the contribution of the right-hand side of equation (1.15), we now have

$$p_{\bar{1}} + p_{\bar{2}} + p_{\bar{3}} = 1, \quad p_{\bar{1}}^2 + p_{\bar{2}}^2 + p_{\bar{3}}^2 = 1 - p_\varphi^2 \,, \tag{4.4}$$

where p_φ^2 is an arbitrary three-dimensional function subject to the restriction $p_\varphi^2 < 1$, to which the energy density of the matter is proportional. In the particular case when the stiff-matter source is realized as a massless scalar field φ, its asymptotic behavior is $\varphi = p_\varphi \ln t$ and this is the formal reason why we use the index φ for the additional exponent p_φ. Thanks to (4.4) and in contrast to the Kasner relations, it is possible for all three Kasner exponents $p_{\bar{\alpha}}$ to be positive. In [16] it has been shown that, even if the contraction of space starts with a Kasner epoch during which one of the exponents $p_{\bar{\alpha}}$ is negative, the asymptotic behavior (1.42)–(1.43) with positive exponents is inevitably established after a finite number of oscillations. Once a regime with positive Kasner exponents is reached, it remains unchanged up to the singularity. Thus, for the equation of state $p = \varepsilon$, the asymptotical behavior to the cosmological singularity is described by a monotonic (but anisotropic) contraction of space along all directions, rather than by the chaotic oscillatory regime.

Reference [18] investigated the general case of stiff-matter perfect fluid and constructed a *general cosmological solution* in the vicinity of the singularity involving eight arbitrary functions of 3-space. For this solution, it was possible not only to derive the form of the main approximation, but also to calculate the subsequent terms of the expansion in powers of time. Such a solution is also of interest from the methodological point of view, since here we deal with the general cosmological solution with a complete set of arbitrary three-dimensional functions for which there exist *explicit analytical expressions* in the vicinity of the singularity. Thus, here we can avoid a number of questions which remain qualitative to a large extent in the description of the properties of the solution with the oscillatory regime, for which no explicit analytic form exists even in the vicinity of the singularity. More on this point in Sections 4.3 and 4.7.

In connection with the analysis of the stiff matter case, we would like to point out that a similar phenomenon, i.e., the disappearance of oscillations and the transition of the general cosmological solution from the oscillatory regime to a monotonic power-law asymptotic behavior in the vicinity of the singularity, was also observed in a space filled with a classical Dirac spinor field, as it was shown in [17]. However, this field was treated in a pure classical way and can be understood only as a one-particle wave function. For this reason, its physical significance is somewhat unclear and it is relegated to Appendix C. Nevertheless, it is worth remarking that its influence near the singularity is as crucial as that of a scalar field; that is, it changes the asymptotics of the general solution from the oscillatory regime to the smooth power law behavior with all three Kasner-like indices positive, as in the case of a perfect fluid with the stiff matter equation of state. This illustrates again possibility (2).

It is important to stress that the disappearance of oscillations for the case of a massless scalar (or spinor) field is a phenomenon specific to this field and can be unstable with respect to inclusion into the right-hand side of the Einstein equations of other kinds of fields. For instance, in the same paper [16], it was shown that if we add a vector to the scalar field, then the endless oscillations reappear. This question is examined in the billiard context in Subsections 7.6.2 and 7.6.3.

(3) The cosmological evolution in the presence of an electromagnetic field may serve as an example of the third possibility. In this case the oscillatory regime in the presence of matter is, as usual, described by an alternation of Kasner epochs, but in this process the energy–momentum tensor plays a role as important as the three-dimensional curvature tensor. This problem has been treated in [17], where it has been shown that in the diagonal projections of the energy–momentum

tensor of the electromagnetic field (as in the components $P_{\bar{\alpha}}^{\bar{\beta}}$ for $\bar{\alpha} = \bar{\beta}$), there appear terms in the course of a Kasner epoch that sooner or later violate the conditions (4.2), i.e., the conditions for the existence of the given epoch. Non-diagonal projections of the energy–momentum tensor are, as before, unimportant in the transition region (i.e., where one of the inequalities (4.2) is violated), and the alternation of epochs qualitatively takes place according to the same laws as in vacuum. Depending on the relative magnitude of the three-dimensional space curvature and the electromagnetic field strength, the alternation of epochs may be caused by the energy-momentum tensor of the electromagnetic field or by the three-dimensional curvature tensor of the gravitational field.

In [18] the problem of the influence of the Yang–Mills fields on the character of the cosmological singularity was also studied. The analysis was restricted to fields corresponding to the gauge group SU(2). The study was performed in a synchronous reference system in the "temporal gauge" where the time components of all three vector fields are equal to zero. It was shown that, in the neighborhood of a cosmological singularity, the behavior of the Yang–Mills fields is similar to the behavior of the electromagnetic field: as in that case, the oscillatory regime described by an alternation of Kasner epochs emerges and is caused either by the three-dimensional curvature or by the energy–momentum tensor. If, in the process of alternation of epochs, it is the energy–momentum tensor of the gauge fields that is dominating, the qualitative behavior of the solution in those epochs and in the corresponding transition region between them is similar to the behavior valid for free Yang–Mills fields (i.e., with abelian group). This does not mean that the nonlinear interaction terms can be neglected completely, but the latter introduce only minor quantitative changes into the picture observed for non-interacting fields. The reason for this lies in the absence of time derivatives of the gauge field strengths in those terms of the equations of motion which describe the interaction. In the neighborhood of the singularity, the dominant role in the equations of motion of matter is played by the terms involving the time derivatives.

4.2 Perfect Fluid

In the vicinity of the cosmological singularity, the duration of the Kasner epochs depends on how small, at the start of each epoch, are those diagonal frame components of the three-dimensional curvature tensor $P_{\bar{\alpha}}^{\bar{\beta}}$ that increase afterward faster than t^{-2} as the time decreases. If the function a^2 increases, an epoch will be the longer, the smaller is the initial value of the first quantity among (1.45):

$$\frac{a^4}{2a^2b^2c^2}\left[\left(l^{\bar{1}}_{\alpha,\beta} - l^{\bar{1}}_{\beta,\alpha}\right) l^{\alpha}_{\bar{2}} l^{\beta}_{\bar{3}}\right]^2 . \tag{4.5}$$

To understand how matter with the hydrodynamic energy–momentum tensor is inserted into the oscillatory regime, let us first study the behavior of a perfect fluid in one Kasner epoch, assuming that this epoch is sufficiently long-lasting, i.e., that at its start the quantity (4.5) is sufficiently small. In this case we may estimate the orders of magnitude of all the quantities, taking into account only that the value of t is small. We can also formally assume that the values of the three-dimensional arbitrary functions and their derivatives which enter the solution are of order one. These conditions just simplify the analysis. The conclusions derived under these assumptions are identical to those obtained through a more thorough investigation.

For a perfect fluid with energy–momentum tensor $T_i^k = (\varepsilon + p)\, u^k u_i + p\delta_i^k$ and with equation of state $p = q\varepsilon$, the right-hand sides of the gravitational equations (1.15)–(1.17) are:

$$T_0^0 - \frac{1}{2}(T_0^0 + T_\mu^\mu) = (1+q)\,\varepsilon u^0 u_0 + \frac{1}{2}\,(1-q)\,\varepsilon\ , \tag{4.6}$$

$$T_\alpha^0 = (1+q)\,\varepsilon u^0 u_\alpha\ , \tag{4.7}$$

$$T_\alpha^\beta - \frac{1}{2}\delta_\alpha^\beta(T_0^0 + T_\mu^\mu) = (1+q)\,\varepsilon u^\beta u_\alpha + \frac{1}{2}\delta_\alpha^\beta\,(1-q)\,\varepsilon\ , \tag{4.8}$$

and the equations of hydrodynamics can be written as

$$\left(\sqrt{g}\varepsilon^{1/(1+q)} u^0\right)^{\cdot} = -\left(\sqrt{g}\varepsilon^{1/(1+q)} u^\alpha\right)_{,\alpha}\ , \tag{4.9}$$

$$\varepsilon^{1/(1+q)} u^0 \left(\varepsilon^{q/(1+q)} u_\alpha\right)^{\cdot} = -\frac{q}{1+q}\varepsilon_{,\alpha} - \frac{q}{1+q} u_\alpha u^\mu \varepsilon_{,\mu} - \varepsilon u^\mu u_{\alpha,\mu} + \frac{1}{2}\varepsilon u^\mu u^\nu g_{\mu\nu,\alpha}\ . \tag{4.10}$$

Let the epoch in question be described by the solution (1.42)–(1.43). According to the spirit of our approach, we shall assume that the main terms in equations (4.9)–(4.10) are the terms with time derivatives. Then, in the first approximation, the solution of these equations is

$$\sqrt{g}\varepsilon^{1/(1+q)} u^0 = c^0\ ,\quad \varepsilon^{q/(1+q)} u_\alpha = c_\alpha\ , \tag{4.11}$$

where c^0 and c_α are arbitrary three-dimensional functions. Besides, we must necessarily take into account the condition $u^k u_k = -1$, which in the synchronous system (where $u_0 = -u^0$) takes the form

$$\left(u^0\right)^2 = 1 + u^\alpha u_\alpha\ . \tag{4.12}$$

What follows depends on the value of the squared velocity $u^\alpha u_\alpha$. If in the given epoch it is large, we may neglect the "1" in the relation (4.12), and then from (1.42)–(1.43) as well as from (4.11)–(4.12) we obtain the following evolution laws for these quantities in the asymptotic region $t \to 0$:

$$u_\alpha \sim t^{\frac{q(1-p_3)}{1-q}}\ ,\quad u^0 \sim t^{\frac{q-p_3}{1-q}}\ ,\quad \varepsilon \sim t^{\frac{(1+q)(p_3-1)}{1-q}}\ . \tag{4.13}$$

Here p_3 is the largest of the three Kasner exponents, varying in the range $2/3 \leqslant p_3 \leqslant 1$. Equations (4.13) are obtained under the assumption that $u^\alpha u_\alpha \gg 1$ or $u^0 \gg 1$, which is the same. Hence, they are actually correct for

$$p_{\bar{3}}\left(x^{\alpha}\right) > q \,, \tag{4.14}$$

since only in this case will the value of u^0 be large at sufficiently small t. Thus the laws (4.13) will be valid in any epoch and everywhere in the three-dimensional space if $q < 2/3$, since in this case the inequality (4.14) is everywhere automatically fulfilled.

If $q > 2/3$, then together with those epochs and regions of the three-dimensional space where the condition (4.14) is still satisfied, there may appear other Kasner epochs and individual regions of the space where the exponent $p_{\bar{3}}$ is smaller than the parameter q. Let us assume that the epoch and the three-dimensional region we are considering are such that $p_{\bar{3}} < q$. It is clear that in that case, the squared velocity $u^{\alpha}u_{\alpha}$ is decreasing at $t \to 0$ and, if the epoch is sufficiently long-lasting, will become in time much smaller than unity. Then it follows from (4.12) that $u^0 \sim 1$, and from (1.42)–(1.43) and (4.11) we get

$$u_{\alpha} \sim t^{q} \,, \quad u^0 \approx 1 \,, \quad \varepsilon \sim t^{-(1+q)} \,, \quad p_{\bar{3}}\left(x^{\alpha}\right) < q \,. \tag{4.15}$$

To confirm the estimates (4.13)–(4.15) we must make sure that the right-hand sides of the hydrodynamic equations (4.9)–(4.10), omitted earlier, are actually small in comparison with the preserved left-hand sides, which involve time derivatives. Besides, it must be checked that the back-reaction of matter upon the metric may also be ignored, i.e., that the inequalities (4.1)–(4.3) are indeed verified. It is evident that both these requirements are satisfied if as $t \to 0$, the two inequalities:

$$\varepsilon u_{\alpha} u^{\beta} \ll t^{-2} \,, \quad \varepsilon \ll t^{-2} \tag{4.16}$$

hold. It is not difficult to verify that, for the case of (4.13)–(4.14) as well as for the asymptotic behavior (4.15), all the conditions (4.16) are fulfilled.

The above analysis, performed for a sufficiently long-lasting Kasner epoch, cannot provide a detailed description of the motion of matter for the oscillatory regime. Yet, this analysis allows one to draw the necessary basic conclusions. Note that, in the oscillatory regime, as the singularity is being approached, a typical state of the metric is the Kasner epoch, during which one of the functions a^2, b^2, c^2 is much smaller than the others. This function decreases in accordance with the fastest law $t^{2p_{\bar{3}}}$ where $p_{\bar{3}}$ is the biggest Kasner exponent. Such epochs largely comprise the whole evolution of the solution in the neighborhood of a singularity.

If the equation of state of the fluid corresponds to the case $q < 2/3$, then in practically all epochs and everywhere in space, the squared velocity $u^{\alpha}u_{\alpha}$ (or u^0) will increase, and in a small vicinity of the singularity the time behavior of the hydrodynamic quantities in the epochs will be in agreement with the laws (4.13). Thus, the values of $\varepsilon u_{\alpha} u^{\beta}$ and ε increase more slowly than t^{-2}, and throughout the evolution the inequalities (4.1)–(4.3) are fulfilled, i.e., matter is inserted into the oscillatory regime with a negligible back influence on the metric.

If $q > 2/3$, the behavior of matter throughout the evolution will not be as simple as in the former case. As has already been pointed out, the numerical values of the Kasner exponents change as the Kasner epochs alternate. Therefore, on approaching a singularity, we shall encounter epochs where the biggest exponent $p_{\bar{3}}$ will be larger than q, as well as epochs where it will be smaller than q. Besides, the relative values of $p_{\bar{3}}(x^\alpha)$ and of the parameter q may also vary from point to point in the three-dimensional space, even within the framework of one epoch. For this reason, we cannot state that the component of u^0 mainly increases as one approaches the singularity. During the cosmological collapse, we shall observe both periods of relativistic motion ($u^\alpha u_\alpha \gg 1$) and nonrelativistic motion of matter ($u^\alpha u_\alpha \ll 1$). The laws governing the behavior of matter in the various epochs must then be sought taking into account the three-dimensional functions that appear in the solution and determine (together with the values of t) the initial data for each epoch. Thus, if the period in which the squared velocity $u^\alpha u_\alpha$ is increasing (when $p_{\bar{3}} > q$) is replaced by a period in which the squared velocity is decreasing (when $p_{\bar{3}} < q$), the evolution of the hydrodynamic quantities in the latter period will still be governed by the relativistic laws (4.13), but not by (4.15), provided the initial value of the square $u^\alpha u_\alpha$ is sufficiently large at the beginning of the period in which it is decreasing. (In this case, the decreasing value of $u^\alpha u_\alpha$ in (4.12) may still be greater than unity.) This, however, relates to specific details of the behavior of the fluid in the process of the oscillatory regime. There are no grounds to doubt that the back-reaction of matter upon the metric is still negligible for $2/3 < q < 1$, since in this case also the increase in the values of ε and $\varepsilon u_\alpha u^\beta$ is slower than t^{-2}.

Hence, if the parameter q lies in the interval $0 \leqslant q < 1$, the influence of matter on the oscillatory regime may be neglected, i.e., the right-hand sides of equations (1.15) and (1.17) may be omitted. This cannot be done in equation (1.16), since the right-hand side $T^0_\alpha \sim \varepsilon u^0 u_\alpha$ and the left-hand side are of the same order, namely $1/\sqrt{g}$, as it is evident from (4.11). In the general case, this circumstance affects only the relations imposed upon the arbitrary three-dimensional functions contained in the solution ("initial data"), and does not influence the dynamical character of the cosmological evolution, which is qualitatively the same as in vacuum.

To avoid misunderstanding, let us stress that the above assertion refers to the general inhomogeneous case. It does not rule out the possibility of the existence of special classes of solutions in which some properties of their dynamics may be largely dependent on the presence of matter.

A homogeneous cosmological model may serve as an example of such a situation. In vacuum, the metric of the model of Bianchi VII type may be taken in the diagonal form of (1.42) with vectors $l^{\bar{a}}_\alpha$ that are fixed once and forever, and taken such that the three defining structure constants (1.47) are $\lambda \neq 0, \mu \neq 0$ but $\nu = 0$ and all other structure constants (2.2) are zero. In this case the oscillatory regime, as is known, will terminate in the epoch where the function c^2 starts increasing. (For $\nu = 0$ the three-dimensional curvature tensor contains no terms capable of stopping this increase and of replacing this epoch by another where c^2 would decrease.) As a result the number of oscillations in this vacuum model appears to be finite and the asymptotic behavior in the vicinity of the singularity turns out to be purely Kasner.

However, when constructing a type VII model for a space containing moving matter ($u_\alpha \neq 0$), we cannot preserve the diagonality of the metric in the fixed triad $l^{\bar{\alpha}}_{\alpha}$; this means that the Kasner axes $L^{\bar{\alpha}}_{\alpha}$ (where the metric is diagonal by definition) do not coincide then with $l^{\bar{\alpha}}_{\alpha}$ but rotate with respect to them from epoch to epoch according to the laws we explained in Section 2.4 on models with rotation of the Kasner axes. Thanks to these rotations, all the three quantities $\left(L^{\bar{1}}_{\alpha,\beta} - L^{\bar{1}}_{\beta,\alpha}\right) L^{\alpha}_{\bar{2}} L^{\beta}_{\bar{3}}$, $\left(L^{\bar{2}}_{\alpha,\beta} - L^{\bar{2}}_{\beta,\alpha}\right) L^{\alpha}_{\bar{3}} L^{\beta}_{\bar{1}}$ and $\left(L^{\bar{3}}_{\alpha,\beta} - L^{\bar{3}}_{\beta,\alpha}\right) L^{\alpha}_{\bar{1}} L^{\beta}_{\bar{2}}$ that play the essential role in the appearance of oscillations, will differ in general from zero throughout the evolution and, consequently, the oscillatory regime will last indefinitely up to the singularity. This phenomenon may be studied also in the fixed triad $l^{\bar{\alpha}}_{\alpha}$ but with non-diagonal projections of the metric. In such a case, in the asymptotic vicinity of the singularity, the solution admits the qualitative description given in Section 2.4. Non-diagonality of the metric creates supplementary potential barriers, so that provided one among the quantities λ, μ, ν differs from zero, the point (the motion of which is described by the solution) indefinitely oscillates in a potential well bounded by three impenetrable walls. Two of the walls appear due to the presence of the non-diagonal projections of the metric tensor in the triad $l^{\bar{\alpha}}_{\alpha}$, i.e., ultimately, due to the presence of matter. This effect was first described in [127, 128] (for the type VII model) and then confirmed in [140] for all homogeneous models (wherever it may be observed) by means of the qualitative theory of differential equations.

In this homogeneous example, the "insertion" of matter leads to a qualitative change in the asymptotic behavior of the solution in the neighborhood of a singularity in comparison with the vacuum case. Yet this phenomenon is typical only of the special class of homogeneous metrics under consideration and is based on the fact that rotations of the Kasner axes appear only in the presence of moving matter. In the general inhomogeneous case, the rotations of the Kasner axes do not depend on the presence of matter and exist already in vacuum. Thus the "insertion" of matter does not bring any new element of principle into the behavior of the general cosmological solution.

4.3 Perfect Fluid of Stiff Matter Equation of State

4.3.1 Main Approximation

Let us now deal with the special case corresponding to the "stiff matter" equation of state $p = \varepsilon$ (i.e., $q = 1$). As has been stated in the introduction to this section, a perfect fluid with such an equation of state cannot be inserted into the vacuum solution of the gravitational equations since the right-hand side of equation (1.15) is then of the same order of magnitude as its left-hand side and cannot be omitted.

Using the results of the paper [16], it is easy to show that for $p = \varepsilon$ the general cosmological solution near the singular point $t = 0$ has a simple power-law asymptotic behavior instead of the oscillatory regime. In the region of sufficiently small values of t, the first terms in the expansion of the hydrodynamic quantities are of the form:

$$u^0 = 1 \ , \quad u_\alpha = -\left(\ln p_\varphi\right)_{,\alpha} t \ln t + c_\alpha t \ , \quad \varepsilon = p_\varphi^2/2t^2 \ , \tag{4.17}$$

where p_φ and c_α are four arbitrary three-dimensional functions. The leading terms of the expansion of the metric tensor are given by equations (1.42)–(1.43) as before, but instead of the Kasner relations (1.21)–(1.22) we get other conditions for the exponents $p_{\bar{\alpha}}$, namely the relations (4.4) *where p_φ is the same three-dimensional function as the one that appears in (4.17)*.

Because the standard Kasner relations are replaced by (4.4), all three exponents $p_{\bar{\alpha}}$ become positive in the vicinity of the singularity and all three metric coefficients a^2, b^2, c^2 decrease monotonically with decreasing t. Then in the asymptotic region $t \to 0$ (as is clear from simple estimates) the three-dimensional curvature tensor P_α^β in equation (1.17) becomes negligible as compared with the first term $(\sqrt{g}\kappa_\alpha^\beta)^{\dot{}}/2\sqrt{g}$ which is of order t^{-2}. Despite the fact that the energy density ε increases as t^{-2}, the right-hand side in (1.17) (thanks to the condition $q = 1$) appears to be small and may be omitted, as is evident from (4.8) and (4.17). This, however, does not happen to the right-hand side of (1.15),

$$T_0^0 - \frac{1}{2}(T_0^0 + T_\mu^\mu) = -2\varepsilon(u^0)^2 = -2\varepsilon - 2\varepsilon u^\alpha u_\alpha \ , \tag{4.18}$$

which thanks to the term -2ε is of order t^{-2}. For this reason, we can observe the change in the relations between the exponents $p_{\bar{\alpha}}$ in comparison with the Kasner case. The second condition of (4.4) results from the fact that equation (1.15) should be satisfied in its leading order t^{-2} with the right-hand side taken into account.

A simple analysis of equation (1.16) shows that the highest terms in the expansion of its left- and right-hand sides are of the order $t^{-1}\ln t$ and in this order the equation in question is identically satisfied thanks to the condition (4.4). In the subsequent order t^{-1}, equation (1.16) provides three relations for the arbitrary three-dimensional functions $l_\alpha^{\bar{\alpha}}, p_{\bar{\alpha}}, p_\varphi, c_\alpha$ (here we assume the factors a_0^2, b_0^2, c_0^2 from (1.43) to be included in $l_\alpha^{\bar{\alpha}}$) of the constructed solution. If we take into account that the synchronous reference system admits only arbitrary transformations of the spatial coordinates containing three arbitrary three-dimensional functions, then it can easily be calculated that *the solution whose asymptotic behavior is described by (1.42), (1.43), (4.4) and (4.17) has eight physically arbitrary three-dimensional functions, i.e., the exact number that should be contained in the general solution of the gravitational equations in a space filled with a perfect fluid. These results provide the unique known type of general solution of the Einstein equations with cosmological singularity for which we have exact analytical expressions for its asymptotics.* It can be shown that all subsequent terms of the expansion in powers of t of the solution under study indeed are negligibly small and all of them are completely determined (i.e., contain no new arbitrary functions) by the leading approximation described above. The respective calculations have been accomplished in [18] and are given in the next subsection (the reader not keen on such details may omit it).

It is worth mentioning that the asymptotic behavior of (1.42), (1.43), (4.4) and (4.17) becomes an exact solution of the Einstein equations if we assume in (4.17) that $c_\alpha = 0$ and that all the remaining three-dimensional functions are constants. Then the velocity of the fluid vanishes and we get an exact homogeneous model of Bianchi type I in the synchronously co-moving reference system.

Let us point out in conclusion that the analysis reported in this subsection loses its validity when the highest Kasner exponent $p_{\bar{3}}$ is close to unity, i.e., when we are in the vicinity of the solution corresponding to the particular set of Kasner exponents $(p_{\bar{1}}, p_{\bar{2}}, p_{\bar{3}}) = (0, 0, 1)$. In [123] it has been shown that, as the singularity is being approached, the probability of the appearance of such particular cases tends to zero, but nevertheless this probability should not be ignored. The character of the oscillatory regime for $p_{\bar{3}} \approx 1$ (when the oscillation amplitudes of the functions a^2 and b^2 are sufficiently close together) has been dwelt upon in [15]. It has also been proved there that in this case the perfect fluid with the ultrarelativistic equation of state ($q = 1/3$) does not bring in any qualitative changes into the vacuum solution of the gravitational equations. Since for $p_{\bar{3}} \approx 1$ it makes sense to study only the case when $p_{\bar{3}} > q$, the absence of any significant effect of matter on the oscillatory regime for $q = 1/3$ (which corresponds to the interval $p_{\bar{3}} > q$) gives reason to conclude that this is correct at all the remaining values of the parameter q.

4.3.2 The Next-Order Terms

In this subsection we shall work out the subsequent terms of the expansion in powers of t (in the vicinity of a singular point $t = 0$) of the general cosmological solution for the case of a perfect fluid with the equation of state $p = \varepsilon$. The first or leading terms of the expansion were just described in the previous subsection.

Such a detailed study is of interest since, as we said already, it concerns a general cosmological solution with a complete set of arbitrary functions, for which it is possible to obtain explicit analytical expressions in the neighborhood of a singularity. Thus we can avoid a number of problems which might be unsolved in the analysis of the solution with the oscillatory regime, since the construction of the latter is possible only by means of matching separate epochs. Explicit expressions for the asymptotics of the general oscillatory solution do not exist even in the vicinity of a singularity.

Thus, as it has been shown, the first terms of the expansion of the general solution for the case $p = \varepsilon$ are of the form:

$$g^{(0)}_{\alpha\beta} = \eta_{\bar{\alpha}\bar{\beta}} l^{\bar{\alpha}}_{\alpha}\, l^{\bar{\beta}}_{\beta}\,, \quad \eta_{\bar{\alpha}\bar{\beta}} = diag(a^2, b^2, c^2). \tag{4.19}$$

$$(a^2, b^2, c^2) = (t^{2p_{\bar{1}}}, t^{2p_{\bar{2}}}, t^{2p_{\bar{3}}}), \tag{4.20}$$

$$u^0_{(0)} = 1\,, \quad u^{(0)}_{\alpha} = -\left(\ln p_{\varphi}\right)_{,\alpha} t \ln t + c_{\alpha} t\,, \quad \varepsilon^{(0)} = p^2_{\varphi}/2t^2, \tag{4.21}$$

where we assume again that the three-dimensional coefficients a^2_0, b^2_0, c^2_0 from (1.43) are included in the frame vectors $l^{\bar{\alpha}}_{\alpha}$. The index 0 in parentheses means that these expressions belong to the leading approximation. The exponents $p_{\bar{1}}, p_{\bar{2}}, p_{\bar{3}}, p_{\varphi}$ satisfy the relations (4.4) and we assume that we are close enough to the singularity, so that all four exponents are positive. For definiteness, the first three of them are assigned the following order:

$$0 < p_{\bar{1}} < p_{\bar{2}} < p_{\bar{3}}\,. \tag{4.22}$$

It is not difficult to show that, for this ordering, the exponents vary over the ranges of

$$0 \leqslant p_{\bar{1}} \leqslant 1/3\,, \quad 0 \leqslant p_{\bar{2}} \leqslant 2/3\,, \quad 1/3 \leqslant p_{\bar{3}} \leqslant 1. \tag{4.23}$$

The next terms of the expansion of this solution may be regarded as small corrections to the metric tensor, velocity and energy density of matter:

$$g_{\alpha\beta} = g^{(0)}_{\alpha\beta} + \delta g_{\alpha\beta}\,, \quad u_{\alpha} = u^{(0)}_{\alpha} + \delta u_{\alpha}\,, \quad \varepsilon = \varepsilon^{(0)} + \delta\varepsilon\,, \tag{4.24}$$

where the quantities with the superscript 0 are given by (4.19)–(4.21).

Substituting these expansions into the Einstein equations (1.15)–(1.17), we should bear in mind that we must retain in them only the variations of those terms which were leading in the first approximation, i.e., which are involved in the derivation of the asymptotic expressions (4.19)–(4.21). As for the other terms (which we have neglected earlier), only their first non-vanishing approximations appear in these equations, i.e., the same terms calculated according to the leading solution (4.19)–(4.21). This procedure of construction of the equations for the corrections is discussed in detail in Section A.1 of Appendix A. Here and elsewhere we shall follow the same method and we shall not provide all the details except where necessary.

First consider equation (1.17), which for $p = \varepsilon$ may be written as follows:

$$\dot{\kappa}^{\beta}_{\alpha} + \frac{1}{2}\kappa\kappa^{\beta}_{\alpha} = -2P^{\beta}_{\alpha} + 4\varepsilon u_{\alpha} u^{\beta}. \tag{4.25}$$

The leading term in this equation is its left-hand side, which vanishes for the zeroth-order solution $g^{(0)}_{\alpha\beta}$. Thus (4.25) yields the following equation for the corrections:

$$(\delta\kappa^{\beta}_{\alpha})^{\cdot} + \frac{1}{2}\kappa^{(0)}\delta\kappa^{\beta}_{\alpha} + \frac{1}{2}\kappa^{(0)\beta}_{\alpha}\delta\kappa = -2P^{(0)\beta}_{\alpha} + 4\varepsilon^{(0)} u^{(0)}_{\alpha} u^{\beta}_{(0)}. \tag{4.26}$$

If we write

$$\delta g_{\alpha\beta} = h_{\alpha\beta} \ , \tag{4.27}$$

then for the variations of the mixed components of the 3-tensor $\kappa_\alpha^\beta = g^{\beta\mu}\kappa_{\mu\alpha}$ we have

$$\delta\kappa_\alpha^\beta = \dot{h}_\alpha^\beta - \kappa_\alpha^{(0)\mu} h_\mu^\beta + \kappa_\mu^{(0)\beta} h_\alpha^\mu \ , \quad \delta\kappa = \dot{h} \ , \tag{4.28}$$

where $h = h_\mu^\mu$ and the tensorial operations on the indices of the corrections are performed by means of the leading metric $g_{\alpha\beta}^{(0)}$. The substitution of (4.28) into (4.26) yields the final system of equations determining the quantities $h_{\alpha\beta}$.

Since these equations contain only derivatives of $h_{\alpha\beta}$ with respect to time, it is convenient to project them along the directions $l_\alpha^{\bar{\alpha}}$. After simple calculations, we get

$$\ddot{h}_{\bar{\alpha}}^{\bar{\beta}} + \frac{1 + 2p_{\bar{\beta}} - 2p_{\bar{\alpha}}}{t}\dot{h}_{\bar{\alpha}}^{\bar{\beta}} + \frac{p_{\bar{\beta}}}{t}\delta_{\bar{\alpha}}^{\bar{\beta}}\dot{h} = -2P_{\bar{\alpha}}^{\bar{\beta}} + 4\varepsilon u_{\bar{\alpha}} u^{\bar{\beta}}, \ \textit{no summation over } \bar{\alpha}, \bar{\beta}. \tag{4.29}$$

Here and elsewhere, we shall omit the index 0 on the quantities of the first approximation in order to simplify the notation.

The computation of $P_{\bar{\alpha}}^{\bar{\beta}}$ in the right-hand side of equation (4.29) proceeds in a way closely analogous to what has been done in Section 1.6 for the computation of the frame components of the three-dimensional Ricci tensor. However, now, to evaluate the asymptotic values of the different terms in $P_{\bar{\alpha}}^{\bar{\beta}}$ and to find which of them are leading should be done on the basis of the inequalities (4.22) and not in accordance with the Kasner relations. The corresponding calculations have been done in [18] and here we show only the order of magnitudes of the highest-order terms (in the limit $t \to 0$) in $P_{\bar{\alpha}}^{\bar{\beta}}$. These are:

$$P_{\bar{1}}^{\bar{1}} \sim P_{\bar{2}}^{\bar{2}} \sim P_{\bar{3}}^{\bar{3}} \sim t^{4p_{\bar{1}}-2}, \tag{4.30}$$

$$P_{\bar{1}}^{\bar{2}} \sim t^{4p_{\bar{1}}-2}\ln t \ , \quad P_{\bar{1}}^{\bar{3}} \sim t^{4p_{\bar{1}}-2}\ln t \ , \quad P_{\bar{2}}^{\bar{3}} \sim t^{-2p_{\bar{3}}}\ln^2 t \ . \tag{4.31}$$

Then, using the solution (4.19)–(4.21) to calculate the projections of the terms $\varepsilon u_\alpha u^\beta$ in the right-hand sides of equation (4.29), we can show that in all components of this equation with the exception of the one corresponding to $\bar{\beta} = \bar{3}$, $\bar{\alpha} = \bar{2}$, these terms are negligible in comparison with the components of the three-dimensional curvature tensor and may therefore be omitted. Only in the aforementioned component alone, the quantity $4\varepsilon u_{\bar{2}} u^{\bar{3}}$ is of the same order in t as $P_{\bar{2}}^{\bar{3}}$ and must be retained.

Equations (4.29) represent a linear inhomogeneous system, and its general solution consists of a particular solution of these inhomogeneous equations to which one adds the general solution of the corresponding homogeneous system. The general solution of the homogeneous system implies, however, no new data for us, since the homogeneous system (generated from (4.29) with the right-hand sides omitted) is a repetition of the equations of the leading approximation

$(\sqrt{g}\kappa_{\alpha}^{\beta})^{\cdot} = 0$, the solution of which is already taken into account in the metric (4.19). Thus, the corrections to the metric tensor are provided only by particular solutions of the system (4.29), which one can find by elementary integrations.

Using the exact asymptotic expressions for $P_{\bar{\alpha}}^{\bar{\beta}}$ (i.e., expressions (4.30)–(4.31) but with the exact three-dimensional coefficient functions in front of the time factors found in [18]), we obtain the following solutions for the diagonal projections of the quantities h_{α}^{β}:

$$h_{\bar{1}}^{\bar{1}} = -\frac{\lambda^2 (1+5p_{\bar{1}})}{16p_{\bar{1}}^2 (1+4p_{\bar{1}})} t^{4p_{\bar{1}}}, \; h_{\bar{2}}^{\bar{2}} = \frac{\lambda^2 (1+4p_{\bar{1}}-p_{\bar{2}})}{16p_{\bar{1}}^2 (1+4p_{\bar{1}})} t^{4p_{\bar{1}}},$$
$$h_{\bar{3}}^{\bar{3}} = \frac{\lambda^2 (1+4p_{\bar{1}}-p_{\bar{3}})}{16p_{\bar{1}}^2 (1+4p_{\bar{1}})} t^{4p_{\bar{1}}}, \tag{4.32}$$

where λ is the quantity introduced earlier in (1.47). Hence, for the contraction $h = h_{\bar{1}}^{\bar{1}} + h_{\bar{2}}^{\bar{2}} + h_{\bar{3}}^{\bar{3}}$, one gets

$$h = \frac{\lambda^2}{4p_{\bar{1}} (1+4p_{\bar{1}})} t^{4p_{\bar{1}}}. \tag{4.33}$$

The expressions for the non-diagonal projections of h_{α}^{β} are more involved and we shall indicate here only their orders of magnitude:

$$h_{\bar{1}\bar{2}} = t^{2p_{\bar{2}}} h_{\bar{1}}^{\bar{2}} \sim t^{4p_{\bar{1}}+2p_{\bar{2}}} \ln t \;, \; h_{\bar{1}\bar{3}} = t^{2p_{\bar{3}}} h_{\bar{1}}^{\bar{3}} \sim t^{4p_{\bar{1}}+2p_{\bar{3}}} \ln t,$$
$$h_{\bar{2}\bar{3}} = t^{2p_{\bar{3}}} h_{\bar{2}}^{\bar{3}} \sim t^2 \ln^2 t \;. \tag{4.34}$$

As is clear from the above formulae, the corrections to the metric tensor found in the region $t \to 0$ are actually small, i.e., satisfy the requirements:

$$h_{\bar{1}}^{\bar{1}} \ll 1, \; h_{\bar{2}}^{\bar{2}} \ll 1, \; h_{\bar{3}}^{\bar{3}} \ll 1, \tag{4.35}$$
$$h_{\bar{1}\bar{2}} \ll \sqrt{g_{\bar{1}\bar{1}} g_{\bar{2}\bar{2}}} = t^{p_{\bar{1}}+p_{\bar{2}}}, \; h_{\bar{1}\bar{3}} \ll \sqrt{g_{\bar{1}\bar{1}} g_{\bar{3}\bar{3}}} = t^{p_{\bar{1}}+p_{\bar{3}}},$$
$$h_{\bar{2}\bar{3}} \ll \sqrt{g_{\bar{2}\bar{2}} g_{\bar{3}\bar{3}}} = t^{p_{\bar{2}}+p_{\bar{3}}} \tag{4.36}$$

The subsequent term of the expansion in the energy density of the matter is then determined from equation (1.15), which is of the form:

$$\dot{\kappa} + \frac{1}{2}\kappa_{\alpha}^{\beta}\kappa_{\beta}^{\alpha} + 4\varepsilon = -4\varepsilon u^{\alpha} u_{\alpha}. \tag{4.37}$$

The leading part of this equation, which is equal to zero for the leading-order solution (4.19)–(4.21), is on the left. Varying it and retaining in the right-hand side the small quantity $-4\varepsilon u^{\alpha} u_{\alpha}$, calculated according to the solution of the leading approximation, we get:

$$\delta\varepsilon = -\frac{1}{4}\left[\ddot{h} + \frac{2}{t}\left(p_{\bar{1}} h_{\bar{1}}^{\bar{1}} + p_{\bar{2}} h_{\bar{2}}^{\bar{2}} + p_{\bar{3}} h_{\bar{3}}^{\bar{3}}\right)\right] - \varepsilon u^{\alpha} u_{\alpha}. \tag{4.38}$$

An estimate of the orders of magnitude of the terms in the right-hand side of this expression proves that the term $\varepsilon u^{\alpha} u_{\alpha}$ may be omitted. Substituting the solution (4.32)–(4.33) into the remaining part, we find:

$$\delta\varepsilon = -\frac{\lambda^2 p_\varphi^2}{8p_{\bar{1}}\left(1+4p_{\bar{1}}\right)} t^{4p_{\bar{1}}-2}. \tag{4.39}$$

The correction to the velocity vector δu_α may now be found from the first variation of equation (1.16). Yet it is simpler to make use of the equation of hydrodynamics (4.10) (setting $q = 1$ and dividing all the terms by $\sqrt{\varepsilon}$):

$$\begin{aligned} u^0 \left(\sqrt{\varepsilon} u_\alpha\right)^{\cdot} + \left(\sqrt{\varepsilon}\right)_{,\alpha} &= -u_\alpha u^\mu \left(\sqrt{\varepsilon}\right)_{,\mu} \\ &- \sqrt{\varepsilon} u^\mu u_{\alpha,\mu} + \frac{1}{2}\sqrt{\varepsilon} u^\mu u^\nu g_{\mu\nu,\alpha} \,. \end{aligned} \tag{4.40}$$

The leading terms of this equation are again inserted in its left-hand side, which becomes equal to zero after the solution (4.21) is substituted into it. The right-hand side of equation (4.40), calculated in terms of the leading approximation (4.19)–(4.21), appears to be small in comparison with the variation of its left-hand side and may be omitted. Among the four terms resulting from the variation in the left-hand side of equation (4.40), the term containing δu^0 (which is found from (4.12) and is of the order of $\delta u^0 \sim t^{2-2p_3} \ln^2 t$) is also small. As a result, for the correction δu_α we have the following simple equation:

$$\left(\sqrt{\varepsilon}\delta u_\alpha\right)^{\cdot} = -\left(u_\alpha \delta\sqrt{\varepsilon}\right)^{\cdot} - \left(\delta\sqrt{\varepsilon}\right)_{,\alpha} \,. \tag{4.41}$$

Substituting here $\delta\varepsilon$ from (4.39), we easily find:

$$\delta u_\alpha = \frac{\chi}{p_\varphi}\left(\ln\frac{p_{\bar{1}}}{p_\varphi}\right)_{,\alpha} t^{4p_{\bar{1}}+1}\ln t + \frac{1}{p_\varphi}\left(\frac{\chi_{,\alpha}}{4p_{\bar{1}}} - \frac{\chi p_{\bar{1},\alpha}}{4p_{\bar{1}}^2} + \chi c_\alpha\right) t^{4p_{\bar{1}}+1} \,, \tag{4.42}$$

where χ denotes:

$$\chi = \lambda^2 p_\varphi \left[8p_{\bar{1}}\left(1+4p_{\bar{1}}\right)\right]^{-1}. \tag{4.43}$$

The supplementary term in (4.42), corresponding to the solution of the homogeneous equation $\left(\sqrt{\varepsilon}\delta u_\alpha\right)^{\cdot} = 0$ and containing three arbitrary three-dimensional functions, is omitted since it has the form $\delta u_\alpha = t f_\alpha\left(x^\beta\right)$ and is related to the small variation of the arbitrary functions c_α in (4.21), i.e., is already taken into account in the leading approximation.

It is evident from (4.39) and (4.42) that the corrections $\delta\varepsilon$ and δu_α to the energy density and velocity in the limit $t \to 0$ are negligibly small in comparison with the leading approximation (4.21) for ε and u_α.

4.4 Yang–Mills and Electromagnetic Fields

4.4.1 Coupled Gravitational and Gauge Fields

In the preceding section we considered the influence of a perfect fluid upon the properties of the general solution of the gravitational equations in the neighborhood of a cosmological singularity and found out that, for the equation of state $p = q\varepsilon$ where $0 \leq q < 1$, the character of the singularity is the same as in vacuum, i.e., the presence of this kind of matter in space does not affect the vacuum oscillatory regime in its leading terms. In contrast to this, the case of a perfect fluid with the equation of state $p = \varepsilon$ (i.e., $q = 1$) is associated with another, simpler situation: the existence of the oscillatory regime in the neighborhood of a cosmological singularity becomes impossible, and is replaced by a simple power-law asymptotic behavior.

Henceforth we shall touch upon the third example of the possibilities mentioned in the Introduction, which is associated with the case when matter does not affect the oscillatory character of the general solution near a singularity, but the influence of the energy–momentum tensor nevertheless cannot be neglected, since its contribution to the formation and alternation of Kasner epochs is as essential as that implied by the three-dimensional curvature tensor P_α^β. Classical Yang–Mills fields and their special case, the electromagnetic field, are examples of such a situation.

For simplicity, we shall confine ourselves to the Yang–Mills fields corresponding to the gauge group $SU(2)$ without charges and currents. In this case we have three vector fields, the potentials of which will be denoted by L_i^A. *Here and elsewhere in Section 4.4, the capital Latin indices A, B, C, D will assume the values of 1,2,3 and are internal $SU(2)$ indices. The position of these indices is of no significance, but these indices if repeated imply a summation.* For the group $SU(2)$, the field strengths are:

$$F_{ik}^A = L_{i,k}^A - L_{k,i}^A + \zeta \varepsilon_{ABD} L_k^B L_i^D \ , \tag{4.44}$$

where ζ a is an arbitrary constant and ε_{ABD} is the unit antisymmetric Levi-Civita symbol ($\varepsilon_{123} = 1$).

The first group of field equations (similar to the first pair of Maxwell equations) is a consequence of the definition of the tensors F_{ik}^A in terms of the potentials L_i^A and may be written down in the form:

$$F_{ik;l}^A + F_{li;k}^A + F_{kl;i}^A = \zeta \varepsilon_{ABD} \left(F_{ik}^B L_l^D + F_{li}^B L_k^D + F_{kl}^B L_i^D \right) . \tag{4.45}$$

In the left-hand side of these equations, the covariant derivatives may be replaced by ordinary derivatives, since all the terms involving the Christoffel symbols cancel each other.

The second group of Yang–Mills equations (i.e., the equations of motion similar to the second pair of Maxwell equations), as well as the Einstein equations, may be obtained by a variational principle, with the action:

$$S = \int \left(R - \frac{1}{4} F_{ik}^A F^{Aik} \right) \sqrt{- \det g_{lm}} \, d^4x \ . \tag{4.46}$$

These equations are of the form:

$$\left(F^{Aik} \right)_{;k} - \zeta \varepsilon_{ABD} L_k^D F^{Bik} = 0, \tag{4.47}$$

$$R_i^k - \frac{1}{2} \delta_i^k R = F_{il}^A F^{Akl} - \frac{1}{4} \delta_i^k F_{lm}^A F^{Alm}. \tag{4.48}$$

We shall use the synchronous reference system with metric given by (1.1)–(1.2) and singularity occurring at $t = 0$. Along with these coordinate conditions, we shall also choose a certain gauge for the potentials L_i^A. In synchronous coordinates, the most convenient gauge conditions are that the time components of all the three vector fields vanish:

$$L_0^A = 0. \tag{4.49}$$

These define the so-called "temporal gauge," and leave the freedom of making time-independent gauge transformations. In the $SU(2)$ case, this corresponds to three arbitrary functions of 3-space.

Let us now introduce the electric and magnetic field strengths of the Yang–Mills fields by analogy with the way the corresponding strengths are introduced in electrodynamics:

$$E^{A\alpha} = F^{A0\alpha} \, , \quad H^{A\alpha} = \frac{1}{2\sqrt{g}} \varepsilon^{\alpha\beta\gamma} F_{\beta\gamma}^A \, , \tag{4.50}$$

where $\varepsilon^{\alpha\beta\gamma}$ denotes again the conventional antisymmetric unit Levi-Civita symbol in three dimensions* already used in (1.25). The definitions (4.50) together with the gauge conditions (4.49) enable one to write down the four-dimensional relations (4.44) in the following three-dimensional form:

$$E_\alpha^A = \dot{L}_\alpha^A \, , \tag{4.51}$$

$$H^{A\alpha} = \frac{1}{2\sqrt{g}} \varepsilon^{\alpha\lambda\mu} \left(L_{\lambda,\mu}^A - L_{\mu,\lambda}^A + \zeta \varepsilon_{ABD} L_\mu^B L_\lambda^D \right) . \tag{4.52}$$

In the above formulae, g denotes the determinant of the three-dimensional metric $g_{\alpha\beta}$ determining all tensorial operations with Greek indices (e.g., $E_\alpha^A = g_{\alpha\mu} E^{A\mu}$).

In the notations that we just introduced, the first group of Yang–Mills equations, namely (4.45), may be easily rewritten in the form:

$$(\sqrt{g} H^{A\alpha})^{\dot{}} = -\varepsilon^{\alpha\mu\lambda} E_{\lambda,\mu}^A + \zeta \varepsilon_{ABD} L_\mu^D \varepsilon^{\alpha\mu\lambda} E_\lambda^B \, , \tag{4.53}$$

$$\frac{1}{\sqrt{g}} (\sqrt{g} H^{A\mu})_{,\mu} - \zeta \varepsilon_{ABD} L_\mu^D H^{B\mu} = 0 \, . \tag{4.54}$$

Equations (4.47) of the second group can be written as

$$(\sqrt{g} E^{A\alpha})^{\dot{}} = \varepsilon^{\alpha\mu\lambda} H_{\lambda,\mu}^A - \zeta \varepsilon_{ABD} L_\mu^D \varepsilon^{\alpha\mu\lambda} H_\lambda^B \, , \tag{4.55}$$

$$\frac{1}{\sqrt{g}} (\sqrt{g} E^{A\mu})_{,\mu} - \zeta \varepsilon_{ABD} L_\mu^D E^{B\mu} = 0 \, , \tag{4.56}$$

* Note that, in a curved three-dimensional space, the covariant components of the unit antisymmetric tensor $e_{\alpha\beta\gamma}$ and its contravariant components $e^{\alpha\beta\gamma}$ are defined as

$$e_{\alpha\beta\gamma} = \sqrt{g} \varepsilon_{\alpha\beta\gamma} \, , \quad e^{\alpha\beta\gamma} = \frac{1}{\sqrt{g}} \varepsilon^{\alpha\beta\gamma} \, ,$$

where g is the determinant of the three-dimensional metric tensor and $\varepsilon_{\alpha\beta\gamma} \equiv \varepsilon^{\alpha\beta\gamma}$ is the unit antisymmetric Levi-Civita symbol in three dimensions, obeying the requirement $\varepsilon_{123} = \varepsilon^{123} = 1$. All the expressions containing the symbol $\varepsilon_{\alpha\beta\gamma}$ (or $\varepsilon^{\alpha\beta\gamma}$) used here and in the sequel are in fact written in the three-dimensional covariant form, containing only the tensor $e_{\alpha\beta\gamma}$ (or its contravariant components $e^{\alpha\beta\gamma}$).

while the Einstein equations (4.48), reduced to the form (1.15)–(1.17), are:

$$\frac{1}{2}\dot{\kappa} + \frac{1}{4}\kappa^{\alpha}_{\beta}\kappa^{\beta}_{\alpha} = -\frac{1}{2}\left(E^{A\mu}E^{A}_{\mu} + H^{A\mu}H^{A}_{\mu}\right) , \tag{4.57}$$

$$\frac{1}{2}(\kappa_{,\alpha} - \kappa^{\beta}_{\alpha;\beta}) = -\sqrt{g}\varepsilon_{\alpha\lambda\mu}E^{A\mu}H^{A\lambda} , \tag{4.58}$$

$$\frac{1}{2\sqrt{g}}(\sqrt{g}\kappa^{\beta}_{\alpha})^{\dot{}} + P^{\beta}_{\alpha} = -E^{A}_{\alpha}E^{A\beta} - H^{A}_{\alpha}H^{A\beta} + \frac{1}{2}\delta^{\beta}_{\alpha}\left(E^{A\mu}E^{A}_{\mu} + H^{A\mu}H^{A}_{\mu}\right). \tag{4.59}$$

4.4.2 Oscillatory Regime in the Presence of Gauge Fields

We shall now show that in the neighborhood of a cosmological singularity, the evolution of the system of fields governed by equations (4.51)–(4.59) is, as before, described by a succession of Kasner epochs. This succession, however, is induced both by the three-dimensional curvature tensor and by the energy momentum tensor of the gauge fields.

The metric tensor $g_{\alpha\beta}$ in each Kasner epoch is written down in the form (1.42). As it has been stated already, the interchange of two Kasner epochs is accompanied by two important phenomena: a change in the form of the metric coefficients a^2, b^2, c^2 and a rotation of the Kasner axes. The most important aspect of the solution is the change of the behavior of the scale factors a^2, b^2, c^2, which is the basic element of the oscillatory regime. This change takes place in the transition region between two epochs when the rotation of the Kasner axes is still a minor effect. Hence, we can simplify the subsequent analysis if we disregard the rotation of the Kasner axes and confine ourselves only to the main part of the problem: the elucidation of the behavior of the coefficients a^2, b^2, c^2. This approximation means that, throughout the interchange of epochs, the metric tensor is written in the form of (1.42) with the same time-independent vectors $l^{\bar{\alpha}}_{\alpha}$.

Let us then derive the equations governing the interchange of Kasner epochs in the above-established approximation. The rules for deriving these equations result from the following analysis: let us assume that, during the Kasner epochs, the leading terms in the Einstein equations and the equations of motion are again the terms containing time derivatives. Neglecting in equations (4.57) and (4.59) the energy–momentum tensor and the three-dimensional Ricci tensor, we obtain the generalized Kasner metric (1.42)–(1.43), (1.21) (we recall that the three-dimensional coefficients a_0^2, b_0^2, c_0^2 in (1.43) can be considered to be equal to unity, since such coefficients can be absorbed in the frame vectors $l^{\bar{\alpha}}_{\alpha}$, and that with this convention, we have in the leading approximation $abc = t$ and, as it follows from (1.25), $\sqrt{g} = t\varepsilon^{\alpha\beta\gamma}l^{\bar{1}}_{\alpha}l^{\bar{2}}_{\beta}l^{\bar{3}}_{\gamma})$. It will also be our assumption, as before, that the exponent $p_{\bar{1}}$ is negative, so that the function a^2 increases as $t \to 0$ while b^2 and c^2 decrease.

From the dynamical equations (4.51), (4.53) and (4.55) (retaining in them only the terms with time derivatives), we have, in the approximation under study:

$$L_{\alpha}^{A} = \beta_{\alpha}^{A}(x^{\mu}) \ , \ E^{A\alpha} = \frac{1}{t}\Phi_{1}^{A\alpha}(x^{\mu}) \, , \ \ H^{A\alpha} = \frac{1}{t}\Phi_{2}^{A\alpha}(x^{\mu}) \, , \qquad (4.60)$$

where $\beta_{\alpha}^{A}(x^{\mu})$, $\Phi_{1}^{A\alpha}(x^{\mu})$, $\Phi_{2}^{A\alpha}(x^{\mu})$ are three-dimensional time-independent functions.

The functions $\Phi_{2}^{A\alpha}(x^{\mu})$ are unambiguously defined by $\beta_{\alpha}^{A}(x^{\mu})$ from (4.52). Then equation (4.54) is automatically fulfilled (recall that equations (4.53) and (4.54) are direct consequences of the definitions of the field strengths (4.51) and (4.52)). The constraint equation (4.56) provides three relations which must be satisfied by the three-dimensional arbitrary functions $\beta_{\alpha}^{A}(x^{\mu})$ and $\Phi_{1}^{A\alpha}(x^{\mu})$. Taking into account the residual gauge freedom of performing time-independent gauge transformations, it is then easy to calculate that the first terms of the expansion of the solution (1.42)–(1.43) and (4.60) contain sixteen arbitrary, physically distinct, three-dimensional functions. This is the exact amount that should be contained in a general solution of the system (4.51)–(4.59) (four for the gravitational field, and four for each of the three vector fields).

Now it is necessary to project equations (4.51), (4.53), (4.55), (4.57) and (4.59) into the triad $l_{\alpha}^{\bar{\alpha}}$, substitute the solution of the leading approximation (1.42)–(1.43) and (4.60) into the projections so obtained, and use them to calculate the earlier-omitted parts of these equations that contain no time derivatives and that we have treated as small. We can omit the terms in these parts which, in the asymptotic region $t \to 0$, are actually small in comparison with those retained. But the terms which increase faster with decreasing t than do the terms with time derivatives will sooner or later exceed them in order of magnitude and therefore violate the validity of our assumptions. We should retain these terms, replace everywhere the explicit asymptotic expressions (1.43) and (4.60) by the (now unknown) functions a^2, b^2, c^2 and $L_{\alpha}^{A}, E^{A\alpha}, H^{A\alpha}$ and include them along with the time derivatives in a new system of "corrected" equations, which will be governing the solution in the region where the approximation (1.43) and (4.60) loses its validity. It turns out that in this new region, the first terms of the expansion of the solution again acquire a form analogous to (1.42)–(1.43) and (4.60) but with different three-dimensional parameters. In other words, the equations constructed as above actually describe the interchange of Kasner epochs.

Let us start the construction of these equations from the simplest relation (4.51). Projecting it on the three directions $l_{\bar{1}}^{\alpha}, l_{\bar{2}}^{\alpha}, l_{\bar{3}}^{\alpha}$ (which are inverse to $l_{\alpha}^{\bar{1}}, l_{\alpha}^{\bar{2}}, l_{\alpha}^{\bar{3}}$) and performing the above analysis, we discover that in the $l_{\bar{2}}^{\alpha}$ and $l_{\bar{3}}^{\alpha}$ projections of (4.51), only the time derivatives should be retained, i.e., throughout the process of epoch alternation these equations are of the form:

$$\dot{L}_{\bar{2}}^{A} = 0 \ , \ \ \dot{L}_{\bar{3}}^{A} = 0 \ , \qquad (4.61)$$

where $L_{\bar{\alpha}}^{A} = L_{\alpha}^{A} l_{\bar{\alpha}}^{\alpha}$. In the $l_{\bar{1}}^{\alpha}$-projection of (4.51) the left-hand side $E_{\bar{1}}^{A} = E_{\alpha}^{A} l_{\bar{1}}^{\alpha}$ increases faster with decreasing t than does the right-hand side (due to the increase of the metric coefficient a^2) and this equation must be taken in its complete form:

$$\dot{L}^A_{\bar{1}} = a^2 E^{A\bar{1}}, \tag{4.62}$$

where $E^{A\bar{\alpha}} = E^{A\alpha} l^{\bar{\alpha}}_\alpha$ and the same notation is used later for the projections of the "magnetic" components: $H^{A\bar{\alpha}} = H^{A\alpha} l^{\bar{\alpha}}_\alpha$. Thus the projections of the potentials $L^A_{\bar{2}}$ and $L^A_{\bar{3}}$ are constant in time not only during the first Kasner epoch but throughout the whole process of interchange:

$$L^A_{\bar{2}} = \beta^A_\alpha l^\alpha_{\bar{2}} \ , \quad L^A_{\bar{3}} = \beta^A_\alpha l^\alpha_{\bar{3}} \ . \tag{4.63}$$

The fact that the projections $L^A_{\bar{2}}$ and $L^A_{\bar{3}}$ are constant immediately simplifies the analysis of equations (4.53) and (4.55) in the region we are interested in. It can easily be shown that, in the $l^{\bar{1}}_\alpha$-projections of these equations (which are the result of their multiplication by $l^{\bar{1}}_\alpha$) in the epoch interchange region, only the following terms should be retained:

$$(abcH^{A\bar{1}})^{\dot{}} = -\lambda a^2 E^{A\bar{1}} \ , \tag{4.64}$$

$$(abcE^{A\bar{1}})^{\dot{}} = \lambda a^2 H^{A\bar{1}} \ , \tag{4.65}$$

(we recall that λ was introduced in (1.47)).

In the $l^{\bar{2}}_\alpha$- and $l^{\bar{3}}_\alpha$-projections of equations (4.53) and (4.55), we must necessarily retain a larger number of terms along with the time derivatives. With (4.63) taken into account, these equations take the form:

$$(abcH^{A\bar{2}})^{\dot{}} = -\left(a^2 E^{A\bar{1}}\right)_{,\alpha} l^\alpha_{\bar{3}} + \gamma a^2 E^{A\bar{1}} + \zeta \varepsilon_{ABD} \beta^D_\alpha l^\alpha_{\bar{3}} a^2 E^{B\bar{1}} \ , \tag{4.66}$$

$$(abcH^{A\bar{3}})^{\dot{}} = \left(a^2 E^{A\bar{1}}\right)_{,\alpha} l^\alpha_{\bar{2}} + \rho a^2 E^{A\bar{1}} - \zeta \varepsilon_{ABD} \beta^D_\alpha l^\alpha_{\bar{2}} a^2 E^{B\bar{1}} \ , \tag{4.67}$$

$$(abcE^{A\bar{2}})^{\dot{}} = \left(a^2 H^{A\bar{1}}\right)_{,\alpha} l^\alpha_{\bar{3}} - \gamma a^2 H^{A\bar{1}} - \zeta \varepsilon_{ABD} \beta^D_\alpha l^\alpha_{\bar{3}} a^2 H^{B\bar{1}} \ , \tag{4.68}$$

$$(abcE^{A\bar{3}})^{\dot{}} = -\left(a^2 H^{A\bar{1}}\right)_{,\alpha} l^\alpha_{\bar{2}} - \rho a^2 H^{A\bar{1}} + \zeta \varepsilon_{ABD} \beta^D_\alpha l^\alpha_{\bar{2}} a^2 H^{B\bar{1}} \ , \tag{4.69}$$

where γ and ρ denote the quantities:

$$\gamma = \left(l^{\bar{1}}_{\alpha,\beta} - l^{\bar{1}}_{\beta,\alpha}\right) l^\alpha_{\bar{3}} l^\beta_{\bar{1}} \ , \quad \rho = \left(l^{\bar{1}}_{\alpha,\beta} - l^{\bar{1}}_{\beta,\alpha}\right) l^\alpha_{\bar{1}} l^\beta_{\bar{2}} \ . \tag{4.70}$$

Then we must consider the Einstein equations (4.57) and (4.59). Note that, with respect to the evolution of the metric tensor in the process of epoch alternation, we have restricted ourselves to the part that is relevant for the behavior of the scale coefficients a^2, b^2, c^2 and have neglected the rotation of the Kasner axes. This means that, out of the projections of (4.59) in the triad $l^{\bar{\alpha}}_\alpha$ we must use only the diagonal projections (since non-diagonal projections of these equations determine the rotation of the Kasner axes which we ignore). It is evident that, in the diagonal projections of equations (4.59), the terms increasing faster than t^{-2} as $t \to 0$ (i.e., faster than the potential increase of the principal term $(\sqrt{g}\kappa^\beta_\alpha)^{\dot{}}$ $/2\sqrt{g})$ are contained both in the three-dimensional Ricci tensor P^β_α and in the energy-momentum tensor. Terms of this kind derived from the three-dimensional curvature tensor P^{β}_{α} are well known: they are the same vacuum terms of the type (4.5) studied earlier in Section 1.7 (see formulae (1.46)):

$$P^{\bar{1}}_{\bar{1}} = \lambda^2 a^4/2a^2b^2c^2,\ P^{\bar{2}}_{\bar{2}} = -\lambda^2 a^4/2a^2b^2c^2,\ P^{\bar{3}}_{\bar{3}} = -\lambda^2 a^4/2a^2b^2c^2. \tag{4.71}$$

As $t \to 0$, these terms increase as a^4t^{-2} during the initial Kasner epoch. Concerning the energy-momentum tensor terms, we note that in the right-hand sides of the diagonal projections of (4.59), as well as in the right-hand side of (4.57), such terms appear in the form of the combination:

$$E^{A\bar{1}}E^{A}_{\bar{1}} + H^{A\bar{1}}H^{A}_{\bar{1}} = a^2\left(E^{A\bar{1}}E^{A\bar{1}} + H^{A\bar{1}}H^{A\bar{1}}\right). \tag{4.72}$$

As $t \to 0$, the terms increase as a^2t^{-2} in the initial epoch (i.e., where the solution is approximately governed by the laws (1.43) and (4.60)).

The final form of the equations involving both groups of terms, is:

$$[abc\,(\ln a)^{\cdot}]^{\cdot}/abc + \lambda^2a^4/2a^2b^2c^2 = -\omega^2a^2/2a^2b^2c^2\ , \tag{4.73}$$

$$[abc\,(\ln b)^{\cdot}]^{\cdot}/abc - \lambda^2a^4/2a^2b^2c^2 = \omega^2a^2/2a^2b^2c^2\ , \tag{4.74}$$

$$[abc\,(\ln c)^{\cdot}]^{\cdot}/abc - \lambda^2a^4/2a^2b^2c^2 = \omega^2a^2/2a^2b^2c^2\ , \tag{4.75}$$

$$[(\ln a)^{\cdot}]^2 + [(\ln b)^{\cdot}]^2 + [(\ln c)^{\cdot}]^2 - [(\ln a)^{\cdot} + [(\ln b)^{\cdot} + [(\ln c)^{\cdot}]^2 + \frac{\lambda^2a^4}{2a^2b^2c^2} + \frac{2\omega^2}{2a^2b^2c^2} = 0\ , \tag{4.76}$$

where ω^2 denotes:

$$\omega^2 = a^2b^2c^2\left(E^{A\bar{1}}E^{A\bar{1}} + H^{A\bar{1}}H^{A\bar{1}}\right). \tag{4.77}$$

Equations (4.73)–(4.75) follow from the diagonal projections of equations (4.59) and their first integral (4.76) results from (4.57) taking into account the sum of (4.73)–(4.75).

The main equations in the resulting system are (4.64), (4.65) and (4.73)–(4.77) containing the functions a^2, b^2, c^2 and the projections $E^{A\bar{1}}, H^{A\bar{1}}$ of the gauge field strengths on the direction $l^{\bar{1}}_{\alpha}$ which increase as t decreases. *These equations can be integrated exactly.* First of all, note that the quantity ω^2 is an integral of the system (4.64)–(4.65), i.e., is constant in time:

$$\omega^2 = \omega^2(x^{\alpha}). \tag{4.78}$$

To verify that this is correct, it suffices to multiply (4.64) by $abcH^{A\bar{1}}$, then (4.65) by $abcE^{A\bar{1}}$, and to add the results. Hence it follows that equations (4.73)–(4.76) form a closed system completely determining the metric coefficients a^2, b^2, c^2. The exact general solution of these equations is not difficult to derive and may be written in the form:

$$a^2 = G\alpha_0^2e^{-2\Lambda p_{\bar{1}}\tau},\ b^2 = G^{-1}\beta_0^2e^{-2\Lambda p_{\bar{2}}\tau},\ c^2 = G^{-1}\gamma_0^2e^{-2\Lambda p_{\bar{3}}\tau}, \tag{4.79}$$

$$G = 8\Lambda^2p_{\bar{1}}^2\left[\left(\alpha_0^2e^{-2\Lambda p_{\bar{1}}\tau} + \omega^2\right)^2 + 4\Lambda^2p_{\bar{1}}^2\lambda^2\right]^{-1}, \tag{4.80}$$

where the quantities $\alpha_0, \beta_0, \gamma_0, \Lambda, p_{\bar{1}}, p_{\bar{2}}, p_{\bar{3}}$ are functions only of the three-dimensional coordinates x^α. The first four functions are arbitrary, while the last three should satisfy the usual Kasner relations: $p_{\bar{1}} + p_{\bar{2}} + p_{\bar{3}} = 1$ and $p_{\bar{1}}^2 + p_{\bar{2}}^2 + p_{\bar{3}}^2 = 1$. The variable τ plays here the role of a new time and, as a function of the synchronous time t, is defined by equation (1.63): $d\tau = -(abc)^{-1}dt$.

Let us consider the variation of the asymptotic form of the functions a^2, b^2, c^2 as the time t goes from ∞ to zero. The corresponding change in the variable τ is from $-\infty$ to $+\infty$ (we put $\Lambda > 0$). It is easy to see that, as $t \to +\infty$, the metric coefficients are governed by the same laws as (1.54): $\left(a^2, b^2, c^2\right) \sim \left(t^{2p_{\bar{1}}}, t^{2p_{\bar{2}}}, t^{2p_{\bar{3}}}\right)$. In the region $t \to 0$, from (4.79)–(4.80) we get $\left(a^2, b^2, c^2\right) \sim \left(t^{2\acute{p}_{\bar{1}}}, t^{2\acute{p}_{\bar{2}}}, t^{2\acute{p}_{\bar{3}}}\right)$, where the new exponents $\acute{p}_{\bar{\alpha}}$ are given by exactly the same relations as (1.56), that is they are again Kasner exponents, but now with $\acute{p}_{\bar{2}} < 0$ so that the function b^2 increases.

Thus we have the same law for the change of Kasner epochs as in vacuum, that is the qualitative character of the variation in the asymptotic form of the functions a^2, b^2, c^2 at the transition from one epoch to another does not depend on the presence or absence of gauge fields. The solution (4.79)–(4.80) is also relevant for the pure vacuum case $\omega^2 = 0$, and implies therefore the opposite possibility that the interchange of epochs is determined only by the energy-momentum tensor of the Yang–Mills fields when the three-dimensional curvature is absent ($\lambda = 0, \omega^2 \neq 0$). In either case, as it has already been stated above, the result will be the same in the leading approximation. The gauge fields will affect only some subtle aspects of the behavior of the coefficients a^2, b^2, c^2 and quantitative characteristics of the rotation of the Kasner axes, which we are not dwelling upon here.

The functions a^2, b^2, c^2 make it possible now to exactly integrate equations (4.64) and (4.65). After substituting into them the explicit expressions for a^2 and abc as well as transforming to the variable τ, we get the following result:

$$abcE^{A\bar{1}} = \left[-4r^A \Lambda p_{\bar{1}} \lambda B + q^A \left(4\Lambda^2 p_{\bar{1}}^2 \lambda^2 - B^2\right)\right] \left(4\Lambda^2 p_{\bar{1}}^2 \lambda^2 + B^2\right)^{-1}, \quad (4.81)$$

$$abcH^{A\bar{1}} = \left[-4q^A \Lambda p_{\bar{1}} \lambda B - r^A \left(4\Lambda^2 p_{\bar{1}}^2 \lambda^2 - B^2\right)\right] \left(4\Lambda^2 p_{\bar{1}}^2 \lambda^2 + B^2\right)^{-1}. \quad (4.82)$$

Here $r^A\left(x^\alpha\right)$ and $q^A\left(x^\alpha\right)$ are new arbitrary three-dimensional functions and B denotes

$$B = \alpha_0^2 e^{-2\Lambda p_{\bar{1}} \tau} + \omega^2. \quad (4.83)$$

Now we can easily calculate the three-dimensional function ω^2, inserting the solution (4.81) and (4.82) into (4.77). It is expressed in terms of the arbitrary three-dimensional parameters r^A and q^A as follows:

$$\omega^2 = r^A r^A + q^A q^A. \quad (4.84)$$

While studying the above solution for a^2, b^2, c^2, we have already pointed out that the Kasner epochs in the given case correspond to the asymptotic regions $\tau \to -\infty$ $(t \to +\infty)$ and $\tau \to +\infty$ $(t \to 0)$. It is clear from (4.81)–(4.83) that the behavior of the fields $E^{A\bar{1}}$ and $H^{A\bar{1}}$ in both epochs is exactly as it has been assumed to be by equations (4.60) at the beginning of this subsection; this confirms the self-consistency of our analysis. Thus, in the first approximation, the quantities $abcE^{A\bar{1}} \sim tE^{A\bar{1}}$ and $abcH^{A\bar{1}} \sim tH^{A\bar{1}}$ are actually time-independent during each epoch and change sharply in the transition region between two epochs.

The $l^{\bar{2}}_{\alpha}$- and $l^{\bar{3}}_{\alpha}$- projections of the gauge fields have identical properties. This becomes evident if equations (4.66)–(4.69) are employed. Their right-hand sides now involve only the known functions $a^2, E^{A\bar{1}}$ and $H^{A\bar{1}}$; therefore the $l^{\bar{2}}_{\alpha}$- and $l^{\bar{3}}_{\alpha}$-projections are found by simple integration. By virtue of (4.64)–(4.65), this integration proves to be trivial. Replacing the quantities $a^2 E^{A\bar{1}}$ and $a^2 H^{A\bar{1}}$ in the right-hand sides of (4.66)–(4.69) with the time derivatives in accordance with (4.64)–(4.65), we immediately get:

$$abcH^{A\bar{2}} = \left(\frac{abc}{\lambda}H^{A\bar{1}}\right)_{,\alpha} l^{\alpha}_{\bar{3}} - \frac{\gamma abc}{\lambda}H^{A\bar{1}} - \frac{\zeta abc}{\lambda}\varepsilon_{ABD}\beta^{D}_{\alpha} l^{\alpha}_{\bar{3}} H^{B\bar{1}} + U^{A\bar{2}}, \quad (4.85)$$

$$abcH^{A\bar{3}} = -\left(\frac{abc}{\lambda}H^{A\bar{1}}\right)_{,\alpha} l^{\alpha}_{\bar{2}} - \frac{\rho abc}{\lambda}H^{A\bar{1}} + \frac{\zeta abc}{\lambda}\varepsilon_{ABD}\beta^{D}_{\alpha} l^{\alpha}_{\bar{2}} H^{B\bar{1}} + U^{A\bar{3}}, \quad (4.86)$$

$$abcE^{A\bar{2}} = \left(\frac{abc}{\lambda}E^{A\bar{1}}\right)_{,\alpha} l^{\alpha}_{\bar{3}} - \frac{\gamma abc}{\lambda}E^{A\bar{1}} - \frac{\zeta abc}{\lambda}\varepsilon_{ABD}\beta^{D}_{\alpha} l^{\alpha}_{\bar{3}} E^{B\bar{1}} + V^{A\bar{2}}, \quad (4.87)$$

$$abcE^{A\bar{3}} = -\left(\frac{abc}{\lambda}E^{A\bar{1}}\right)_{,\alpha} l^{\alpha}_{\bar{2}} - \frac{\rho abc}{\lambda}E^{A\bar{1}} + \frac{\zeta abc}{\lambda}\varepsilon_{ABD}\beta^{D}_{\alpha} l^{\alpha}_{\bar{2}} E^{B\bar{1}} + V^{A\bar{3}}. \quad (4.88)$$

The last terms in the right-hand sides of these formulae denote arbitrary three-dimensional functions. Since the quantities $abcE^{A\bar{1}}$ and $abcH^{A\bar{1}}$ are, as has been discovered before, constant during each separate Kasner epoch, it follows from (4.85)–(4.88) that the projections of the quantities $abcE^{A\alpha}$ and $abcH^{A\alpha}$ on the directions $l^{\bar{2}}_{\alpha}$ and $l^{\bar{3}}_{\alpha}$ are also constant in these regions, so that actually all the components of $abcE^{A\alpha}$ and $abcH^{A\alpha}$ are constant, as it has been our assumption from the very beginning.

This character of the behavior of the fields enables us to conclude that the projections of the potentials $L^{A}_{\bar{2}}$ and $L^{A}_{\bar{3}}$ are unchanged in the transition region between two epochs. This is not valid for the $l^{\bar{1}}_{\alpha}$-projection of L^{A}_{α}, satisfying equation (4.62), which is integrated as simply as is the system (4.66)–(4.69). Taking into account (4.64), we obtain:

$$L^{A}_{\bar{1}} = -\frac{abc}{\lambda}H^{A\bar{1}} + W^{A}_{\bar{1}}\,, \quad (4.89)$$

where $W^{A}_{\bar{1}}$ are some three-dimensional functions. Hence, it becomes clear that the projections of the potentials $L^{A}_{\bar{1}}$ behave in time in the same way as the quantities $abcH^{A\bar{1}}$: during Kasner epochs they can be treated as constants, but in the transition region they change sharply.

The influence of an electromagnetic field on the character of the cosmological singularity need not a separate study because all information on this case can directly be extracted from the foregoing analysis of the behavior of the gauge fields. To get the results for the electromagnetic field, it is enough to put the constant ζ to zero and consider the case when the index A takes only one value. One reaches exactly the same conclusions, confirming that the non-abelian structure of the gauge field does not play a significant role at leading order as one goes to the singularity.

4.5 Scalar Field

The influence of a scalar field (called "dilaton" in the billiard context developed in Part II) on the character of the cosmological singularity does not need any new investigation of the corresponding scalar-tensor gravitational equations, because these are equivalent to the equations for the gravitational field in the presence of a stiff fluid, analyzed already above in Section 4.3.

This is because, near the singularity, a massive field becomes effectively massless since in this region the field starts to be ultrarelativistic and the influence of the mass-terms ceases to be of any significance. Furthermore, the energy–momentum tensor of a massless scalar field can be represented as the energy–momentum tensor of a perfect fluid with equation of state $p = \varepsilon$ [16]. Indeed, consider the action:

$$S = \int (R + \mathcal{L}_\varphi) \sqrt{-\det g_{ik}}\, d^4x. \tag{4.90}$$

The simplest form of the scalar–tensor theory corresponds to the Lagrangian:

$$\mathcal{L}_\varphi = -\varphi_{;k}\varphi^{;k}. \tag{4.91}$$

We stress that there is no loss of generality in choosing this form of the Lagrangian since any other scalar–tensor theory known in the literature (e.g., Brans–Dicke theory) can be derived from this form by a conformal transformation of the metric accompanied by an appropriate transformation of the scalar field [16].

The Lagrangian (4.91) corresponds to the energy-momentum tensor

$$T_{ik} = \varphi_{;i}\varphi_{;k} - \frac{1}{2} g_{ik}\varphi_{;l}\varphi^{;l}\ , \tag{4.92}$$

and to the following system of field equations:

$$\varphi^{;k}_{;k} = 0, \tag{4.93}$$

$$R_{ik} = \varphi_{;i}\varphi_{;k}. \tag{4.94}$$

If we define the "velocity" u_i and "pressure" p as

$$u_i = \frac{\varphi_{;i}}{\sqrt{-\varphi_{;l}\varphi^{;l}}},\ p = -\frac{1}{2}\varphi_{;l}\varphi^{;l}, \tag{4.95}$$

then it is easy to see that the energy–momentum tensor (4.92) takes the form $T_{ik} = 2pu_iu_k + pg_{ik}$, that is the form of a fluid with stiff equation of state $\varepsilon = p$.

Consequently, we can directly apply to the scalar case all the results found in Section 4.3. The first approximation to the general solution of equations (4.93)–(4.94) near the singularity is described by the expressions (1.42)–(1.43), where all three exponents $p_{\bar{\alpha}}$ are positive and satisfy the conditions (4.4). The leading term in the solution for the scalar field φ follows from (4.17) in which one substitutes the "hydrodynamic" quantities using (4.95). The result is:

$$\varphi = p_\varphi \ln t + \varphi_0 \ , \tag{4.96}$$

where φ_0 is an arbitrary three-dimensional function. All subsequent terms in the expansion of the solution can easily be calculated in the same way, using the results presented in Subsection 4.3.2. All these corrections are completely determined by the main approximation and do not contain any new arbitrariness. In the leading approximation (1.42)–(1.43), (4.4) and (4.96) we have nine arbitrary three-dimensional functions represented by the vectors $l^{\bar{\alpha}}_\alpha$ (considering the factors a_0^2, b_0^2, c_0^2 to be included in the vectors $l^{\bar{\alpha}}_\alpha$), two such functions among the exponents $p_{\bar{\alpha}}, p_\varphi$ and also φ_0 in (4.96). This amounts to a total of twelve, but three of these arbitrary parameters represent the non-physical gauge freedom existing in the synchronous system and another three are fixed by the initial conditions equation (1.16). We thus get in the end six physical arbitrary parameters, which is exactly the number necessary for the solution to be general.

It is worth making the following remarks. If in (1.42)–(1.43), (4.4) and (4.96) we take all three-dimensional functions $l^{\bar{\alpha}}_\alpha, p_{\bar{\alpha}}, p_\varphi, \varphi_0$ to be constants then this first approximation becomes an exact solution of the coupled Einstein and Klein–Gordon equations (4.93)–(4.94). Therefore, in the presence of a scalar field, the basic structure of the general solution arises already from the homogeneous case when the metric and the scalar field depend only on time. To get the asymptotic form of the general solution, it is enough to simply replace all the arbitrary constants of the general homogeneous model by three-dimensional functions (subject to equation (1.16)). This means stability of this homogeneous regime (in contrast to the Kasner-like evolution).

As we already mentioned at the end of the introduction to this chapter, the disappearance of the oscillations (in conventional four-dimensional gravity) due to the presence of a scalar field is an effect inherent only to the case when no other tensor fields is present. If we add also such fields, the endless oscillations generically reappear.

4.6 Pure Gravity in Higher Dimensions

A change in the space-time dimension does not invalidate the general qualitative approach to the description of the singularity, which can still be expressed in terms of Kasner free flights interrupted by collisions. However, it turns out that for space-time dimensions greater than or equal to eleven, the Kasner free flight motion is an asymptotic solution of the vacuum Einstein equations – and hence is uninterrupted as $t \to 0$ – for values of the Kasner parameters lying in an *open* region [63]. Therefore, the existence or nonexistence of the oscillatory behavior depends also on the space-time dimension, just as it does on the matter content.

That the space-time dimension plays a role is not surprising, since the dimensional reduction of a pure gravitational system from D space-time dimensions to four space-time dimensions yields the Einstein theory coupled to electromagnetic and scalar fields. We have seen that, while the vector fields preserve the oscillatory behavior, a scalar field alone does not. The combination of the two may or may not preserve chaos depending on the scalar couplings. These couplings, and the number of fields, depend on the dimension from which one descends.

We consider the case of pure gravity to exhibit in the simplest context the dependence on the space-time dimension. We treat the system directly in D dimensions without making any dimensional reduction.

4.6.1 Generalized Kasner Metric in D Space-Time Dimensions

As in four space-time dimensions, the basic building block is the generalized Kasner metric, obtained by letting the integration constants in the exact Kasner solution depend on the spatial coordinates. We work again in a synchronous coordinate system adapted to the singularity,

$$- ds^2 = -dt^2 + g_{\alpha\beta}(t, x^\gamma) dx^\alpha dx^\beta, \tag{4.97}$$

which we take to be located at $t = 0$. The indices α, β, γ, ... (as well as their bared counterparts) take now the $d = D - 1$ spatial values $1, 2, \cdots, d$.

The generalized Kasner metric is explicitly given by

$$g_{\alpha\beta} = \eta_{\bar{\alpha}\bar{\beta}} l^{\bar{\alpha}}_{\alpha} \, l^{\bar{\beta}}_{\beta} \, , \quad \eta_{\bar{\alpha}\bar{\beta}} = diag(t^{2p_{\bar{1}}}, t^{2p_{\bar{2}}}, \cdots, t^{2p_{\bar{d}}}). \tag{4.98}$$

Here, the "Kasner exponents" $p_{\bar{\alpha}}(x^1, \cdots, x^d)$ should satisfy the restriction

$$\sum_{\bar{\alpha}=1}^{d} p_{\bar{\alpha}} = 1 \tag{4.99}$$

($g \sim t^2$), as well as

$$\sum_{\bar{\alpha}=1}^{d} p^2_{\bar{\alpha}} = 1. \tag{4.100}$$

The functions $l^{\bar{\alpha}}_{\alpha}$ depend also on the spatial coordinates and define an invertible change of frame $(\det(l^{\bar{\alpha}}_{\alpha}) \neq 0)$ but are otherwise arbitrary at this stage. Just as in three spatial dimensions, the Kasner conditions (4.99) and (4.100) imply that there are positive and negative Kasner exponents $p_{\bar{\alpha}}$, so that some directions expand while some others contract as one goes towards the singularity. The

$(d-2)$-dimensional surface in the space of the ps given by the intersection of the plane (4.99) with the $(d-1)$-sphere (4.100) is called thereafter the "Kasner sphere."

The Kasner solution is spatially homogenous and corresponds to Kasner exponents $p_{\bar{\alpha}}$ and frame components $l^{\bar{\alpha}}_{\alpha}$ that are constants. It is an exact solution of the vacuum Einstein equations, the Kasner condition (4.100) originating from the constraint $G_{00} = 0$. The above-generalized Kasner metric is not a solution of the Einstein equations when the Kasner exponents and frame components are inhomogeneous, however. But it is an approximate solution as $t \to 0$ provided that the time derivatives indeed dominate the spatial gradients as $t \to 0$, so that the terms coming from the spatial inhomogeneities in the Einstein equations can be dropped in the limit.

We thus investigate when the spatial gradient terms are subdominant with respect to the time derivative terms in the vacuum Einstein equations. A computation that parallels the same computation performed in four space-time dimensions shows that the time derivative terms for the metric (4.98) are of order t^{-2} in the components ${}^{(D)}G^0_0$ and ${}^{(D)}G^{\alpha}_{\beta}$ of the space-time Einstein tensor. On the other hand, the spatial gradients (which are zero for the exact Kasner metric) enter ${}^{(D)}G^0_0$ and ${}^{(D)}G^{\alpha}_{\beta}$ through the d-dimensional Ricci tensor ${}^{(d)}P^{\alpha}_{\beta}$. Hence we can neglect the spatial gradients with respect to the time derivatives in the Einstein equations ${}^{(D)}G^0_0 = 0$ and ${}^{(D)}G^{\alpha}_{\beta} = 0$ if and only if ${}^{(d)}P^{\alpha}_{\beta}$ can be neglected with respect to t^{-2}, i.e.,

$$\lim_{t \to 0} t^2 \, {}^{(d)}P^{\alpha}_{\beta} = 0 \tag{4.101}$$

Now, many powers of t appear in the expansion of $t^2 \, {}^{(d)}P^{\alpha}_{\beta}$. However, one can directly verify that the only "potentially dangerous" terms (i.e., the only terms that might violate (4.101)) are of the form

$$t^{2\alpha_{\bar{\alpha}\bar{\beta}\bar{\gamma}}} \tag{4.102}$$

where $\bar{\alpha}, \bar{\beta}, \bar{\gamma}$ are distinct and where the exponents $\alpha_{\bar{\alpha}\bar{\beta}\bar{\gamma}}$ are related to the Kasner exponents through

$$\alpha_{\bar{\alpha}\bar{\beta}\bar{\gamma}} = 2p_{\bar{\alpha}} + \sum_{\bar{\delta} \neq \bar{\alpha}, \bar{\beta}, \bar{\gamma}} p_{\bar{\delta}} \qquad (\bar{\alpha} \neq \bar{\beta}, \bar{\alpha} \neq \bar{\gamma}, \bar{\beta} \neq \bar{\gamma}). \tag{4.103}$$

The potentially dangerous powers (4.102) are multiplied in $t^2 \, {}^{(d)}P^{\alpha}_{\beta}$ by coefficients $A_{\bar{\alpha}\bar{\beta}\bar{\gamma}}$ that involve the spatial functions $l^{\bar{\alpha}}_{\alpha}$, $p_{\bar{\alpha}}$ ("initial data") and their spatial derivatives and that are generically non-vanishing.

The computations leading to this result are somewhat cumbersome but otherwise straightforward. They are in fact the same as the computations of the spatial curvature walls made in the next two chapters to derive the billiard picture and are therefore left to the reader.

The actual dangerous terms leading to a violation of the Einstein equations by the metric (4.98) are those for which $\alpha_{\bar{\alpha}\bar{\beta}\bar{\gamma}} \leq 0$, since it is only for those that the corresponding terms in $t^2\ {}^{(d)}P^\alpha_\beta$ blow up or remain of order $O(1)$. These terms must therefore be absent if (4.98) is to be an asymptotic solution of the vacuum Einstein equations as $t \to 0$.

This can occur in two different ways:

- Either there is an open region on the Kasner sphere where all the $\alpha_{\bar{\alpha}\bar{\beta}\bar{\gamma}}$ are positive, $\alpha_{\bar{\alpha}\bar{\beta}\bar{\gamma}} > 0 \quad \forall \bar{\alpha}, \bar{\beta}, \bar{\gamma}$ and the Kasner exponents lie in that region. There is then no dangerous term. As we shall prove in the next subsection, this alternative is absent for $D < 11$, but present for $D \geq 11$.
- Or some of the $\alpha_{\bar{\alpha}\bar{\beta}\bar{\gamma}}$ are negative, and one must then impose on the "initial data" $l^{\bar{\alpha}}_\alpha$ and $p_{\bar{\alpha}}$ the requirement that they fulfill the conditions $A_{\bar{\alpha}\bar{\beta}\bar{\gamma}} = 0$ for those $\alpha_{\bar{\alpha}\bar{\beta}\bar{\gamma}} < 0$, ensuring that the corresponding dangerous terms are absent in $t^2\ {}^{(d)}P^\alpha_\beta$.

In both cases, the generalized Kasner metric is an asymptotic solution. However, in the second case, there are less freely specified functions among the initial conditions than in the first case. Of course, the initial data are also subject to the constraints $G^0{}_\alpha = 0$, but these are easily checked to be independent from $A_{\bar{\alpha}\bar{\beta}\bar{\gamma}} = 0$, and furthermore must be imposed whether the first or the second situation holds.

In four dimensions, the exponent $\alpha_{\bar{\alpha}\bar{\beta}\bar{\gamma}}$ reduces to $2p_{\bar{\alpha}}$. This matches what we derived in Chapter 2, where we showed that the components $t^2\ {}^{(3)}P^\alpha_\beta$ indeed contained terms that behaved like $t^{4p_{\bar{\alpha}}}$. Since one of the Kasner exponents is necessarily negative ($p_{\bar{1}}$ if they are ordered), the first alternative cannot hold: there is no region on the Kasner sphere where all $a_{\bar{\alpha}\bar{\beta}\bar{\gamma}}$s are positive. This forces one to impose on the initial conditions the extra condition that the coefficient of the "dangerous terms" $t^{4p_{\bar{1}}}$ be absent if the generalized Kasner metric is to be an asymptotic solution. There is nothing wrong if this extra condition is not imposed, but then the metric (4.98) is not a solution all the way to $t = 0$ and the oscillatory behavior prevails.

In higher dimensions, it is *not* the Kasner exponents themselves that control the relative growth of the spatial gradients with respect to the time derivatives in the Einstein equations, but rather their linear combinations $\alpha_{\bar{\alpha}\bar{\beta}\bar{\gamma}}$ [63]. This has the profound implication that the first possibility becomes available for $D = d + 1 \geq 11$. Before establishing this fact, we first do the counting of the number of arbitrary functions involved in the generalized Kasner solutions.

The generalized Kasner metric (4.98) contains d^2 (number of components of $l^{\bar{\alpha}}_\alpha$) + d (number of Kasner exponents) $-d$ (number of constraints $G^0{}_\alpha = 0$) -2 (number of Kasner conditions) $= d^2 - 2$ functions of space. These functions are redundant since there is the residual symmetry of making spatial diffeomorphisms (d functions). This leaves $d^2 - d - 2$ functions of space. This is precisely the number of physically distinct arbitrary functions of space necessary to describe a "general solution" of the vacuum gravitational field equations in $D = d + 1$ space-time dimensions. This can be seen by direct counting from the Hamiltonian formulation, or by recalling that a symmetric traceless tensor for the little group $SO(d-1)$ contains $\frac{1}{2}(d^2-d-2)$ components, to be multiplied by 2 to get the number of physically distinct independent initial data since the equations are of second order. Therefore, if the functions $l^{\bar{\alpha}}_{\alpha}$ and the Kasner exponents can be chosen freely (apart from the Kasner conditions and the constraints $G^0{}_\alpha = 0$), without conflicting with the Einstein equations, the metric is a "general solution." If, on the contrary, one must impose an extra condition of the type $A_{\bar{\alpha}\bar{\beta}\bar{\gamma}} = 0$ on the initial data to guarantee that the Einstein equations hold asymptotically, then, the generalized Kasner solution does not contain sufficiently many arbitrary functions to accommodate an open set of admissible initial data. It is not a general solution.

We stress that if the conditions $\alpha_{\bar{\alpha}\bar{\beta}\bar{\gamma}} > 0$ can be fulfilled on the Kasner sphere, they define an open region and therefore do not invalidate the above counting since there are still $d-2$ independent Kasner exponents even after these strict inequalities are imposed – the independent Kasner exponents have restricted range but are otherwise arbitrary.

The question then is whether the conditions $\alpha_{\bar{\alpha}\bar{\beta}\bar{\gamma}} > 0$ can be fulfilled on an open region of the Kasner sphere. It is to this question that we now turn.

4.6.2 Critical Dimension

It was shown in [63], where the analysis was carried out for the first time, that the conditions $\alpha_{\bar{\alpha}\bar{\beta}\bar{\gamma}} > 0$ have no solution for space-time dimensions $D < 11$.

To that end, one observes that one may restrict the Kasner exponents, which play a symmetrical role, to the region on the Kasner sphere defined by the inequalities

$$p_{\bar{1}} \leq p_{\bar{2}} \leq p_{\bar{3}} \leq \cdots \leq p_{\bar{d}}. \tag{4.104}$$

This is a closed – and hence compact – region of the Kasner sphere. The condition that the exponents $\alpha_{\bar{\alpha}\bar{\beta}\bar{\gamma}}$ be positive for all $\bar{\alpha}, \bar{\beta}, \bar{\gamma}$ is equivalent to the condition that the smallest exponent, namely $\alpha_{\bar{1}\overline{d-1}\bar{d}}$ in the region under consideration, be positive,

$$\alpha_{\bar{1}\overline{d-1}\bar{d}} = 2p_{\bar{1}} + p_{\bar{2}} + \cdots + p_{\overline{d-2}} > 0. \tag{4.105}$$

The analysis proceeds then as follows.

- First, one notes that if the inequalities $\alpha_{\bar{\alpha}\bar{\beta}\bar{\gamma}} > 0$ can all be fulfilled in d spatial dimensions, they can also be all fulfilled in $d+1$ spatial dimensions. Indeed, let $\{p_{\bar{\alpha}}\}$ be a set of ordered Kasner exponents that fulfill $\alpha_{\bar{\alpha}\bar{\beta}\bar{\gamma}} > 0$ in d spatial dimensions. Set $p_{\overline{d+1}} = 0$. Then the set $\{p_{\bar{\alpha}}, p_{\overline{d+1}}\}$ defines a point on the Kasner sphere. It is not ordered because $p_{\overline{d+1}} = 0 < p_{\bar{d}}$, and we call $p'_{\bar{\alpha}}$ the exponents obtained by reordering. One has $p'_{\bar{1}} = p_{\bar{1}}$ and, because $p_{\overline{d-2}} > 0$ (otherwise, $\alpha_{\bar{1}\bar{d}\,\overline{d-1}}$ could not be positive), one has also $p'_{\overline{d-1}} = p_{\overline{d-2}}$, $p'_{\bar{d}} = p_{\overline{d-1}}$, $p'_{\overline{d+1}} = p_{\bar{d}}$, the value zero being somewhere between p_1 and $p_{\overline{d-2}}$. For this ordered set of $p'_{\bar{\alpha}}$, one has, for the smallest $\alpha'_{\bar{\alpha}\bar{\beta}\bar{\gamma}}$,

$$\alpha'_{\bar{1}\bar{d}\,\overline{d+1}} = 2p'_{\bar{1}} + p'_{\bar{2}} + \cdots + p'_{\overline{d-1}} = \alpha_{\bar{1}\,\overline{d-1}\,\bar{d}} > 0 \tag{4.106}$$

and hence $\alpha'_{\bar{\alpha}\bar{\beta}\bar{\gamma}} > 0$ for all $\alpha'_{\bar{\alpha}\bar{\beta}\bar{\gamma}}$s.

- Second, we note that for $d = 9$, there is no point on the Kasner sphere that fulfills simultaneously $\alpha_{\bar{\alpha}\bar{\beta}\bar{\gamma}} > 0$ for all $\alpha_{\bar{\alpha}\bar{\beta}\bar{\gamma}}$s. This is shown by extremization of the smallest exponent $\alpha_{\bar{1}\,\overline{d-1}\,\bar{d}}$ in the compact region (4.104) of the Kasner sphere. The maximum of that function is easily verified to be achieved at the boundary of the region, at the point

$$p_{\bar{1}} = p_{\bar{2}} = p_{\bar{3}} = -\frac{1}{3}, \quad p_{\bar{4}} = p_{\bar{5}} = \cdots = p_{\bar{9}} = \frac{1}{3}. \tag{4.107}$$

  One has there $\alpha_{\bar{1}\bar{8}\bar{9}} = 0$ and this is the best one can do. Hence, the inequality $\alpha_{\bar{1}\bar{8}\bar{9}} > 0$ is violated.

- Third, one observes that in $d = 10$ spatial dimensions ($D = 11$), the exponents $\alpha_{\bar{\alpha}\bar{\beta}\bar{\gamma}}$ are all strictly positive at the point

$$p_{\bar{1}} = p_{\bar{2}} = p_{\bar{3}} = \frac{1-\sqrt{21}}{10}, \quad p_{\bar{4}} = p_{\bar{5}} = \cdots = p_{\overline{10}} = \frac{7+3\sqrt{21}}{70} \tag{4.108}$$

  and in its vicinity: the smallest exponent $\alpha_{\bar{1}\bar{9}\,\overline{10}}$ is equal to $\frac{63-13\sqrt{21}}{70}$ and is strictly positive.

We thus conclude that for pure gravity in $D \geq 11$ space-time dimensions, the region $\alpha_{\bar{\alpha}\bar{\beta}\bar{\gamma}} > 0$ defines a nonempty open subset of the Kasner sphere and that the generalized Kasner metric with exponents in that region is a general asymptotic solution containing sufficiently many arbitrary functions of space. This general solution does not exhibit the oscillatory behavior, since the uniform Kasner solution holds all the way to the singularity $t = 0$.

One can ask what happens if one starts with Kasner exponents that do not lie in the open region $\alpha_{\bar{\alpha}\bar{\beta}\bar{\gamma}} > 0$. One then has, just as in four dimensions, a "collision" due to the curvature term that cannot be neglected, inducing a transition to a new Kasner regime with new Kasner exponents. It has been argued in [62] that after a finite number of collisions, the system generically settles in a Kasner regime with exponents fulfilling $\alpha_{\bar{\alpha}\bar{\beta}\bar{\gamma}} > 0$, which remains valid up to $t = 0$. The never-ending oscillatory behavior therefore ceases. The collision law relating the new Kasner exponents to the old ones was also derived in [62] and will be given in the next chapter in the billiard context.

We end this section with four comments: (i) First, matter may re-establish the chaotic behavior and render the Kasner solution unstable. This notably happens for $D = 11$ supergravity, where the 3-form present in the model precisely produces this effect though its energy density, which forces new types of collisions leading to the never-ending oscillatory behavior [45, 46]. (ii) Second, some of the general features discussed here can be exhibited in the simpler context of higher-dimensional homogeneous models [61]. It is interesting to observe that in those homogeneous cases, rotations of the Kasner axes is important to generate chaos for $4 < D < 11$, just as in the Bianchi model of type VII filled with a perfect fluid discussed at the end of Section 4.2 above. (iii) Third, there is a beautiful connection between the critical value $d = 9$ found here and the theory of hyperbolic Coxeter groups and Kac–Moody algebras, which will be explained in Chapter 7. (iii) Finally, some of the stochastic properties of the vacuum gravitational system have been investigated in higher dimensions in [70, 71].

4.7 Generalized Kasner Solutions: Rigorous Results

We have established in this chapter that, in general, matter fields and modification of the space-time dimension do not change the description of the asymptotic dynamics of the scale factors in terms of Kasner free flight motions interrupted by collisions, the net effect of which is to replace the Kasner exponents by new ones. For many menus of matter fields and space-time dimension, the collisions never end and the system undergoes the oscillatory regime with an infinite number of oscillations as $t \to 0$. However, for some special menus of matter fields and space-time dimensions, there is only a finite number of collisions and the system settles in a final Kasner regime as $t \to 0$.

Anticipating the results of the next chapter, where we shall develop the billiard description, the first case corresponds to a billiard table with finite volume, exhibiting strong chaotic properties, while the second case corresponds to a billiard table with infinite volume.

When there are only finitely many collisions, the dynamics in the vicinity of the singularity is particularly simple. After the last collision, the asymptotic dynamics are dominated by the time derivatives of the scale factors (among which we include the scalar fields – "dilatons" – if any). This case where spatial gradients terms can be neglected, has been called "asymptotically velocity-dominated" (or "AVD") in [66].* The first example of such a situation encountered above is given by a perfect fluid with stiff equation of state (Section 4.3) or equivalently the pure coupled scalar-Einstein system (Section 4.5). The second example is given by pure gravity in space-time dimensions $D \geq 11$ (Section 4.6).

* Strictly speaking, it is not sufficient to impose that time derivatives dominate spatial gradients in order to get a last uninterrupted Kasner regime holding all the way to the singularity. Subdominance of the spatial gradients is only one necessary condition in order to insure the existence of a final Kasner regime. For instance, as we have seen, purely electric fields, which correspond to the time derivatives of the electromagnetic vector potential, induce never-ending collisions when present. Such fields – more generally p-form fields – should be absent or negligible in order to get a final Kasner regime. This is implicitly assumed in the definition of AVD singularities, which were originally studied only in the case were the system settles in a final generalized Kasner solution. Perhaps the terminology "Quiescent cosmological singularity," introduced in [3], is less ambiguous and more appropriate for this reason.

Just as in the stiff case, AVD singularities allow a rigorous analysis of their asymptotic dynamics by means of Fuchsian techniques [3, 52]. The idea is that one can write an explicit analytic form of the metric valid asymptotically, namely the generalized Kasner metric with exponents in the appropriate range. One can then show analytically that the deviation with respect to this asymptotic behavior is subdominant as $t \to 0$, along the lines of Section 4.3. Reference [3] deals with gravity coupled to a scalar field; reference [52] deals with pure gravity in dimension $D \geq 11$. Both models are velocity-dominated.

By contrast, as we pointed out in the introduction, rigorous results are much more difficult to obtain for the case of infinitely many collisions, because the asymptotic solution is then much more complicated. There exists however analytic advances [144] as well as a lot of numerical support [25, 23, 76, 24, 77, 82].

4.8 On the Influence of Viscous Matter

The previous analysis showed that there exist some special cases when the influence of matter can change the asymptotics of the general solution near a cosmological singularity in an essential way, namely, instead of the oscillatory regime, the behavior of the fields becomes of the smooth power law character. However, this behavior still remains anisotropic since the scale factors corresponding to different space directions behave differently.

At the same time, observations show that the early Universe in a good approximation is isotropic. A number of authors [9, 139, 81] expressed the point of view that also the initial cosmological singularity might be in conformity with these properties, that is, it might be of the Friedmann isotropic type. But from all our preceding considerations and directly from the seminal paper of E. Lifshitz [122], it follows that the isotropic singularity for the conventional types of matter is unstable, which means that space-time cannot start an isotropic expansion unless an artificial fine tuning of unknown origin appears. This instability is due to the sharp anisotropy which develops unavoidably near the generic cosmological singularity for the standard "perfect" types of matter. Nevertheless, an intuitive understanding suggests that anisotropy can be damped down in the presence of dissipative effects, first of all due to the shear viscosity which, if taken into account, might result in a *generic solution with isotropic Big Bang*. In the present section we show that such a case can indeed be realized.

4.8.1 Relativistic Dissipative Fluid

To search for an analytical realization of a generic isotropic asymptotical behavior in the presence of shear viscosity, it would be inappropriate to use just the Eckart [67] or Landau–Lifshitz [121] approaches to relativistic hydrodynamics with dissipative processes. These theories are valid provided the characteristic times of the macroscopic motions of matter are much bigger than the time of relaxation of the medium to the equilibrium state. It might happen that this is not so near the cosmological singularity since all characteristic macroscopic times in this region tend to zero, in which case one needs a theory which takes into account Maxwell's relaxation times. In a literal sense, such a theory does not exist. However, it can be constructed in an approximate form for the cases when a medium does not deviate too much from equilibrium and the relaxation times do not exceed noticeably the characteristic macroscopic times. It is reasonable to expect that these conditions will be satisfied (and in a model described below, this is indeed the case) for a generic solution near the isotropic singularity describing the beginning of the Friedmann Universe accompanied by arbitrary infinitesimally small corrections.

The main target of the efforts by many authors (starting from the first idea of Cattaneo [32] up to the final formulation of the generalized relativistic theory by Israel and Stewart [98, 99]) was to bring the theory in line with relativistic causality, that is to eliminate the supraluminal propagation of the thermal and viscous excitations. The existence of such supraluminal effects was the main stumbling-block for the Eckart's and Landau–Lifshitz's descriptions of dissipative fluids. One of the first applications of the Israel–Stewart theory to the problems of cosmological singularity was undertaken in the article [13]. Already in this paper the stability of the Friedmann models under the influence of the shear viscosity was investigated and it was found that relativistic causality and stability of the Friedmann singularity are in contradiction with each other. However, it was shown in [12] that this "no go" conclusion was the result of too restricted a range for the dependence of the shear viscosity coefficient on the energy density. As usual, in the vicinity of the singularity where the energy density ε diverges, we approximate the coefficient of viscosity η by the power law asymptotics $\eta \sim \varepsilon^{\nu}$ with some exponent ν. In the article [13], the values of this exponent was chosen from the region $\nu > 1/2$. For such values of ν, the negative result of the paper [13] remains correct, but in [12] it was found that the boundary value $\nu = 1/2$ leads to a dramatic change of the state of affairs. It turns out that, for this case, there exists a window in the space of the free parameters of the theory in which the Friedmann singularity becomes stable and at the same time *no supraluminal signals* exist in its vicinity.

It is worth adding that the case $\nu < 1/2$ was also analyzed in the article [12], but this is of no relevance here since it leads to a strong instability of the isotropic singularity, independently of the question of relativistic causality.

Also it is necessary to stress that we consider only the standard models for a physical fluid for which the pressure is nonnegative and is less than the energy density.

4.8.2 On the Basic Equations and Their Solution

Shear stresses generate an extra term S_{ik} that must be added to the standard energy–momentum tensor of a fluid:

$$T_{ik} = (\varepsilon + p)\, u_i u_k + p g_{ik} + S_{ik} \ , \tag{4.109}$$

and this additional term has to satisfy the following constraints [121]:

$$S_{ik} = S_{ki} \ , \ S_k^k = 0 \ , \ u^i S_{ik} = 0 \ . \tag{4.110}$$

Besides, we have the usual normalization condition for the 4-velocity:

$$u_i u^i = -1. \tag{4.111}$$

If Maxwell's relaxation time τ of the stresses is not zero, then there does not exist any closed expression for S_{ik} in terms of the viscosity coefficient η and the 4-gradients of the 4-velocity. Instead the stresses S_{ik} should be defined from the following differential equations [98]:

$$\begin{aligned} & S_{ik} + \tau \left(\delta_i^m + u_i u^m\right)\left(\delta_k^n + u_k u^n\right) S_{mn;l} u^l \\ & = -\eta \left(u_{i;k} + u_{k;i} + u^l u_k u_{i;l} + u^l u_i u_{k;l}\right) + \frac{2}{3}\eta \left(g_{ik} + u_i u_k\right) u^l{}_{;l} \ , \end{aligned} \tag{4.112}$$

which, due to the normalization condition for the velocity, is compatible with the constraints (4.110). In the case $\tau = 0$, the expression for S_{ik} following from this equation coincides with the one introduced by Landau and Lifshitz [121]. If the equations of state $p = p(\varepsilon)\, ,\ \eta = \eta(\varepsilon)\, ,\ \tau = \tau(\varepsilon)$ are fixed, then the Einstein equations

$$R_{ik} = T_{ik} - \frac{1}{2} g_{ik} T_l^l \tag{4.113}$$

together with equation (4.112) for the stresses give a closed system from which from all quantities of interest, that is $g_{ik}, u_i, \varepsilon, S_{ik}$ can be found.

As we already said, in the vicinity of the cosmological singularity the viscosity coefficient η can approximated by a power law asymptotics $\eta = const \cdot \varepsilon^\nu$ with some constant exponent ν. However, the only truly interesting value of this exponent here is $\nu = 1/2$:

$$\eta = C_0 \varepsilon^{1/2} \ , \tag{4.114}$$

where C_0 is some dimensionless constant. As for the relaxation time τ the choice is more or less definite. It is known that $\eta/\varepsilon\tau$ represents a measure of velocity of propagation of the shear excitations. Then we can model this ratio by a positive constant f (in a more accurate theory f can be a slowly varying function of time but in any case this function should be bounded in order to exclude the appearance of supraluminal signals). Consequently we choose the following model for the relation between the relaxation time and the viscosity coefficient:

$$\eta = f \varepsilon \tau, \ f = const. \tag{4.115}$$

For the dependence $p = p(\varepsilon)$, we follow the standard approximation with constant parameter γ:

$$p = (\gamma - 1)\, \varepsilon \ , \ 1 \leqslant \gamma < 2. \tag{4.116}$$

Now the system of all equations is closed and we can search for the asymptotic behavior of its solution in the vicinity of the cosmological singularity. It is convenient to work in the synchronous reference system (1.2). Our task is to take the standard Friedmann metric in this system as background and to find the asymptotic (near the singularity) solution of equations (4.109)–(4.116) for the linear perturbations around this background in the same synchronous system. The background solution is:*

$$-ds^2 = -dt^2 + R^2\left[\left(dx^1\right)^2 + \left(dx^2\right)^2 + \left(dx^3\right)^2\right], \; R = (t/t_c)^{2/3\gamma}, \tag{4.117}$$

$$\varepsilon = 4\left(3\gamma^2 t^2\right)^{-1}, \; u_0 = -1, \; u_\alpha = 0, \; S_{ik} = 0, \tag{4.118}$$

where $t > 0$ and t_c is some arbitrary positive constant (it is worth remarking that in the comoving and at the same time synchronous system, the right-hand side of equation (4.112) is identically zero, then the background value $S_{ik} = 0$ indeed satisfies this equation). We have to deal with the following linear perturbations (as usual any quantity X is written as $X = X_{(0)} + \delta X$ where $X_{(0)}$ represents the background value of X):

$$\delta g_{\alpha\beta}, \delta u_\alpha, \delta\varepsilon, \delta S_{\alpha\beta}. \tag{4.119}$$

In the linearized version of the system (4.109)–(4.116) around the Friedmann solution (4.117)–(4.118), only these variations will appear. The variations δu_0 and δS_{0k} cannot be of first (linear) order because of the exact relations $u_i u^i = -1$ and $u^i S_{ik} = 0$ and the properties (4.118) of the background. The variations $\delta\tau$ and $\delta\eta$ of the relaxation time and viscosity coefficient exist as first-order quantities, but disappear from the linear approximation since they enter only as factors in front of terms that vanish for the isotropic Friedmann seed.

The exact form of the linear equations for the perturbations (4.119) and the technique for their integration can be found in the paper [12]. Here we describe only the results. First of all, it is easy to derive the conditions of the absence of supraluminal excitations. From the equations of motion for the scalar, vector and tensor type excitations (in accordance with Lifshitz classification [122]), taking the limit of the short wavelengths, one can obtain the following propagation velocities:

$$v^2_{scalar} = \gamma - 1 + \frac{4f}{3\gamma}, \; v^2_{vector} = \frac{f}{\gamma}, \; v^2_{tensor} = 1\,, \tag{4.120}$$

(the typical behavior of excitations in this limit is going like $A(t)\exp(ivkT)$ where A is a slowly varying amplitude, k is the wave vector and T is the time in the Friedmann metric (4.117) written in the form conformal to the Minkowski interval, that is $dT = dt/R$). This result has been obtained in [13] and it shows

* It is enough to analyze the flat Friedmann model. As was indicated in [124], flatness essentially simplifies the calculations and at the same time the generalization to the closed or open models contributes nothing fundamentally new to the behavior of the perturbations.

that gravitational waves propagate with velocity of light but in order to exclude supraluminal signals for the other two types of perturbations it is necessary to demand $v^2_{scalar} < 1$ and $v^2_{vector} < 1$. Both of these conditions will be satisfied in the region $1 \leqslant \gamma < 2$ if

$$f < \frac{3}{4}\gamma\,(2-\gamma)\,. \tag{4.121}$$

The main result is that in the space of parameters C_0, f, γ, there exists an open set of finite volume where the Friedmann Big Bang (4.117)–(4.118) is stable. That is, near the singularity, all perturbations to the Friedmann solution are negligibly small and they contain the full set of arbitrary three-dimensional functions that are necessary to match any initial conditions. This stability means that for viscous matter, there exists a *generic* solution of the Einstein equations with isotropic Friedmann-like initial cosmological singularity. The asymptotics of the metric near the singularity $t = 0$ in this generic solution has the following structure:

$$g_{\alpha\beta} = R^2\left(a_{\alpha\beta} + t^{s_1}b^{(1)}_{\alpha\beta} + t^{s_2}b^{(2)}_{\alpha\beta} + t^{s_3}b^{(3)}_{\alpha\beta} + \dots\right)\,, \tag{4.122}$$

where $R = (t/t_c)^{2/3\gamma}$ and the exponents s_1, s_2 and s_3 are defined in terms of *positive* constants C_0, f, γ by the relations:

$$\begin{aligned} s_1 &= \frac{3\gamma-\gamma\beta-2}{2\gamma} + \frac{1}{2\gamma}\sqrt{(\gamma-\gamma\beta+2)^2 - \frac{32f}{3}}\,, \\ s_2 &= \frac{3\gamma-\gamma\beta-2}{2\gamma} - \frac{1}{2\gamma}\sqrt{(\gamma-\gamma\beta+2)^2 - \frac{32f}{3}}\,, \\ s_3 &= \frac{2\,(3\gamma-2)}{3\gamma}\,, \end{aligned} \tag{4.123}$$

where $\beta = \frac{2f}{\sqrt{3\gamma C_0}}$. The coefficients $a_{\alpha\beta}, b^{(1)}_{\alpha\beta}, b^{(2)}_{\alpha\beta}, b^{(3)}_{\alpha\beta}$ in (4.122) are some functions depending on the space coordinate x^1, x^2, x^3.

The crucial fact is that in the space of arbitrary parameters C_0, f, γ there exists a region (its configuration was displayed in [12]) where all three exponents s_1, s_2, s_3 are either positive or two of them (s_1 and s_2) are complex conjugate to each other but with positive real part (s_3 is always positive). In case of complex conjugate s_1 and s_2, the components $b^{(1)}_{\alpha\beta}$ and $b^{(2)}_{\alpha\beta}$ are also complex, but in such a way that the metric tensor is real. The additional terms denoted by the triple dots are higher-order corrections which contain terms of the orders $t^{2s_3}, t^{s_1+s_3}, t^{s_2+s_3}$ as well as all their powers and cross products. The main term $a_{\alpha\beta}$ represents six arbitrary three-dimensional functions. Each tensor $b^{(1)}_{\alpha\beta}$ and $b^{(2)}_{\alpha\beta}$ consists of six three-dimensional functions subject to the restrictions $a^{\alpha\beta}b^{(1)}_{\alpha\beta} = 0$ and $a^{\alpha\beta}b^{(2)}_{\alpha\beta} = 0$ (here $a^{\alpha\beta}$ is inverse to $a_{\alpha\beta}$), consequently $b^{(1)}_{\alpha\beta}$ and $b^{(2)}_{\alpha\beta}$ contain another ten arbitrary three-dimensional functions. The last term $b^{(3)}_{\alpha\beta}$ and all the corrections denoted by the triple dots in the expansion (4.122) are expressible in terms of the $a_{\alpha\beta}, b^{(1)}_{\alpha\beta}, b^{(2)}_{\alpha\beta}$ and their derivatives; accordingly they do not contain any new arbitrariness. The shear stresses, velocity and energy density follow from the exact Einstein equations in terms of the metric tensor (4.122) and its derivatives. Consequently all these quantities contain only those arbitrary three-dimensional functions that already appeared in metric tensor (4.122) (the behavior in time of the velocities and energy density are schematically like $u_\alpha \sim t^{1+s_1} + t^{1+s_2} + t^{1+s_3} + \ldots$ and $\varepsilon \sim t^{-2} + t^{-2+s_3} + \ldots$).

It follows that the solution contains sixteen arbitrary three-dimensional functions, three of which represent the gauge freedom due to the possibility of making an arbitrary three-dimensional coordinate transformation. Then the physical freedom in the solution corresponds to thirteen arbitrary functions as it should be for the generic solution in the presence of shear viscosity. Indeed, in this case we should have four functions for the gravitational field, one for the energy density, three for the velocity and five for the viscous stresses (five because in the synchronous system, due to equations (4.110)–(4.112), the six components $S_{\alpha\beta}$ of the stresses follow from six differential equations of first order in time with one additional constraint, while the components S_{00} and $S_{0\alpha}$ are expressible in terms of $S_{\alpha\beta}, g_{\alpha\beta}, u^0, u^\alpha$ algebraically and do not contain any proper arbitrariness).

Therefore, the results just described show that the viscoelastic matter with shear viscosity coefficient $\eta \sim \sqrt{\varepsilon}$ can stabilize the Friedmann cosmological singularity and that the corresponding *generic solution of the Einstein equations for the viscous fluid possessing the isotropic Big Bang (or Big Crunch) exists.* Depending on the free parameters C_0, f, γ of the theory such a solution can be either of smooth power law asymptotics near the singularity (when both exponents s_1 and s_2 are real and positive) or it can have the character of damped oscillations (in the limit $t \to 0$) (when s_1 and s_2 are complex conjugate with positive real part). The last possibility reveals itself as a weak trace of the chaotic oscillatory regime which is characteristic for the most general asymptotics near the cosmological singularity and which cannot be described in closed analytical form. The present case show that the shear viscosity can smooth such a chaotic behavior up to the quiet oscillations which have simple asymptotic expressions in terms of the elementary functions of the type $t^{\Re s} \sin[(\Im s) \ln t]$ and $t^{\Re s} \cos[(\Im s) \ln t]$, where $\Re$ and $\Im$ denote the real and imaginary parts, respectively.

Part II

Cosmological Billiards

5

The Billiard of Four-Dimensional Vacuum Gravity

In this second part of the book, we develop the billiard description of the BKL behavior. We show how, in the BKL limit, the equations of motion of systems containing gravity can be recast at each spatial point as equations of motion for a billiard ball moving in a region of hyperbolic space. The emerging billiards that capture the motion near a spacelike singularity are called "cosmological billiards." We start with pure gravity in four space-time dimensions. The inclusion of matter fields, and the extension to higher dimensions, are considered in Chapter 6.

In most papers devoted to cosmological billiards, index conventions different from those used so far are adopted. These different conventions are that Greek indices are space-time indices while Latin indices are spatial indices. Thus, in four space-time dimensions, $\alpha, \beta, \cdots = 0, 1, 2, 3$, and $a, b, \cdots = 1, 2, 3$. Furthermore, indices in non-coordinate frames are not distinguished with a bar from coordinate indices. In order to ease the reading of the cosmological billiard literature, we shall switch to these different conventions. No confusion should arise as the context is always clear.

The derivation of the billiard description of the BKL limit given in this second part of the book was originally worked out in a collaboration of the second author with Thibault Damour and Hermann Nicolai [45, 46, 47, 49, 51]. One of its main ingredients is the use of the Iwasawa decomposition of the metric for handling the influence of the off-diagonal metric components on the dynamics of the scale factors. Chapters 5 and 6 are largely based on the ideas developed in those works, and in particular draw very much from the review [51], from which we adopt in particular the notations.

5.1 Hamiltonian Form of the Action

The cosmological billiard picture is based on the Hamiltonian formalism, which we first briefly review. For more information, we refer the reader to the original papers [65, 5] as well as to the book [133], Chapter 21.

The starting point is the Einstein–Hilbert action

$$S[g_{\alpha\beta}] = \int d^4x \sqrt{-{}^{(4)}g}\, R \tag{5.1}$$

where units are now chosen such that $16\pi G = 1$ (where G is Newton's constant). We impose no isometry conditions on the metric components $g_{\alpha\beta}$ which are thus arbitrary functions of both space and time.

The coordinate x^0 defines a (local) foliation of space-time by a family of three-dimensional hypersurfaces, namely, the hypersurfaces of constant x^0. In order to follow how the gravitational variables change as one moves from one hypersurface to the next, it is convenient to decompose the vector $\frac{\partial}{\partial x^0}$ tangent to the x^0-coordinate lines as

$$\frac{\partial}{\partial x^0} = N\mathbf{n} + N^k \frac{\partial}{\partial x^k}$$

where $\mathbf{n}$ is the unit vector normal to the hypersurfaces x^0 = const., and where $\frac{\partial}{\partial x^k}$ are the vectors tangent to the x^k-coordinate lines. One has

$$\mathbf{n}\cdot\mathbf{n} = -1, \qquad \mathbf{n}\cdot\frac{\partial}{\partial x^k} = 0.$$

The function N, which indicates how much one moves along the normal as one goes from one hypersurface in the family to the next, is called the "lapse." The 3-vector N^k, which parametrizes the tangential displacement, is called the "shift."

The space-time metric can be expressed in terms of the lapse, the shift and the spatial metric g_{ij} as

$$ds^2 = \left(-N^2 + N^k N_k\right)\left(dx^0\right)^2 + 2N_k dx^0 dx^k + g_{ij}dx^i dx^j \tag{5.2}$$

where $N_k \equiv g_{kj}N^j$. In the Hamiltonian description of the time evolution, one makes the change of field variables $(g_{\alpha\beta}) \to (g_{ij}, N, N^k)$, because the lapse, the shift and the spatial components of the metric play different dynamical roles [65, 5, 133]: the g_{ij}s are canonical variables with conjugate momenta denoted π^{ij}, while the lapse and the shift are Lagrange multipliers for constraints that the dynamical variables must fulfill. If the shift is equal to zero, the line element reduces to $ds^2 = -N^2\left(dx^0\right)^2$ along the coordinate lines x^k = const., and thus x^0 is $\pm$ the proper time t along these lines when the lapse is equal to one, $N = 1$. In general, $dt = \pm N dx^0$. We take $N > 0$. Then the different choices of sign correspond to common orientation for t and x^0 (+ sign) or opposite orientations (− sign).

The Hamiltonian action corresponding to (5.1) reads

$$S\left[g_{ij}, \pi^{ij}, \tilde{N}, N^k\right] = \int dx^0 \int d^3x \left(\pi^{ij}\dot{g}_{ij} - \tilde{N}\mathcal{H} - N^j\mathcal{H}_j\right) \tag{5.3}$$

where

$$\mathcal{H} = \mathcal{K} + \mathcal{M} \tag{5.4}$$

$$\mathcal{K} = \pi^{ij}\pi_{ij} - \frac{1}{2}\pi^i{}_i\pi^j{}_j, \quad \mathcal{M} = -gR \tag{5.5}$$

$$\mathcal{H}_i = -2\pi^j{}_{i|j} \tag{5.6}$$

and again $\dot{F} \equiv \frac{\partial F}{\partial x^0}$. Here R is the spatial curvature scalar built out of the spatial metric, the subscript $|j$ stands for the spatial covariant derivative, and we have rescaled the lapse function as

$$\tilde{N} \equiv N/\sqrt{g} \tag{5.7}$$

where $g \equiv \det g_{ij}$ in order to absorb a factor $1/\sqrt{g}$ in $\mathcal{K}$. As we shall see below, this leads to a simpler description of the null geodesics of the geometry of the space of the scale factors (Section 5.4). We shall often refer to $\tilde{N}$ just as the "lapse," omitting the "rescaled," when no confusion can arise.

The equations of motion are derived by varying the action (5.3) with respect to the spatial metric components, their conjugate momenta, the lapse and the shift. Variation with respect to the lapse yields the "Hamiltonian constraint" on the dynamical variables,

$$\mathcal{H} \approx 0 \quad (\text{"Hamiltonian constraint"}) \tag{5.8}$$

while variation with respect to the shift yields the "momentum constraint"

$$\mathcal{H}_i \approx 0 \quad (\text{"momentum constraint"}). \tag{5.9}$$

5.2 Supermetric

The Hamiltonian constraint $\mathcal{H}(x)$ is the sum of a "kinetic term" $\mathcal{K}(x)$ and a "potential term" $\mathcal{M}(x)$. While the potential term $\mathcal{M}(x)$ involves the dynamical variables and their spatial derivatives at x up to second order, the kinetic term is "ultralocal," in the sense that $\mathcal{K}$ involves only the dynamical variables at x and not their spatial derivatives. There is complete decoupling of the spatial points in the kinetic term.

Because of this crucial property, the kinetic term, which is quadratic in the momenta,

$$\mathcal{K} = G_{ijmn}\pi^{ij}\pi^{mn}, \quad G_{ijmn} = \frac{1}{2}\left(g_{im}g_{jn} + g_{in}g_{jm}\right) - \frac{1}{2}g_{ij}g_{mn}$$

defines a metric in the finite dimensional space of the spatial metric coefficients g_{ij}. This space, sometimes called the "superspace of spatial metrics," is six-dimensional and isomorphic to the homogeneous space $GL(3,\mathbb{R})/O(3)$ since any two positive definite symmetric forms can be related by a $GL(3,\mathbb{R})$ transformation, while the stability group of any such quadratic form is the orthogonal group $O(3)$. The "supermetric" in that "superspace" is simply obtained by inverting G_{ijmn},

$$G_{ijmn}G^{mnpq} = \frac{1}{2}\left(\delta_i^p \delta_j^q + \delta_i^q \delta_j^p\right) ,$$

and was introduced by De Witt [64]. It reads explicitly

$$G^{ijmn} = \frac{1}{2}\left(g^{im}g^{jn} + g^{in}g^{jm}\right) - g^{ij}g^{mn}.$$

To be precise, the supermetric used here differs by the factor $\sqrt{g}$ from the supermetric of [64], because we have rescaled the lapse. This simplifies some of its properties (see below). The line element $\frac{1}{4}G^{ijmn}dg_{ij}dg_{mn}$ in the superspace of metrics, which we denote by $d\sigma^2$ to distinguish it from the line element ds^2 in physical space-time, can be rewritten

$$d\sigma^2 = \tfrac{1}{4}\left[\mathrm{tr}\,(\mathrm{g}^{-1}d\mathrm{g})^2 - (\mathrm{tr}\,\mathrm{g}^{-1}d\mathrm{g})^2\right]. \tag{5.10}$$

We have normalized $d\sigma^2$ by an overall factor of $1/4$ for later convenience and, as in [51], we have adopted a matrix notation where g stands for the 3×3 symmetric matrix (g_{ij}) representing the spatial components of the metric at each spatial point.

Logarithmic Scale Factors

We first explore the properties of the supermetric in the subspace of diagonal metrics.

For diagonal metrics,

$$\mathrm{g}_D = \exp\left[\mathrm{diag}(-2\beta)\right] \quad\Longleftrightarrow\quad g^D_{ij} = \exp(-2\beta^i)\delta_{ij}\,, \tag{5.11}$$

the supermetric (5.10) reduces to

$$\begin{aligned} d\sigma^2 &= \mathrm{tr}\,d\beta^2 - (\mathrm{tr}\,d\beta)^2 \\ &= \sum_{i=1}^{3}(d\beta^i)^2 - (\sum_{i=1}^{3} d\beta^i)^2 \equiv G_{ij}d\beta^i d\beta^j. \end{aligned} \tag{5.12}$$

where the explicit form of the metric G_{ij} in the three-dimensional space of the β^is can be read off directly from (5.12) and is

$$G_{ij} = \begin{pmatrix} 0 & -1 & -1 \\ -1 & 0 & -1 \\ -1 & -1 & 0 \end{pmatrix}. \tag{5.13}$$

Note that the determinant g of g_{ij} ("volume") is given by

$$g = \exp(-2\sum_i \beta^i). \tag{5.14}$$

The variables β^i that parametrize the diagonal metrics are called "logarithmic scale factors," or even, more simply, just "scale factors."

The metric G_{ij} does not involve the βs. *Hence it is flat.* It is actually flat because we have absorbed the factor $\sqrt{g}$ in the definition of the supermetric. This is the reason why we did so.

The flat metric induced in the space of the logarithmic scale factors has Minkowskian signature $(-,+,+)$. Indeed, the transformation

$$\beta^1 = \frac{1}{\sqrt{6}}\left(\bar{\beta}^1 - \bar{\beta}^2 - \sqrt{3}\bar{\beta}^3\right), \; \beta^2 = \frac{1}{\sqrt{6}}\left(\bar{\beta}^1 - \bar{\beta}^2 + \sqrt{3}\bar{\beta}^{\bar{3}}\right), \tag{5.15}$$
$$\beta^3 = \frac{1}{\sqrt{6}}\left(\bar{\beta}^1 + 2\bar{\beta}^2\right),$$

brings the metric to the form

$$d\sigma^2 = -(d\bar{\beta}^1)^2 + (d\bar{\beta}^2)^2 + (d\bar{\beta}^3)^2. \tag{5.16}$$

The timelike deformation $d\bar{\beta}^1 = \sqrt{6}\epsilon$, $d\bar{\beta}^2 = d\bar{\beta}^3 = 0$ corresponds to the rescaling $d\beta^i = \epsilon$, while the spacelike deformations $d\bar{\beta}^1 = 0$, $d\bar{\beta}^2$ and $d\bar{\beta}^3$ arbitrary, define volume-preserving deformations ($\sum_i d\beta^i = 0$). For an arbitrary vector v^i in the space of the (logarithmic) scale factors, the first component in the Minkowskian coordinates $\bar{\beta}^i$ is equal to $\sqrt{6}\sum_i v^i$.

The Lorentzian signature of the metric in field space is a distinguishing feature of gravity. It is associated with changes in the volume as we have just demonstrated. All other gravitational variables and, as we shall see, all matter variables when matter is included, contribute with + signs to the signature of this metric. This property holds in any space-time dimension.

Because the metric in the space of the scale factors has Lorentzian signature, one can consider the light cone through any point in that space. We define the time-orientation to be such that future-pointing vectors v^i have $\sum_i v^i > 0$. Small volumes (small g) are associated with large positive values of $\sum_i \beta^i$, while large volumes (large g) mean large negative values of $\sum_i \beta^i$. Thus, within the light cone, the small volume limit, i.e., $\sum_i \beta^i \to +\infty$, corresponds to going to future infinity.

Iwasawa Parametrization of Off-Diagonal Components

We now introduce a parametrization of the metric which includes also the off-diagonal components. There are many such possible parametrizations, of course, but the one adopted here has the advantage to lead to a simple asymptotical behavior as one goes to the singularity, in which the off-diagonal variables freeze.

As we have seen, the space of symmetric matrices g_{ij} of Euclidean signature can be identified with $GL(3,\mathbb{R})/O(3)$ since $GL(3,\mathbb{R})$ acts transitively on the space of such matrices, and $O(3)$ is the isotropy subgroup. The space $GL(3,\mathbb{R})/O(3)$

itself is isomorphic to $SL(3,\mathbb{R})/SO(3) \times \mathbb{R}^+$, and the space $SL(3,\mathbb{R})/SO(3)$ is a symmetric space [84], for which there exist classical parametrizations. The one to be adopted here is the Iwasawa decomposition, which consists in diagonalizing the matrix g_{ij} through the Gram–Schmidt procedure, by performing an appropriate upper triangular change of basis.

Explicitly, one sets

$$\mathrm{g} = \mathcal{N}^T \mathcal{A}^2 \mathcal{N} \tag{5.17}$$

where $\mathcal{N}$ is an upper triangular matrix with 1s on the diagonal and $\mathcal{A}$ is a diagonal matrix with positive entries, which we parametrize as in the diagonal case (where g reduces to $\mathcal{A}^2$) as

$$\mathcal{A} = \exp(-\beta)\ , \quad \beta = \mathrm{diag}\,(\beta^1, \beta^2, \beta^3), \tag{5.18}$$

i.e.,

$$\mathcal{A} = \begin{pmatrix} \exp(-\beta^1) & 0 & 0 \\ 0 & \exp(-\beta^2) & 0 \\ 0 & 0 & \exp(-\beta^3) \end{pmatrix}, \tag{5.19}$$

The component form of (5.17) reads

$$g_{ij} = \sum_{a=1}^{3} e^{-2\beta^a} \mathcal{N}^a{}_i \mathcal{N}^a{}_j. \tag{5.20}$$

The Iwasawa decomposition amounts to orthogonalize the coordinate coframe $\{dx^i\}$ to a new frame $\{\theta^a\}$

$$\theta^a = \mathcal{N}^a{}_i\, dx^i \tag{5.21}$$

where the metric is diagonal,

$$g_{ij} dx^i dx^j = \sum_{a=1}^{3} e^{-2\beta^a}\, \theta^a \otimes \theta^a \tag{5.22}$$

by performing the triangular Gram–Schmidt orthogonalization process. We do not request the new coframe $\{\theta^a\}$ to be orthonormal. We request instead that the first term (with the order prescribed below) in the definition of the θ's come with a coefficient equal to 1. The associated orthonormal frame is $\{e^{-\beta^a}\theta^a\}$.

With the choice (5.17) where $\mathcal{N}$ is upper triangular, one starts by setting $\theta^3 = dx^3$. One next successively constructs θ^2 by adding to dx^2 a term proportional to dx^3 and then θ^1 by adding to dx^1 a linear combination of θ^2 and θ^3, in such a way that θ^2 is orthogonal to θ^3 and θ^1 is orthogonal to both θ^3 and θ^2. With

$$\mathcal{N} = \begin{pmatrix} 1 & n_1 & n_2 \\ 0 & 1 & n_3 \\ 0 & 0 & 1 \end{pmatrix} \tag{5.23}$$

this yields

$$\begin{aligned}\theta^3 &= dx^3\\ \theta^2 &= dx^2 + n_3\, dx^3,\\ \theta^1 &= dx^1 + n_1\, dx^2 + n_2\, dx^3.\end{aligned} \tag{5.24}$$

The component $\mathcal{N}^a{}_i$ vanishes for $a > i$, is equal to one for $a = i$, and is nontrivial only for $a < i$. Because the matrix $\mathcal{N}$ is upper triangular with 1s on its diagonal, its determinant is equal to 1. Therefore, as in the diagonal case, the sum of the βs is directly related to the metric determinant: $g = \det \mathrm{g} = \det \mathcal{A}^2 = \exp\big(-2\Sigma_a \beta^a\big)$.

One refers to the β^as as the (logarithmic) scale factors in the Iwasawa frame, and to the n_is as the off-diagonal components. The space of the βs is called "β-space." When the matrix $\mathcal{N}$ is equal to the unit matrix, the metric is diagonal and reduces to (5.11).

The change of variables from the metric to the Iwasawa variables and its inverse are respectively given by [91]

$$\begin{aligned}&g_{11} = e^{-2\beta^1}, \quad g_{12} = n_1 e^{-2\beta^1}, \quad g_{13} = n_2 e^{-2\beta^1},\\ &g_{22} = n_1^2 e^{-2\beta^1} + e^{-2\beta^2}, \quad g_{23} = n_1 n_2 e^{-2\beta^1} + n_3 e^{-2\beta^2},\\ &g_{33} = n_2^2 e^{-2\beta^1} + n_3^2 e^{-2\beta^2} + e^{-2\beta^3}\end{aligned} \tag{5.25}$$

and

$$\begin{aligned}&\beta^1 = -\frac{1}{2}\ln g_{11}, \quad \beta^2 = -\frac{1}{2}\ln\left[\frac{g_{11}g_{22} - g_{12}^2}{g_{11}}\right],\\ &\beta^3 = -\frac{1}{2}\ln\left[\frac{g}{g_{11}g_{22} - g_{12}^2}\right], \quad n_1 = \frac{g_{12}}{g_{11}},\\ &n_2 = \frac{g_{13}}{g_{11}}, \quad n_3 = \frac{g_{23}g_{11} - g_{12}g_{13}}{g_{11}g_{22} - g_{12}^2}.\end{aligned} \tag{5.26}$$

We shall also need below the vector basis $\{e_a\}$ dual to the basis of the θ^as. One has:

$$e_a = \mathcal{N}^i{}_a \frac{\partial}{\partial x^i}, \tag{5.27}$$

where the matrix $\mathcal{N}^i{}_a$ is the inverse of $\mathcal{N}^a{}_i$, i.e., $\mathcal{N}^a{}_i \mathcal{N}^i{}_b = \delta^a_b$. It is given by

$$\left(\mathcal{N}^i{}_a\right) = \begin{pmatrix} 1 & -n_1 & -n_2 + n_1 n_3\\ 0 & 1 & -n_3\\ 0 & 0 & 1 \end{pmatrix} \tag{5.28}$$

and is again an upper triangular matrix with 1s on the diagonal.

In the Iwasawa variables, the supermetric reads

$$\begin{aligned}d\sigma^2 &= \mathrm{tr}\, d\beta^2 - (\mathrm{tr}\, d\beta)^2\\ &\quad + \frac{1}{2}\mathrm{tr}\left[\mathcal{A}^2 (d\mathcal{N}\mathcal{N}^{-1}) \mathcal{A}^{-2} (d\mathcal{N}\mathcal{N}^{-1})^T\right]\end{aligned} \tag{5.29}$$

i.e.,

$$
\begin{aligned}
d\sigma^2 &= \sum_{a=1}^{3}(d\beta^a)^2 - \left(\sum_{a=1}^{3} d\beta^a\right)^2 \\
&\quad + \frac{1}{2}\sum_{a<b} e^{2(\beta^b-\beta^a)}\left(d\mathcal{N}^a{}_i\,\mathcal{N}^i{}_b\right)^2 \\
&= \sum_{a=1}^{3}(d\beta^a)^2 - \left(\sum_{a=1}^{3} d\beta^a\right)^2 \\
&\quad + \frac{1}{2}e^{2(\beta^2-\beta^1)}\left(dn_1\right)^2 + \frac{1}{2}e^{2(\beta^3-\beta^1)}\left(dn_2 - n_2 dn_1\right)^2 \\
&\quad + \frac{1}{2}e^{2(\beta^3-\beta^2)}\left(dn_3\right)^2 .
\end{aligned} \tag{5.30}
$$

5.3 More on Hyperbolic Space $\mathbb{H}_2$

Polar Coordinates

It will be useful when investigating the dynamics to pass to hyperbolic coordinates (ρ, γ^i) in the space of the logarithmic scale factors β^i. We define these coordinates in the case when β^i lies in the future light cone, the case relevant for analyzing the BKL limit. Thus, we assume $\beta^i\beta_i < 0$ and $\sum_i \beta^i > 0$.

One defines the radial (timelike) variable ρ through

$$\rho^2 \equiv -\beta^i\beta_i > 0 . \tag{5.31}$$

One then sets

$$\beta^i = \rho\gamma^i, \tag{5.32}$$

where γ^i are coordinates on the future sheet of the unit hyperboloid, which are constrained by

$$\gamma^i\gamma_i = -1 . \tag{5.33}$$

The upper sheet of the unit hyperboloid ("γ-space") is a well-known realization of the two-dimensional hyperbolic (Lobachevsky) space $\mathbb{H}_2$. An explicit choice of the γs in terms of two independent variables r, ϕ $(0 \le r < \infty,\ 0 \le \phi \le 2\pi)$ is

$$
\begin{aligned}
\gamma^1 &= \frac{1}{\sqrt{6}}\cosh r - \frac{1}{\sqrt{6}}\left(\sin\phi + \sqrt{3}\cos\phi\right)\sinh r, \\
\gamma^2 &= \frac{1}{\sqrt{6}}\cosh r - \frac{1}{\sqrt{6}}\left(\sin\phi - \sqrt{3}\cos\phi\right)\sinh r, \\
\gamma^3 &= \frac{1}{\sqrt{6}}\cosh r + \frac{2}{\sqrt{6}}\sin\phi\sinh r.
\end{aligned} \tag{5.34}
$$

In terms of the "polar" coordinates just defined, the metric in β-space becomes

$$d\sigma^2 = -d\rho^2 + \rho^2 d\Sigma^2 \tag{5.35}$$

where $d\S^2$ is the metric on the unit hyperboloid $\mathbb{H}_2$,

$$d\Sigma^2 = dr^2 + \sinh^2 r d\phi^2. \tag{5.36}$$

The relationship between the Minkowskian coordinates $\bar{\beta}^1, \bar{\beta}^2, \bar{\beta}^3$ and the polar coordinates ρ, r, ϕ is the familiar one,

$$\bar{\beta}^1 = \rho \cosh r, \;\; \bar{\beta}^2 = \rho \sinh r \sin\phi, \;\; \bar{\beta}^3 = \rho \sinh r \cos\phi. \tag{5.37}$$

The radial projection of the motion of the logarithmic scale factors on the Lobachevsky space $\mathbb{H}_2$ is a central tool in developing the billiard approach to the BKL behavior, and for that reason, we review now some relevant properties of $\mathbb{H}_2$.

Vector Model

The description of the Lobachevsky plane as the upper sheet of the unit spacelike hyperboloid in three-dimensional Minkowki space is the so-called vector model of hyperbolic space.

In Minkowskian coordinates $(\bar{\beta}^1, \bar{\beta}^2, \bar{\beta}^3) \equiv (x^0, x^1, x^2)$ the vector model is defined by

$$(x^0)^2 - (x^1)^1 - (x^2)^2 = 1, \quad x^0 > 0 \tag{5.38}$$

and the Lobachevsky distance $d(\mathbf{a}, \mathbf{b})$ between two points $\mathbf{a}$ and $\mathbf{b}$ of the unit hyperboloid is

$$\cosh d(\mathbf{a}, \mathbf{b}) = -\mathbf{a} \cdot \mathbf{b} \tag{5.39}$$

where $\mathbf{a} \cdot \mathbf{b} = -a^0 b^0 + a^1 b^1 + a^2 b^2$ is the Minkowskian scalar product. One can take x^1 and x^2 as coordinates on the unit hyperboloid instead of the polar coordinates (r, ϕ). In these Cartesian-like coordinates, the metric of $\mathbb{H}_2$ reads

$$d\Sigma^2 = \left(dx^1\right)^2 + \left(dx^2\right)^2 - \frac{\left(x^1 dx^1 + x^2 dx^2\right)^2}{1 + \left(x^1\right)^2 + \left(x^2\right)^2}. \tag{5.40}$$

A key feature of the vector model is that geodesics in hyperbolic space are the intersections of timelike planes of Minkowski space containing the origin with the upper sheet of the unit hyperboloid. This is the hyperbolic analog of the well-known spherical fact that great circles on the sphere are intersections of the sphere with planes containing the origin.

This property can be seen as follows. Any straight line in flat Minkowski space (including a "light-like" one) can be described in polar coordinates (ρ, r, ϕ) by the equations:

$$\begin{aligned} \rho \cosh r &= A_1 \zeta + B_1, \\ \rho \sinh r \sin\phi &= A_2 \zeta + B_2, \\ \rho \sinh r \cos\phi &= A_3 \zeta + B_3, \end{aligned} \tag{5.41}$$

where ζ is some parameter along the trajectory and A_i, B_i are constants. Eliminating from (5.41) the parameter ζ and the radial coordinate ρ, we obtain an equation connecting only the variables r and ϕ:

$$\coth r = \frac{A_1B_2 - A_2B_1}{A_3B_2 - A_2B_3}\cos\phi - \frac{A_1B_3 - A_3B_1}{A_3B_2 - A_2B_3}\sin\phi. \tag{5.42}$$

This is the equation of the radial projection of the trajectory (5.41) into the Lobachevsky plane $\rho = 1$, or what is the same, the equation of the intersection of the plane through the origin containing the geodesic with the upper sheet of the unit hyperboloid. On the other hand, we can derive independently the equation of the geodesics in the two-dimensional Lobachevsky space (5.36). This simple exercise gives:

$$\coth r = r_0 \cos\phi_0 \cos\phi + r_0 \sin\phi_0 \sin\phi, \tag{5.43}$$

where r_0 and ϕ_0 are arbitrary constants. It is therefore clear that the projection (5.42) indeed yields a geodesic curve of $\mathbb{H}_2$. If the parameters A_i, B_i are given, the corresponding constants r_0 and ϕ_0 follows from the relations:

$$\tan\phi_0 = -\frac{A_1B_3 - A_3B_1}{A_1B_2 - A_2B_1}, \qquad r_0^2 = \frac{(A_1B_2 - A_2B_1)^2 + (A_1B_3 - A_3B_1)^2}{(A_3B_2 - A_2B_3)^2}. \tag{5.44}$$

Another important property that can easily be verified is that the angle between two geodesics in $\mathbb{H}_2$ is just the dihedral angle between the planes through the origin containing them. This property is again the translation in hyperbolic geometry of a similar property holding in spherical geometry. It follows from the fact that the tangent vectors to the geodesics are orthogonal to the intersection line of the corresponding planes.

Klein Model

There exist other models of hyperbolic space that turn out to be useful in order to understand the billiard dynamics. One model is the Klein projective model, which is the unit disk in the plane with metric

$$d\Sigma^2 = \frac{dx^2 + dy^2}{1 - (x^2 + y^2)} + \frac{(xdx + ydy)^2}{(1 - (x^2 + y^2))^2}. \tag{5.45}$$

The points on the boundary of the unit disk are the points at infinity. The Klein model is obtained by radially projecting the unit hyperboloid on the tangent plane $x^0 = 1$ at the point with Minkowskian coordinates $(1, 0, 0)$. One has

$$x^0 = \frac{1}{(1 - x^2 - y^2)^{\frac{1}{2}}}, \quad x^1 = \frac{x}{(1 - x^2 - y^2)^{\frac{1}{2}}}, \quad x^2 = \frac{y}{(1 - x^2 - y^2)^{\frac{1}{2}}}. \tag{5.46}$$

A useful property of the Klein model is that geodesics are just chords of the boundary unit circle, i.e., Euclidean straight line segments joining two points on the unit circle. This follows immediately from the projective construction of the Klein model. The Klein model is not conformal, however, in that it distorts the angles: the angles in the hyperbolic geometry are not equal to the Euclidean angles in the (x, y)-plane.

Let P and Q be two distinct points within the unit disk. The geodesic joining them is the chord AB where A and B are the intersections of the line PQ with the unit circle – called "ideal points." We take A to be on the side of P and B to be on the side of Q so that the Euclidean distances $|AP|$, $|AQ|$, $|PB|$, $|QB|$ fulfill $|AP| < |AQ|$ and $|PB| > |QB|$. Integration of the line element gives then the following formula for the hyperbolic distance between P and Q,

$$d(P, Q) = \frac{1}{2} \log \frac{|AQ|\,|PB|}{|AP|\,|QB|}. \tag{5.47}$$

Consider a geodesic in $\mathbb{H}_2$, i.e., a chord AB where A and B are the ideal points where the chord meets the boundary circle. Let l_A and l_B be the tangents to the unit circle at A and B, respectively. One defines the pole of AB to be the intersection point of the lines l_A and l_B. This is not a point in the unit disk. The pole of AB has the important property that every line through it that intersects the disk defines a geodesic orthogonal to AB.

Poincaré Disk

Another useful model of the Lobachevsky plane is the Poincaré disk, obtained by projecting through the point $(-1, 0, 0)$ the upper sheet $\mathbb{H}_2$ of the unit hyperboloid on the plane $x^0 = 0$. The relationship between Minkowskian coordinates and Euclidean coordinates (X, Y) on the Poincaré disk is

$$x^0 = \frac{1 + X^2 + Y^2}{1 - X^2 - Y^2}, \quad x^1 = \frac{2X}{1 - X^2 - Y^2}, \quad x^2 = \frac{2Y}{1 - X^2 - Y^2} \tag{5.48}$$

and the metric is

$$d\Sigma^2 = \frac{4\left(dX^2 + dY^2\right)}{\left(1 - X^2 - Y^2\right)^2}. \tag{5.49}$$

The Poincaré disk representation is a conformal representation preserving the angles. The boundary (infinity) is the unit circle. The geodesics are arcs of Euclidean circles orthogonal to the boundary. In the limiting case when they go through the origin, they are diameters (infinite radius). One may go from the Poincaré coordinates to the above polar coordinates by setting

$$X = \left(\frac{\cosh r - 1}{\sinh r}\right) \sin \phi, \quad Y = \left(\frac{\cosh r - 1}{\sinh r}\right) \cos \phi, \tag{5.50}$$

in which the metric becomes indeed $d\Sigma^2 = dr^2 + \sinh^2 r\, d\phi^2$ as it should.

Upper Half Plane

Yet another useful model is the Poincaré upper half plane model. One gets it from the Poincaré disk representation through the holomorphic change of variables

$$z = \frac{Z+i}{iZ+1} \tag{5.51}$$

i.e., with $Z = X + iY$, and $z = u + iv$,

$$u = \frac{2X}{X^2 + (1-Y)^2}, \qquad v = \frac{1 - X^2 - Y^2}{X^2 + (1-Y)^2}. \tag{5.52}$$

The Lobachevsky metric reads, in Poincaré coordinates (u, v),

$$d\Sigma^2 = \frac{du^2 + dv^2}{v^2}, \qquad v > 0. \tag{5.53}$$

This is also a conformal representation, as it was guaranteed by the fact that it is obtained from the Poincaré disk model by a holomorphic transformation. The boundary (infinity) is now the x-axis $y = 0$ as well as the point at infinity in directions parallel to the y-axis ("vertical directions"). The geodesics are arcs of circles orthogonal to the x-axis as well as vertical straight lines – also orthogonal to the x-axis.

Classification of Motions (Orientation-Preserving Isometries) of Hyperbolic Space $\mathbb{H}_2$

The motions of hyperbolic space can be of different types. This can be easily analyzed in the upper half plane model. Any motion of $\mathbb{H}_2$ is, in this model, of the form

$$z \mapsto \frac{az+b}{cz+d}, \qquad a, b, c, d \in \mathbb{R}, \quad ad - bc = 1. \tag{5.54}$$

The classification of the motions is based on the number of fixed points. When $c \neq 0$, the fixed point condition $z(cz + d) = az + b$ is quadratic and such that if z_0 is a solution, so is also its complex conjugate $\bar{z}_0$. Three cases can arise.

- The roots of the fixed point equation have a non-vanishing imaginary part, i.e., are of the form $u \pm iv$ with $v \neq 0$. We can assume $v > 0$ so that $u + iv \in \mathbb{H}_2$. The motion has one fixed point in $\mathbb{H}_2$, namely $u + iv$ and is called "elliptic." It is a rotation about $u + iv$. For instance, a rotation around i takes the form

$$z \mapsto \frac{z\cos\alpha - \sin\alpha}{z\sin\alpha + \cos\alpha}.$$

  In terms of the vector model, an elliptic motion is just an ordinary rotation in $\mathbb{R}^{2,1}$ around the timelike straight line joining the origin to the fixed point on the hyperboloid ($SO(2) \subset SO^+(2,1)$).

- The roots are real and distinct. There is then no fixed point in $\mathbb{H}_2$. The line L of hyperbolic space joining the two roots is invariant. The motion is called *hyperbolic*. This case is conjugate to the case $c = 0$, $a \neq 1$ as can be seen by performing a motion that maps one of the roots on the point at infinity. For instance, the hyperbolic motion leaving the imaginary axis invariant is $z' = a^2 z$ with $a \neq 1$, corresponding to $c = b = 0$ and $d = 1/a \neq 1$. The hyperbolic motion with invariant line L is also called a parallel displacement along L. In terms of the vector model, this is a boost in $\mathbb{R}^{2,1}$ in the plane joining L to the origin, leaving the orthogonal spacelike straight line through the origin pointwise fixed.
- There is a single, degenerate root. There is again no fixed point in $\mathbb{H}_2$ but there is no invariant line either. The motion is called *parabolic*. This case is conjugate to the case $c = 0$, $a = 1$ as can be seen by performing a motion that maps the root on the point at infinity. A typical parabolic motion is $z \mapsto z + b$ ($a = d = 1$, $c = 0$). Parabolic motions are also called parabolic translations, or parabolic rotations about the point at infinity. In terms of the vector model, this is a null Lorentz transformation leaving a lightlike line pointwise invariant.

5.4 Kasner Solution Revisited

Kasner Solution and Null Geodesics in Superspace

It turns out that the Kasner solution – the spatially homogeneous diagonal solution of the vacuum Einstein equations depending only on time that plays, as we have seen, such an important role in the description of the BKL behavior – describes just a null geodesic of the supermetric defined in the space of metrics, as we shall now show.

For metrics that do not depend on the spatial coordinates, the Einstein–Hilbert action (5.1) reduces to

$$S[g_{ij}, \tilde{N}] = \int dx^0 \tilde{N}^{-1} \left[\frac{1}{4} \left(\mathrm{tr}\, (\mathrm{g}^{-1}\dot{\mathrm{g}})^2 - (\mathrm{tr}\, \mathrm{g}^{-1}\dot{\mathrm{g}})^2 \right) \right] , \tag{5.55}$$

up to a spatial volume factor that we have dropped for notational simplicity. When g_{ij} is diagonal, this becomes

$$S[\beta^i, \tilde{N}] = \int dx^0 \, \tilde{N}^{-1} G_{ij} \dot{\beta}^i \dot{\beta}^j \, . \tag{5.56}$$

The action (5.56) is the (quadratic-in-velocities) action for a massless free particle with coordinates (β^i) moving in the space of the scale factors with metric G_{ij}, the rescaled lapse $\tilde{N}$ playing the role of a Lagrange multiplier enforcing the "zero-mass constraint"

$$G_{ij} \dot{\beta}^i \dot{\beta}^j = 0. \tag{5.57}$$

The equations of motion are $\frac{d(\tilde{N}^{-1}\dot{\beta}^i)}{dx^0} = 0$. An affine parameter along those geodesics is $d\tau = +\tilde{N}dx^0 = -dt/\sqrt{g}$, since with this evolution parameter, the equations are simply $\ddot{\beta}^i = 0$. Here, t is the space-time proper time. We take the minus sign because we assume, as before, that the singularity of the Kasner solution is in the past at $t = 0$ and that the other time coordinates *increase* as $t \to 0$. Thus, in the gauge $\tilde{N} = 1$, the time coordinate x^0 is an affine parameter for the geodesics in the space of the scale factors.

The solutions to the equations of motion are explicitly

$$\beta^i = v^i \, \tau + \beta_0^i \,, \tag{5.58}$$

where v^i and β_0^i are constants of the motion. In particular, v^i is the τ-parameter velocity, $v^i \equiv d\beta^i/d\tau$. Because of the constraint, these integration constants are subject to the "zero-mass condition"

$$G_{ij}v^i v^j = 0 \,, \tag{5.59}$$

enforcing that the geodesic is null.

One can rewrite the solution (5.58) in the above proper time Kasner form by performing the corresponding change of evolution parameter. Since $dt = -\sqrt{g}d\tau$, with $\sqrt{g} = \exp(-\Sigma_i\beta^i)$, one finds $t \propto \exp\big(-(\Sigma_i v^i)\tau\big)$, or

$$\tau = -\frac{1}{\sum_i v^i} \ln t + const. \tag{5.60}$$

The condition $\Sigma_i v^i > 0$, which guarantees that the geodesic goes towards the future, is consistent with our convention that τ increases (in fact to $+\infty$) as $t \to 0^+$ near the singularity. Eliminating τ in favor of t in the space-time metric leads then to

$$ds^2 = -dt^2 + \sum_{i=1}^{d} A_i^2(t)(dx^i)^2 \,, \quad A_i(t) = b_i t^{p_i} \,, \tag{5.61}$$

where $b_i \equiv \exp(-\beta_0^i)$. Here, the Kasner exponents p_i are given in terms of the affine velocities by

$$p_i = \frac{v^i}{\sum_j v^j}. \tag{5.62}$$

As they should be, they are subject to the quadratic constraint

$$\sum_{i=1}^{d} p_i^2 - \left(\sum_{i=1}^{d} p_i\right)^2 = 0. \tag{5.63}$$

coming from the "zero-mass condition," and to the linear constraint

$$\sum_{i=1}^{d} p_i = 1 \tag{5.64}$$

coming from their definition in terms of the velocities.

By rescaling the spatial coordinates, one can set $b_i = 1$ and obtain the standard proper time form of the Kasner metric with $g \sim t^2$, which was extensively used in the previous chapters.

Thus, the Kasner solution, which extremizes the Einstein–Hilbert action in the space of diagonal metrics depending only on time, is given by a null geodesic of the supermetric (5.12).

Hamiltonian for Diagonal, Space-Independent Metrics

The Hamiltonian form of the action (5.3) reduces, for diagonal metrics depending only on time, to

$$S[\beta^i, \pi_i, \tilde{N}] = \int dx^0 \left[\pi_i \dot{\beta}^i - H_0\right] \tag{5.65}$$

with

$$H_0 = \tfrac{1}{4}\tilde{N} G^{ij} \pi_i \pi_j \,, \tag{5.66}$$

where π_i are the momenta conjugate to β^i, i.e.,

$$\pi_i = 2\tilde{N}^{-1} G_{ij} \dot{\beta}^j = 2G_{ij} \frac{d\beta^j}{d\tau} \equiv 2G_{ij} v^j \tag{5.67}$$

in terms of the τ-parameter velocities $v^i \equiv d\beta^i/d\tau$. Here, G^{ij} is the inverse of G_{ij} ($G^{ij} G_{jk} = \delta^i_k$). Explicitly*

$$G^{ij} = \frac{1}{2} \begin{pmatrix} 1 & -1 & -1 \\ -1 & 1 & -1 \\ -1 & -1 & 1 \end{pmatrix} . \tag{5.68}$$

and

$$G^{ij} \pi_i \pi_j = \sum_{i=1}^{3} \pi_i^2 - \frac{1}{2} \left(\sum_{i=1}^{3} \pi_i \right)^2 . \tag{5.69}$$

The extra index 0 on the Hamiltonian H is to emphasize the fact that H_0 describes only the kinetic terms associated to the logarithmic scale factors β^i. In the general description given below, H_0 will be only a piece – an essential piece, but an incomplete one – of the total Hamiltonian.

Radial Projection of the Kasner Motion

We now radially project the Kasner dynamics on hyperbolic space $\mathbb{H}_2$. This will be particularly useful for the billiard description.

* When dealing with tensors in the flat space of the logarithmic scale factors – such as the velocity of a trajectory, or the wall forms encountered below – we shall systematically lower or raise indices with the metric G_{ij} and its inverse G^{ij}.

One can assume, by shifting if necessary the origin in β-space, that $v_i\beta_0^i < 0$. Since $\beta^i\beta_i = 2v_i\beta_0^i\tau + \beta_{0\,i}\beta_0^i$, this ensures that the β^i trajectories (5.58) will, for large enough values of τ, get inside the future light cone of the origin in superspace, i.e.,

$$\beta^i\beta_i = 2v_i\beta_0^i\tau + \beta_{0\,i}\beta_0^i < 0\,. \tag{5.70}$$

Once the motion is inside the future light cone of the origin, one can radially project it on the future sheet of the unit hyperboloid. It is this part of the motion that is relevant for studying the asymptotics $\tau \to \infty$, and it is this part of the motion that will be from now on of interest to us.

In terms of the hyperbolic polar coordinates (ρ, γ^i), the Hamiltonian H_0 reads

$$H_0 = \frac{\tilde{N}}{4}\left[-\pi_\rho^2 + \frac{1}{\rho^2}\pi_\gamma^2\right] \tag{5.71}$$

where π_ρ is the radial momentum and where π_γ are the momenta conjugate to the coordinates γ^i on hyperbolic space. More precisely, in terms of the independent hyperbolic variables r, ϕ, H_0 is explicitly given by

$$H_0 = \frac{\tilde{N}}{4}\left[-\pi_\rho^2 + \frac{1}{\rho^2}\left(\pi_r^2 + \frac{1}{\sinh^2 r}\pi_\phi^2\right)\right]. \tag{5.72}$$

The description becomes simpler if instead of ρ, one uses the timelike configuration variable λ defined by

$$\lambda \equiv \ln\rho \equiv \tfrac{1}{2}\ln\left(-G_{ij}\beta^i\beta^j\right) \tag{5.73}$$

with conjugate momentum π_λ. The Hamiltonian becomes then

$$H_0 = \frac{\tilde{N}}{4\rho^2}\left[-\pi_\lambda^2 + \pi_\gamma^2\right]. \tag{5.74}$$

The gauge

$$\tilde{N} = \rho^2 \tag{5.75}$$

reduces the Hamiltonian (5.74) to a free Hamiltonian on the pseudo-Riemannian space with metric $-d\lambda^2 + d\S^2$, $H_0 = \frac{1}{4}\left[-\pi_\lambda^2 + \pi_\gamma^2\right]$. The coordinate time associated with the special gauge defined by equation (5.75) will be denoted by T, i.e.,

$$dT = -\frac{dt}{\rho^2\sqrt{g}}. \tag{5.76}$$

In the gauge (5.75), the Hamiltonian H_0 is a sum of two decoupled parts, one for the logarithmic radial coordinate λ and one for the $\mathbb{H}_2$- coordinates. The logarithmic radial momentum $\pi_\lambda = \rho\pi_\rho$ is a constant of the motion, say $\pi_\lambda = C$. Similarly, $\frac{1}{4}\pi_\gamma^2$ is also a constant of the motion. The constraint $H_0 = 0$ implies that the two constants of integration are not independent, but rather, $\frac{1}{4}\pi_\gamma^2 = \frac{1}{4}C^2$.

One can solve the constraint and get rid of λ and π_λ by using λ as an intrinsic time. Because π_λ is constant, λ is linear in the time T. In the λ-time, the Hamiltonian for the remaining degrees of freedom is $-\pi_\lambda = \sqrt{\pi_\gamma^2}$ and describes the geodesic motion in hyperbolic space.

Conclusion: Radially projecting the dynamics on the future sheet of the hyperboloid is a way to implement the Hamiltonian constraint. The unconstrained degrees of freedom are the coordinates of hyperbolic space $\mathbb{H}_2$ and their conjugate momenta. Their motion is the free (geodesic) motion on $\mathbb{H}_2$. One can thus view the Kasner dynamics as describing the motion of a "ball" moving freely on $\mathbb{H}_2$.

Relationship Between Different Time Variables

It is useful to collect the relations between the different time variables introduced along the way of the analysis, namely,

- the proper time t between successive equal time hypersurfaces in space-time;
- the affine parameter τ related to the proper time t through $d\tau = -\frac{dt}{\sqrt{g}}$;
- the time T;
- the time λ.

In terms of the affine parameter τ along the null geodesics in β-space, one has for the time T

$$d\tau = -\frac{dt}{\sqrt{g}} = \rho^2 dT. \tag{5.77}$$

From (5.70) one sees that ρ^2 varies linearly with τ:

$$\rho^2 = -\beta_i\beta^i = -2v_i\beta_0^i\tau - \beta_0^i\beta_{0i} \tag{5.78}$$

which implies

$$T = -\frac{1}{2v_i\beta_0^i}\ln\tau + const. \tag{5.79}$$

Recalling that τ varies logarithmically with the proper time t, we see that $T \propto \ln|\ln t|$. Finally, for the configuration variable λ used as an intrinsic time variable, we have

$$\lambda = \tfrac{1}{2}\ln\tau + const. = \tfrac{1}{2}\ln|\ln t| + const. \tag{5.80}$$

in view of (5.78).

At the singularity the proper time t decreases toward 0^+. By contrast, the coordinates λ, T and τ both increase toward $+\infty$, as ensured by the minus sign in $dt = -Ndx^0$ and (5.77). Irrespective of the choice of coordinates, the spatial volume density g collapses to zero in this limit.

Inclusion of Off-Diagonal Metric Components

The previous discussion dealt with diagonal, homogeneous solutions. One can easily include the off-diagonal components.

The Iwasawa change of variables (5.20) can be extended to phase space as a point canonical transformation. The new momenta are obtained in the usual manner through the formula

$$\pi^{ij}\dot{g}_{ij} \equiv \sum_a \pi_a \dot{\beta}^a + \sum_a P^i{}_a \dot{\mathcal{N}}^a{}_i. \tag{5.81}$$

The momenta

$$P^i{}_a = \sum_b e^{2(\beta^b - \beta^a)} \dot{\mathcal{N}}^a{}_j \mathcal{N}^j{}_b \mathcal{N}^i{}_b \tag{5.82}$$

conjugate to the non-constant off-diagonal Iwasawa components $\mathcal{N}^a{}_i$ are obviously only defined for $a < i$.

The kinetic term becomes, in the Iwasawa variables,

$$\mathcal{K} = \mathcal{H}_0 + \mathcal{V}_S \tag{5.83}$$

where $\mathcal{H}_0$ is the kinetic term for the (logarithmic) scale factors derived above,

$$\mathcal{H}_0 = \frac{1}{4} G^{ab} \pi_a \pi_b \tag{5.84}$$

and where $\mathcal{V}_S$ is the kinetic term for the off-diagonal variables derived by mere substitution of the Iwasawa change of variables in the original form of $\mathcal{K}$,

$$\mathcal{V}_S = \frac{1}{2} \sum_{a<b} e^{-2(\beta^b - \beta^a)} \left(P^j{}_a \mathcal{N}^b{}_j\right)^2, \tag{5.85}$$

The term $\mathcal{V}_S$ is called "centrifugal term" in analogy with a similar decomposition in classical mechanics, or "symmetry walls term" (hence the subscript "S") for reasons that will become clear in the subsequent paragraphs. In (5.85), the sum over i is restricted by the fact that $a < j$.

The easiest way to understand the role of the off-diagonal terms is to observe that the non-diagonal vacuum homogeneous metrics can always be diagonalized by a time-independent change of basis. Indeed, at any time, one can simultaneously diagonalize the metric and its first-order time derivative. The vacuum Einstein equations guarantee then that the metric remains diagonal at all times. One can therefore derive the general non-diagonal homogeneous solution by (i) performing the constant change of basis that makes the metric diagonal at all times, i.e., go to the "Kasner frame"; (ii) solve the equations of motion for the diagonal metric, which is the kown Kasner solution; (iii) go back to the original frame where the metric is non-diagonal and study in that parametrization the influence of the off-diagonal terms.

This procedure yields

$$g_{ij} = e^{-2p_1\tau} l_i l_j + e^{-2p_2\tau} m_i m_j + e^{-2p_3\tau} r_i r_j, \tag{5.86}$$

where the frame $\{l_i, m_i, r_i\}$ in which the metric is diagonal is not necessarily the original coordinate frame. The metric (5.86) is obtained by making the change of frame induced by the matrix

$$L = \begin{pmatrix} l_1 & l_2 & l_3 \\ m_1 & m_2 & m_3 \\ r_1 & r_2 & r_3 \end{pmatrix}. \tag{5.87}$$

To display as clearly as possible the role of the centrifugal terms, we assume that only one of them is present and take $r_1 = r_2 = 0 = l_3 = m_3$ and $r_3 = 1$, so that L reduces to

$$L = \begin{pmatrix} l_1 & l_2 & 0 \\ m_1 & m_2 & 0 \\ 0 & 0 & 1 \end{pmatrix}. \tag{5.88}$$

This very special choice (of "measure zero" in the space of the Ls) amounts to keep only the potential term involving $\exp -2(\beta^2 - \beta^1)$ (see below). The general case will be covered below when we discuss the billiard picture in full generality. We also assume, as before, that $p_1 < p_2 < p_3$.

Using the relationship between the metric and the Iwasawa variables, one gets that the scale factors and the off-diagonal components in the Iwasawa frame behave as

$$\begin{aligned} \beta^1 &= -\frac{1}{2}\ln X, \quad \beta^2 = -\frac{1}{2}\ln\left(\frac{Y}{X}\right) \\ \beta^3 &= p_3\tau \\ n_1 &= \frac{e^{-2p_1\tau}l_1 l_2 + e^{-2p_2\tau}m_1 m_2}{X}, \quad n_2 = 0, \quad n_3 = 0 \end{aligned}$$

with

$$\begin{aligned} X &= e^{-2p_1\tau}(l_1)^2 + e^{-2p_2\tau}(m_1)^2 \\ Y &= e^{-2(p_1+p_2)\tau}\,(\det L)^2 . \end{aligned}$$

Because $n_2 = n_3 = 0$, the only centrifugal term present here is indeed the one involving $\exp -2(\beta^2 - \beta^1)$, so that the solution exhibits just the effect of that potential term on the free motion.

One easily finds from the above formulas that for $\tau \to -\infty$ (well before the collision with the potential), the Iwasawa variables behave as (to leading order)

$$\beta^1 \to p_2\tau, \quad \beta^2 \to p_1\tau, \quad n_1 \to \frac{m_2}{m_1} \quad (\tau \to -\infty)$$

(to leading order), while for $\tau \to +\infty$ (well after the collision with the potential), the Iwasawa variables behave as

$$\beta^1 \to p_1\tau, \quad \beta^2 \to p_2\tau, \quad n_1 \to \frac{l_2}{l_1} \quad (\tau \to +\infty)$$

(again to leading order).

This leads to the following behavior of the metric components:

- For very negative τs, one has $\beta^1 << \beta^2$ because $p_1 < p_2$ and $\tau << 0$.
- As τ increases, β^1 increases faster than β^2 and ultimately catches it. At the time τ_0 such that $Y = X^2$, the two scale factors become equal, $\beta^1 = \beta^2$.
- However, a "collision" takes place when β^1 and β^2 become comparable, which has the effect of permuting the velocities of the scale factors β^1 and β^2.
- After the collision, it is β^2 that increases faster than β^1 and therefore, well after the collision, for large $\tau >> 0$, one has again $\beta^1 << \beta^2$.
- At the same time, the off-diagonal Iwasawa variable n_1 evolves with time, but asymptotically tends to a constant in the limit of large $|\tau|$. If the matrix elements of L are of the same order, the asymptotic values of n_1 before and after the collision are of the same order so that n_1 does not change significantly.

Thus, the effect of the collision with the term in $\mathcal{V}_S$ involving the exponential $\exp -2(\beta^2 - \beta^1)$ is just to permute the velocities v^1 and v^2. Had we included the other potential terms in $\mathcal{V}_S$, we would have found that the third component v^3 of the velocity would also participate in the permutations, so that the full symmetry group S_3 is generated by the potential $\mathcal{V}_S$ – hence its name.

That the full permutation group emerges will be shown in Section 5.6 below where we derive the BKL limit in full generality. It can also be established by considering a change of basis L more general than (5.88). We refer the interested reader to [50], where the corresponding computations are carried out for generic (invertible) matrices L.

The effect of the centrifugal term described here, leading to a permutation of the Kasner exponents between the directions 1 and 2, is exactly the effect uncovered in Chapter 2 in the context of homogeneous cosmological models. More precisely, in Section 2.4 where the rotation of the Kasner axes was analyzed, it was shown that the growth of terms such as B^2/A^2 or C^2/B^2 (in the notations of Section 2.4) led indeed to the exchange of the Kasner exponents that we have exhibited here; see formula (2.63).

5.5 Hamiltonian in Pseudo-Gaussian Gauge

Pseudo-Gaussian Gauge

We now return to the general, inhomogeneous case.

The Hamiltonian formulation was developed in Section 5.1 for generic space-time and generic slicings by spacelike hypersurfaces. To incorporate the information that there is a spacelike singularity and study the behavior of the fields in its vicinity by Hamiltonian techniques, we conform to the analysis performed in the previous chapters and take a space-time slicing such that the singularity "occurs" on a constant time slice. We assume that the singularity is in the past, at $t \to 0^+$ in terms of proper time. That is, proper time decreases from positive values to zero as one goes to the singularity. We shall also choose the orientation of x^0 and of the subsequent time variables such that they *increase* as one tends to the singularity. Thus, we take again

$$dt = -N dx^0 = -\tilde{N}\sqrt{g}dx^0. \tag{5.89}$$

(with the minus sign and $N > 0$).

The slicing is built from $t = 0$ by imposing "pseudo-Gaussian" coordinate conditions. Pseudo-Gaussian coordinates are such that the shift vanishes, $N^i = 0$. This means geometrically that one marches orthogonally to the slices of constant time as one moves from one slice to the next along the time coordinate lines. However, one does not restrict the lapse to be equal to unity and allows it to be more general, so that the proper time between neighboring slices can vary.

The metric thus reads

$$ds^2 = -\left(N(x^0, x^i)dx^0\right)^2 + g_{ij}(x^0, x^i)dx^i dx^j. \tag{5.90}$$

The slicing is completely fixed by choosing the lapse N – or, equivalently, the rescaled lapse $\tilde{N}$. Once N (or $\tilde{N}$) is fixed, and given that the singularity is at $t = 0$, the only coordinate freedom left is that of making time-independent changes of spatial coordinates $x^i \to x'^i = f^i(x^j)$, exactly as for the synchronous case corresponding to $N = 1$.

Keeping the freedom to adjust the lapse to values different from 1 is motivated by the analysis of the homogeneous case, which has shown that it is useful to keep this flexibility.

The Hamiltonian action in any pseudo-Gaussian gauge reads

$$S\left[g_{ij}, \pi^{ij}, \tilde{N}\right] = \int dx^0 \int d^3x \left(\pi^{ij} \dot{g}_{ij} - H\right) \tag{5.91}$$

where the Hamiltonian density H is given by

$$H \equiv \tilde{N}\mathcal{H} \tag{5.92}$$

with $\mathcal{H}$ given by (5.4), $\mathcal{H} = \mathcal{K} + \mathcal{M}$.

Decomposition of Hamiltonian in Iwasawa Variables

We have seen in Section 5.4 that the scale factors and the off-diagonal variables play very different role in the BKL limit. For that reason, we perform at each spatial point the Iwasawa decomposition of the spatial metric components and their conjugate momenta,

$$(g_{ij}, \pi^{ij}) \quad \longrightarrow \quad (\beta^i, \mathcal{N}^a{}_j, \pi_i, P^i{}_a)$$

and use as new variables the logarithmic scale factors, the off-diagonal variables, and their conjugate momenta. Because the kinetic term is ultralocal, everything that was said above about it in the context of homogeneous models remain valid.

The change of variables is exactly the same because $\mathcal{K}$ does not involve spatial derivatives.

The Hamiltonian density $\mathcal{H}$ is then split into two parts. The first one $\mathcal{H}_0$, is the kinetic term for the local (logarithmic) scale factors β^i. The other, denoted by $\mathcal{V}$, involves all the other contributions to $\mathcal{H}$, that is, the kinetic terms of the off-diagonal metric components and the curvature term. This term $\mathcal{V}$ turns out to act as a potential for the dynamics of the scale factors.

Explicitly, we have

$$\mathcal{H} = \mathcal{H}_0 + \mathcal{V} \tag{5.93}$$

where $\mathcal{H}_0$ is the kinetic term $\mathcal{H}_0$ for the β variables and where the total (weight 2) potential density,

$$\mathcal{V} = \mathcal{V}_S + \mathcal{V}_G, \tag{5.94}$$

is the sum of the centrifugal contribution coming from the kinetic energy of the off-diagonal components,

$$\mathcal{V}_S = \frac{1}{2}\sum_{a<b} e^{-2w^S_{ab}} \left(P^j{}_b \mathcal{N}^a{}_j\right)^2, \tag{5.95}$$

with

$$w^S_{ab} = \beta^b - \beta^a \quad a < b \tag{5.96}$$

and a "gravitational" contribution

$$\mathcal{V}_G = -gR, \tag{5.97}$$

given by the spatial curvature.

Each linear form w^S_{ab} defines through the equation $w^S_{ab} = 0$ a hyperplane H^S_{ab} in β-space. The hyperplane H^S_{ab} is timelike because the normal to it, which has components $(w^S_{ab})^i = G^{ij}(w^S_{ab})_j$ is spacelike,

$$G_{ij}(w^S_{ab})^i(w^S_{ab})^j = G^{ij}(w^S_{ab})_i(w^S_{ab})_j = +2.$$

Here, the $(w^S_{ab})_i$ are the coefficients of the β's in the linear form w^S_{ab},

$$w^S_{ab} = (w^S_{ab})_i\beta^i,$$

i.e., $(w^S_{ab})_i = \delta^b_i - \delta^a_i$. Because it is timelike, the hyperplane H^S_{ab} intersects the upper sheet of the unit hyperboloid and defines a hyperplane in hyperbolic space $\mathbb{H}_2$, which we also denote by H^S_{ab}.

Computation of Spatial Curvature

To complete the Iwasawa reformulation, we need to express the spatial curvature in terms of the Iwasawa variables. This is a standard computation using the Cartan formalism. We follow closely [51], sometimes almost verbatim. We use the short-hand notation $A_a \equiv e^{-\beta^a}$ for the Iwasawa scale factors.

Let $C^a{}_{bc}(x)$ be the structure functions of the Iwasawa basis $\{\theta^a\}$, i.e.,

$$d\theta^a = -\frac{1}{2}C^a{}_{bc}\theta^b \wedge \theta^c \tag{5.98}$$

where d is the spatial exterior differential. The structure functions obviously depend only on the off-diagonal components $\mathcal{N}^a{}_i$, but not on the scale factors. Using the Cartan formulas for the connection one-form $\omega^a{}_b$,

$$d\theta^a + \sum_b \omega^a{}_b \wedge \theta^b = 0 \tag{5.99}$$

$$d\gamma_{ab} = \omega_{ab} + \omega_{ba} \tag{5.100}$$

where $\omega_{ab} \equiv \gamma_{ac}\omega^c{}_b$, and

$$\gamma_{ab} = \delta_{ab}A_a^2 \equiv \exp(-2\beta^a)\delta_{ab} \tag{5.101}$$

is the metric in the frame $\{\theta^a\}$, one finds

$$\begin{aligned}\omega^c{}_d = &\sum_b \frac{1}{2}\left(C^b{}_{cd}\frac{A_b^2}{A_c^2} + C^d{}_{cb}\frac{A_d^2}{A_c^2} - C^c{}_{db}\frac{A_c^2}{A_d^2}\right)\theta^b \\ &+\sum_b \frac{1}{2A_c^2}\left[\delta_{cd}(A_c^2)_{,b} + \delta_{cb}(A_c^2)_{,d} - \delta_{db}(A_d^2)_{,c}\right]\theta^b.\end{aligned} \tag{5.102}$$

In the last bracket above, the commas denote the frame derivatives $\partial_a \equiv \mathcal{N}^i{}_a\partial_i$.

The Riemann tensor $R^c{}_{def}$, the Ricci tensor R_{de} and the scalar curvature R are obtained through

$$\Omega^a{}_b = d\omega^a{}_b + \sum_c \omega^a{}_c \wedge \omega^c{}_b \tag{5.103}$$

$$= \frac{1}{2}\sum_{e,f} R^a{}_{bef}\theta^e \wedge \theta^f \tag{5.104}$$

where $\Omega^a{}_b$ is the curvature 2-form and

$$R_{ab} = \sum_c R^c{}_{acb}, \quad R = \sum_a \frac{1}{A_a^2}R_{aa}. \tag{5.105}$$

Direct, but somewhat cumbersome, computations yield

$$R = -\frac{1}{4}\sum_{a,b,c}\frac{A_a^2}{A_b^2A_c^2}(C^a{}_{bc})^2 + \sum_a \frac{1}{A_a^2}F_a(\partial^2\beta, \partial\beta, \partial C, C) \tag{5.106}$$

where F_a is some complicated function of its arguments whose explicit form will not be needed here. The only property of F_a that will be of importance is that it is a polynomial of degree two in the derivatives $\partial\beta$ and of degree one in $\partial^2\beta$. Thus, the exponential dependence on the βs which determines the asymptotic behavior in the BKL limit, occurs only through the A_a^2-terms written explicitly in (5.106).

In the first sum on the right-hand side of (5.106) one obviously has $b \neq c$ because the structure functions $C^a{}_{bc}$ are antisymmetric in the pair $[bc]$. In addition to this restriction, we can assume, without loss of generality, that $a \neq b, c$. Indeed, the terms with either $a = b$ or $a = c$ can be absorbed into a redefinition of F_a. We can thus write the gravitational potential term as

$$\mathcal{V}_G \equiv -gR = \frac{1}{4}\sum_{a,b,c}{}' e^{-2\alpha_a(\beta)}(C^a{}_{bc})^2 - \sum_a e^{-2\mu_a(\beta)} F_a \tag{5.107}$$

where the prime on $\sum$ indicates that the sum is to be performed only over unequal indices, i.e., $a \neq b, b \neq c, c \neq a$, and where the linear forms $\alpha_a(\beta)$ and $\mu_a(\beta)$ are given by

$$\alpha_a(\beta) = 2\beta^a \tag{5.108}$$

and

$$\mu_a(\beta) = \sum_{c \neq a} \beta^c. \tag{5.109}$$

respectively.

The linear forms α_a define hyperplanes H_a^G in β-space through the equations $\alpha_a = 0$. These hyperplanes H_a^G are timelike because their normals, which have components $(\alpha_a)^i = G^{ij}(\alpha_a)_j$ are all spacelike, $G^{ij}(\alpha_a)_i(\alpha_a)_j = +2$, with $\alpha_a = (\alpha_a)_i\beta^i$, i.e., $(\alpha_a)_i = 2\delta_i^a$. Because they are timelike, the hyperplanes H_a^G intersect the upper sheet of the unit hyperboloid and define hyperplanes in hyperbolic space $\mathbb{H}_2$, which we also denote by H_a^G.

By contrast, the linear forms μ_a defines hyperplanes that are lightlike, $G^{ij}(\mu_a)_i(\mu_a)_j = 0$ and which therefore do not intersect the upper sheet of the unit hyperboloid.

5.6 BKL Limit and Emergence of Billiard Description

We are now ready to investigate the limit $t \to 0^+$ of the dynamics. To that end, we pass to hyperbolic coordinates in β-space and we work in the slicing of space-time defined by the gauge condition (5.75), i.e., $\tilde{N} = \rho^2$, taken everywhere in the region of space near the singularity under consideration, with corresponding time coordinate T (see (5.76)). We thus assume that the hyperbolic radial coordinate ρ used to define the gauge can actually be meaningfully introduced everywhere in that given region of space. The limit to the singularity is $t \to 0^+$, and we have in that limit $\sqrt{g} \to 0$, $\rho \to +\infty$, $T \to +\infty$ with β^a going to infinity inside the future light cone of some origin (which may be redefined by a spatially dependent transformation).

Again, because the kinetic term is ultralocal, all the transformations performed in the homogeneous case remain valid and take the same form at each point in space. The various expressions for $\mathcal{K}$ worked out above hold unchanged.

Our aim is thus to study the asymptotic behavior of all the dynamical variables $\beta^a(T), \mathcal{N}^a{}_j(T)$ and their momenta as $T \to +\infty$. This is the Hamiltonian formulation of the BKL limit.

In the gauge (5.75) and in terms of the variables λ, γ^i adapted to the projection on hyperbolic space, the Hamiltonian takes at each spatial point the form

$$H(\lambda, \pi_\lambda, \gamma, \pi_\gamma, \mathcal{N}^a{}_j, P^i{}_a) = \tilde{N}\mathcal{H} \qquad (5.110)$$
$$= \frac{1}{4}\left[-\pi_\lambda^2 + \pi_\gamma^2\right] + \rho^2 \sum_A c_A \exp\big(- 2\rho w_A(\gamma)\big)$$

where $\rho \equiv e^l$ and where we are using the same conventions for the kinetic energy π_γ^2 of a particle moving on H_2 as in (5.71) and (5.72). We have seen that the terms in the potential $\mathcal{V}$ all take the characteristic exponential form of (5.110). The sum over A collectively denotes a sum over all of them. The coefficients c_A involve the undifferentiated variables $\mathcal{N}^a{}_j, P^i{}_a$ and only them for $\mathcal{V}_S$. In the case of $\mathcal{V}_G$, the coefficients c_A involve the variables $\mathcal{N}^a{}_j$ and their spatial derivatives, as well as the spatial derivatives of first and second degree of the scale factors β^a, but these derivatives appear only polynomially as we pointed out.

Now in the BKL limit $\rho \to +\infty$, each term $\rho^2 \exp\big(- 2\rho w_A(\gamma)\big)$ either goes to zero when $w_A(\gamma) > 0$, or explodes to $+\infty$ if $w_A(\gamma) < 0$. In other words, one can replace $\rho^2 \exp\big(- 2\rho w_A(\gamma)\big)$ by $\Theta(-2w_A(\gamma))$, where the sharp wall Θ-function is defined by $\Theta(x) = 0$ if $x < 0$ and $\Theta(x) = +\infty$ if $x > 0$. Of course, $\Theta(-2w_A(\gamma)) = \Theta(-w_A(\gamma))$, but, as in [47, 51], we keep the extra factor of 2. This is because normalization questions are relevant for the Kac–Moody developments. We thus want to keep track of the original normalization of the arguments of the exponentials, from which the Θ-functions originate.

A basic formal property of the sharp wall Θ-function is its invariance under multiplication by a positive quantity, $k\Theta(x) = \Theta(x)$ for $k > 0$. Thus, for the exponential terms in the potential $\mathcal{V}$ that come with a positive prefactor c_A, one gets

$$\lim_{\rho\to\infty} \Big[c_A\rho^2 \exp\big(- 2\rho w_A(\gamma)\Big] = c_A\,\Theta\big(- 2w_A(\gamma)\big) \equiv \Theta\big(- 2w_A(\gamma)\big) \qquad (5.111)$$

while for the exponential terms in the potential $\mathcal{V}$ that come with a negative prefactor c_A, one gets

$$\lim_{\rho\to\infty} \Big[c_A\rho^2 \exp\big(- 2\rho w_A(\gamma)\Big] = c_A\,\Theta\big(- 2w_A(\gamma)\big) \equiv -\Theta\big(- 2w_A(\gamma)\big). \qquad (5.112)$$

It follows that the limiting Hamiltonian density H_∞ describing the dynamics in the limit $\rho \to \infty$ reduces at each spatial point to a function of the conjugate momenta π_λ and π_γ and to the hyperbolic space variables γ only, given by

$$H_\infty(\pi_\lambda, \gamma, \pi_\gamma) = \frac{1}{4}\left[-\pi_\lambda^2 + \pi_\gamma^2\right] + \sum_B \Theta\big(- 2w_B(\gamma)\big) - \sum_C \Theta\big(- 2w_C(\gamma).$$

Here, the sum over B refers to the exponentials that come with a positive prefactor, while the sum over C refers to the exponentials that come with a negative prefactor. The terms originating from the potential $\mathcal{V}_S$ and from the α-terms in the potential $\mathcal{V}_G$ clearly belong to the first set because the corresponding prefactors are manifestly positive.* This is not the case for the μ-terms in the potential $\mathcal{V}_G$, for which the sign of the prefactor can vary.

Dropping Lightlike Walls

Fortunately, these μ-terms can be neglected in the BKL limit no matter what the sign of their prefactor is.

Indeed, the linear μ-forms associated with the potentially "dangerous" terms read explicitly:

$$\mu_1 = \gamma^2 + \gamma^3, \quad \mu_2 = \gamma^1 + \gamma^3, \quad \mu_3 = \gamma^1 + \gamma^2.$$

In the BKL limit, one has $\mu_a > 0$ for all as, so that the argument of the corresponding sharp wall Θ-functions is negative and these therefore vanish. This is because the γ^as lie on the positive sheet of the unit hyperboloid in the space of the scale factors, i.e., are constrained by the conditions $G_{ab}\gamma^a\gamma^b = -1 \Leftrightarrow 2\gamma^1\gamma^2 + 2\gamma^2\gamma^3 + 2\gamma^3\gamma^1 = 1$. Together with $\gamma^1 + \gamma^2 + \gamma^3 > 0$, this implies that the linear forms μ_a cannot vanish and have thus a definite sign, which is easily verified to be positive. [If μ_3, say, were to vanish, one would have $2\gamma_1\gamma_2 = 1$ which, together with $\gamma_2 = -\gamma_1$, implies the contradiction $-2(\gamma_1)^2 = 1$. On the other hand, the solution $\gamma_1 = \gamma_2 = \gamma_3 = \left(\sqrt{6}\right)^{-1}$ makes all three μ_as positive.] This reflects, of course, the property derived above that the hyperplanes $\mu_a(\beta) = 0$ are lightlike and do not intersect the unit hyperboloid. The μ-terms in the gravitational potential $\mathcal{V}_G$ can thus be asymptotically neglected. The gravitational potential consequently becomes, in the BKL limit, a positive sum of sharp wall potentials.

Once the μ-terms are dropped, the Hamiltonian becomes

$$H_\infty(\pi_\lambda, \gamma, \pi_\gamma) = \frac{1}{4}\left[-\pi_\lambda^2 + \pi_\gamma^2\right] + \sum_B \Theta\big(-2w_B(\gamma)\big) \tag{5.113}$$

where only the terms from $\mathcal{V}_S$ and the α-terms from $\mathcal{V}_G$ appear in the sum $\sum_B \Theta\big(-2w_B(\gamma)\big)$. These are the sharp walls potentials associated with the

* We should say "generically positive." Indeed, all that can be affirmed is that the prefactor is nonnegative. It can actually vanish but this happens for special, "non-generic" choices of initial data. For instance, in the case of vacuum homogeneous cosmological models, the spatial metric can be taken to be diagonal, eliminating the potential $\mathcal{V}_S$. But any small amount of matter or inhomogeneity will introduce non-diagonal terms, so that generic spatial metrics in the synchronous gauges are non-diagonal, and $\mathcal{V}_S$ does not vanish. It is only for particular solutions with symmetries that some of the nonnegative prefactors are zero.

timelike hyperplanes H^S_{ab} and H^G_a in β-space, which intersect the upper sheet of the unit hyperboloid along the hyperplanes $w_B(\gamma) = 0$.

The fact that the μ-terms in $\mathcal{V}_G$ have a prefactor with a sign that is not universal was observed in the previous chapter on homogeneous cosmological models, where these signs were found not to be the same for types VIII and IX – while the prefactors of the other terms have always the same sign. The fact that they can be neglected was also observed there.

Because the limiting Hamiltonian (5.113) no longer depends on λ, $\mathcal{N}^a{}_i$ and $P^i{}_a$, the asymptotic Hamiltonian equations of motion for λ, $\mathcal{N}^a{}_i$ and $P^i{}_a$ read $\dot{\pi}_\lambda = 0$, $\dot{\mathcal{N}}^a_i = 0$ and $\dot{P}^i_a = 0$. The variables π_λ, $\mathcal{N}^a{}_i$ and $P^i{}_a$ are asymptotic constants of the motion, i.e., freeze in the limit $\lambda \to +\infty$. This freezing phenomenon was observed before in the case of homogeneous models.

The limiting value of π_λ is determined by the limiting constraint $H_\infty = 0$. In the chosen gauge (5.75), the variable λ evolves asymptotically according to $d\lambda/dT = -\frac{1}{2}\pi_\lambda$ and hence becomes a linear function of T as in the Kasner solution. The nontrivial dynamical variables are (γ, π_γ) – i.e., r, π_r, ϕ, π_ϕ . Their evolution is governed by the sum of a free (non-relativistic) kinetic term $\pi_\gamma^2/4$ and a sum of sharp wall potentials. Therefore, the resulting motion of the γs indeed constitutes a billiard motion, namely, portions of geodesic motion on the two-dimensonial hyperbolic space $\mathbb{H}_2$ interrupted by reflections on the walls defined by $w_B(\gamma) = 0$. These walls are hyperbolic hyperplanes because they are given by the intersection of the unit hyperboloid $\beta^a\beta_a = -1$ with the Minkowskian timelike hyperplanes $w_B(\beta) = 0$.

Dominant Walls and Billiard Table

In fact, one may restrict the sum in (5.113) over the so-called *dominant walls*,

$$H_\infty(\pi_\lambda, \gamma, \pi_\gamma) = \frac{1}{4}\left[-\pi_\lambda^2 + \pi_\gamma^2\right] + \sum_{A'} \Theta(-2w_{A'}(\gamma)), \tag{5.114}$$

which are such that the restricted set of inequalities $\{w_{A'}(\gamma) \geq 0\}$ imply the full set $\{w_B(\gamma) \geq 0\}$.

In the case of pure gravity in four space-time dimensions analyzed here, the walls are the boundaries of the regions

$$\gamma^2 - \gamma^1 \geq 0, \quad \gamma^3 - \gamma^2 \geq 0, \quad \gamma^3 - \gamma^1 \geq 0$$

("symmetry walls" from $\mathcal{V}_S$) and

$$2\gamma^1 \geq 0, \quad 2\gamma^2 \geq 0 \quad 2\gamma^3 \geq 0$$

("gravitational walls" from $\mathcal{V}_G$). In terms of the unconstrained (r, ϕ) parametrization of $\mathbb{H}_2$, the symmetry walls are

$$\cos\phi \geq 0, \quad -\cos\left(\phi + \frac{\pi}{3}\right) \geq 0, \quad \cos\left(\phi - \frac{\pi}{3}\right) \geq 0 \tag{5.115}$$

while the gravitational walls are

$$\coth r - 2\sin(\phi + \frac{\pi}{3}) = 0, \quad \coth r - 2\sin(\phi - \frac{\pi}{3}) = 0, \quad \coth r + 2\sin\phi = 0. \quad (5.116)$$

These are all geodesics – i.e., hyperplanes in two-dimensional hyperbolic plane.

It is clear that the inequalities

$$\gamma^2 - \gamma^1 \geq 0, \quad \gamma^3 - \gamma^2 \geq 0, \quad 2\gamma^1 \geq 0$$

imply all the other ones and are independent. Accordingly, these are the dominant walls, or "cushions," that define the billiard table. In the r, ϕ parametrization, the billiard table is thus defined by the following three independent inequalities

$$\cos\phi \geq 0, \quad -\cos\left(\phi + \frac{\pi}{3}\right) \geq 0, \quad \coth r - 2\sin(\phi + \frac{\pi}{3}) \geq 0 \qquad (5.117)$$

which corresponds to the triangle in hyperbolic space given in Figure 5.1.

The billiard table found here in the general inhomogeneous case is precisely the billiard table alluded to in Chapter 2 for Bianchi IX models with rotating Kasner axes. The major difference, however, is that here *we did not make any assumption on homogeneity. Instead of a single billiard for the whole space, one finds one independent billiard per space point.* The independence of the billiards at each point is sometimes summarized by saying that the dynamics becomes asymptotically "ultralocal."

Further consideration on the self-consistency of the "BKL limit" leading to the billiard description may be found in reference [51].

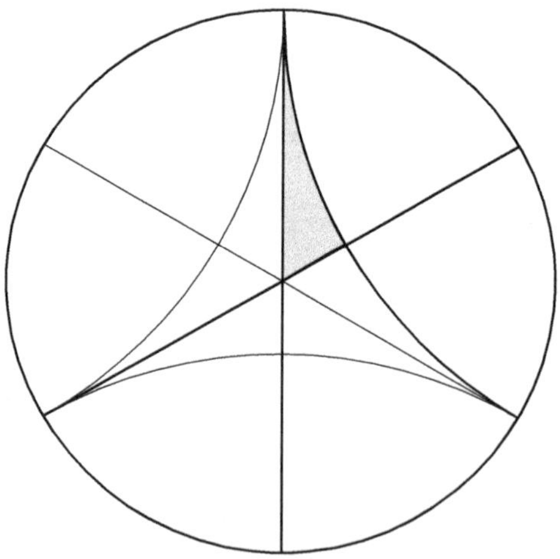

Figure 5.1: The billiard table for four-dimensional pure gravity (one billiard system at each point in space) in the Poincaré disk model. Straight lines are symmetry walls. Arc of circles are gravitationnal walls. Thick lines indicate the dominant walls. The billiard table, which lies on the positive side of each wall, is indicated in grey. It forms a triangle with angles 0, $\frac{\pi}{3}$, $\frac{\pi}{2}$.

Remarks

1. Although the dynamical evolution becomes ultralocal in the BKL limit and described by ODEs with respect to time, spatial gradients, which plays an effective crucial role in the gravitational potential, also appear explicitly in the momentum constraints $\mathcal{H}_k = 0$. As we saw in Chapter 1, these constraints restrict the initial data and need only be imposed at one time, since they are preserved by the dynamical equations of motion. It is important to note that the restrictions on the initial data imposed by the momentum constraints do not bring "dangerous" conditions on the coefficients of the dominant walls in the sense that these coefficients may all take nonzero values, even when the constraints $\mathcal{H}_k = 0$ are imposed. In some non-generic contexts, however, the constraints could force some of the wall coefficients to be zero; the corresponding walls would thus be absent. For instance, as we pointed out in Chapter 2 on homogeneous cosmological models, the momentum constraints force the symmetry wall coefficients to vanish in the vacuum, homogeneous Bianchi IX or VIII models. But this is peculiar to the homogeneous vacuum case and disappears, as we also pointed out, by introducing a small amount of inhomogeneity or generic matter.
2. The coefficients of the dominant exponentials involve only the undifferentiated structure functions $C^a{}_{bc}$. This explains why one can model the dynamical effect of the gravitational potential in the context of spatially homogeneous cosmological models, which have constant structure functions. The condition for these models to correctly represent the gravitational walls is that none of the coefficients of the α-exponentials vanish. This is a condition on the structure constants of the group, which selects the Bianchi types VIII and IX models.
3. As pointed out in [51] which we follow again almost verbatim, the computations in this subsection involve only the Cartan formulas for the curvature in general frames. They remain valid if in (5.21) we replace the one-forms dx^i by some anholonomic frame $f^i = f^i{}_j(x)dx^j$. This modifies the Iwasawa frame $\{\theta^a\}$, which has anyway no intrinsic geometrical meaning. The structure functions $C^a{}_{bc}(x)$ of the new frame get extra contributions from the spatial derivatives acting on f^i. In fact, for $f^i = dx^i$ not all gravitational walls α_a appear because we then find $C^3{}_{bc} = 0$ from (5.24) for the top component (since $\theta^3 = dx^3$), and $C^2{}_{bc} = 0$ unless b or c is equal to 3. Hence, the corresponding gravitational walls are absent. To get all the gravitational walls, one therefore needs an anholonomic frame f^i. However, the dominant gravitational wall α_1 is present, and this is the only gravitational wall relevant for the billiard.

   One may also consider different diagonalization procedures of the spatial metric, e.g., one may diagonalize the metric through a rotation as in Chapter 2. The same BKL picture and the same billiard picture emerge as $t \to 0^+$.

4. The various links between the different time variables were derived above only for the exact Kasner solution. The similar time variables introduced at each spatial point in the general inhomogeneous case fulfill similar relations asymptotically. It is amazing to note that in this general case, since the gauge conditions for τ and T involve the density $\sqrt{g}$ and are thus not invariant under spatial coordinate transformations, such changes of coordinates have the unusual feature of also changing the slicing of space-time by the "equal time" spacelike hypersurfaces.
5. The cosmological constant was assumed to be zero in the precedent discussion. Its effect can easily be taken into account because the cosmological constant contribution to the potential (in its weight-2 form) simply reads

$$\mathcal{V}_\Lambda = \Lambda g = \Lambda \exp\Big[-2\sum_a \beta^a\Big] \tag{5.118}$$

and is thus also exponential in the βs. The wall in β-space associated with the cosmological constant is spacelike. When Λ is positive (de Sitter sign) the spacelike wall (5.118) is repulsive and tend to prevent collapse. Whether collapse does or does not occur is determined by the initial conditions. In the case when the initial conditions are such that collapse does not occur, the system does not reach the BKL small volume regime. The spacelike wall associated with the positive cosmological constant acts as a "time barrier" to the motion of the billiard ball. The reflection against it forces the billiard ball to run "backwards in time" in β-space, i.e., to move in the direction of *increasing* spatial volume (this is inhomogeneous generalization of the familiar bounce of the (global) de Sitter solution, which is a hyperboloid – a sphere that first contracts and then expands). In the second case where collapse is not prevented by the cosmological constant wall, the system enters the BKL regime, which is not modified: the presence of the spacelike wall has only a subdominant effect on its motion because the volume term Λg goes to zero faster than the other terms in $\mathcal{V}$. When Λ is negative (anti-de Sitter sign) the spacelike wall (5.118) is attractive and tends to favor collapse. But again, it quickly becomes negligible in the BKL limit for the same reason that the volume term Λg goes to zero faster than the other terms in $\mathcal{V}$.
The role of the walls that are not timelike can thus be summarized as follows. (i) Lightlike walls (like the μ-type gravitational walls) can never cause reflections because in order to hit them the billiard ball would have to move at superluminal speeds in violation of the Hamiltonian constraint. (ii) Spacelike walls (like the cosmological constant wall) are either irrelevant (if they are "behind the motion"), or otherwise they reverse the time-orientation, inducing a motion towards increasing spatial volume ("bounce") and preventing the system from reaching the BKL small volume regime.
6. Although the μ-type lightlike walls are negligible, they need a more careful examination when β is close to the lightlike direction defined by them. This

is the case of "small oscillations" considered by BKL, where two scale factors are comparable. This case was alluded to at the end of Subsection 4.3.1 and verified in [15] not to alter qualitatively the asymptotic dynamics controlled by the gravitational walls.

5.7 Collision Law

Billiard Dynamics in β-Space

The derivation of the billiard description given in the previous section relied on the projection of the motion on hyperbolic space $\mathbb{H}_2$. This approach is technically useful in that it represents the walls as being located at an asymptotically fixed position in hyperbolic space, namely $w_A(\gamma) = 0$.

Once the BKL billiard behavior has been established, it is convenient to reexpress it in terms of the original scale factors β^a, because these variables run over a linear (Minkowski) space. This provides a simple mathematical representation of the dynamics.

In β-space, the dynamics of the scale factors at each point of space is given, in the gauge $\tilde{N} = 1$ corresponding to the time coordinate τ, by the Hamiltonian

$$H_\infty(\beta^a, \pi_a) = \tfrac{1}{4} G^{ab} \pi_a \pi_b + \sum_{A'} \Theta\big(- 2 w_{A'}(\beta)\big) \tag{5.119}$$

where the sum is over the dominant walls. We have used the property $w_A(\beta) = \rho w_A(\gamma)$ that follows from the linearity of the wall forms.

The asymptotic motion thus takes place at each spatial point in a "polywedge" contained within the future light cone, bounded by the hyperplanes $w_{A'}(\beta) = 0$ defining the dominant walls, which are all timelike. These walls are here the hyperplanes H_1^G, H_{12}^S and H_{23}^S. Any hyperplane H_ω defined as the zero locus $\omega = 0$ of a linear form $\omega(\beta)$ divides the β-space into a positive half-space H_ω^+ where $\omega \geq 0$ and and a negative half-space H_ω^- where $\omega \leq 0$. The polywedge where the motion takes place is thus given by the convex cone

$$H_{12}^{S+} \cap H_1^{G+} \cap H_{23}^{S+} . \tag{5.120}$$

The corresponding one-forms in β-space are the linear forms

$$w_1(\beta) = \beta^2 - \beta^1, \;\; w_2(\beta) = 2\beta^1, \;\; w_3(\beta) = \beta_3 - \beta_2. \tag{5.121}$$

When taking equal time slices of this polywedge (e.g., slices on which $\Sigma_i \beta^i$ is constant), it is clear that with increasing time (i.e., increasing $\Sigma_i \beta^i$, or increasing τ or ρ) the walls recede from the observer, while they are fixed in the hyperbolic space picture on $\mathbb{H}_2$ of the previous chapter.

One can equivalently define the polywedge where the motion takes place in terms of the spacelike vectors associated with the forms $w_{A'}$, which we denote by the same letter ($w^a_{A'} = G^{ab} w_{A'b}$),

$$w_{A'} \cdot \beta \equiv G_{ab} w^a_{A'} \beta^b = w_{A'}(\beta) \geq 0. \tag{5.122}$$

The vector $w_{A'}$ is orthogonal to the billiard wall $H_{A'}$ ($H_{A'} = \{\beta^a : w_{A'}(\beta) = 0\}$ and points inside the positive half-space $H^+_{A'}$ ($w_{A'} \cdot w_{A'} > 0$).

Off the walls, the motion is governed by the free Hamiltonian $\frac{1}{4}G^{ab}\pi_a\pi_b$ with the constraint $\frac{1}{4}G^{ab}\pi_a\pi_b = 0$ and is thus a null straight line motion. The billiard motion is then a zigzag lightlike motion consisting of free motions of the scale factors β^a on straight lightlike lines within this polywedge, which are interrupted by reflections off the walls. As shown later in this section, these reflections are exactly of mirror type.

In the Lorentzian coordinates $\bar{\beta}^1, \bar{\beta}^2, \bar{\beta}^3$ related to the original variables $\beta^1, \beta^2, \beta^3$ by the transformation (5.15), the interval in the space of the logarithmic scale factors is $-(d\bar{\beta}^1)^2 + (d\bar{\beta}^2)^2 + (d\bar{\beta}^3)^2$ and $\bar{\beta}^1$ is timelike. The dominant gravitational wall $2\beta^1 = 0$ is $\frac{\sqrt{6}}{3}(\bar{\beta}^1 - \bar{\beta}^2) - \sqrt{2}\bar{\beta}^3 = 0$, while the dominant symmetry walls $\beta^2 - \beta^1 = 0$ and $\beta^3 - \beta^2 = 0$ are $\sqrt{2}\bar{\beta}^3 = 0$ and $\frac{\sqrt{6}}{2}\bar{\beta}^2 - \frac{\sqrt{2}}{2}\bar{\beta}^3 = 0$.

Mirror Reflection Derived

To compute the effect of a collision on a particular wall $w_{A'}(\beta)$ is particularly simple in the β-space picture. One just solves (5.119), with only one term retained in the sum, namely, the term corresponding to the wall $w_{A'}(\beta)$.

The equations of motion can be solved exactly by decomposing the lightlike motion of the β-particle into its component parallel to the timelike wall hyperplane under consideration, and its orthogonal component. The parallel motion is clearly left unperturbed by the presence of the wall, while the orthogonal one-dimensional motion undergoes a reflection, wich changes the sign of the orthogonal velocity, from ingoing to outgoing.

This yields the usual formula for a geometric reflection in the hyperplane $w_{A'}(\beta) = 0$:

$$v'^a = v^a - 2\frac{(w_{A'} \cdot v) w^a_{A'}}{(w_{A'} \cdot w_{A'})} \tag{5.123}$$

("mirror reflection"). Because the vector $w_{A'}$ orthogonal to the wall hyperplane is spacelike, (5.123) is an element of the orthochronous Lorentz group $O^+(2,1)$: the reflection in the timelike hyperplane $H_{A'}$ preserves the norm and the time-orientation. This implies that the velocity vector after reflection is also null and future-oriented. When re-expressed in terms of Kasner exponents, the reflection law against a gravitational wall coincides with the change of Kasner exponents discussed in Chapter 1, while the reflection against a symmetry wall simply interchanges p_1 with p_2, or p_2 with p_3.

Example: Reflection in a Gravitational Wall

We illustrate the method for a reflection in the gravitational wall H^G_1 : $w_2(\beta) = 2\beta^1 = 0$ and verify explicitly that it reproduces the change in the Kasner exponents due to the growth of the spatial curvature derived in Chapter 1.

Before the reflection, we have

$$\begin{aligned} \beta^1 &= v^1\tau, \\ \beta^2 &= v^2\tau + \beta_0^2, \\ \beta^3 &= v^3\tau + \beta_0^3, \end{aligned} \tag{5.124}$$

for some constants β_0^2, β_0^3. The velocity fulfills

$$(v^1)^2 + (v^2)^2 + (v^3)^2 = (v^1 + v^2 + v^3)^2$$

since it is a null vector. Without loss of generality, we have assumed that the reflection instant is $\tau = 0$. The velocity v^1 is negative and $v^1\tau > 0$ as it should before the reflection ($\tau < 0$).

The one-form defining the wall H_1^G has covariant components $(2, 0, 0)$ and has norm squared equal to 2. The vector orthogonal to the wall has contravariant components obtained by raising the indices with G^{ab}, i.e., $(1, -1, -1)$. Hence, the scalar product of the velocity (v^1, v^2, v^3) with the normal to the wall H_1^G is equal to $2v^1$. Formula (5.123) yields for the velocity after collision:

$$v'^1 = -v^1, \quad v'^2 = v^2 + 2v^1, \quad v'^3 = v^3 + 2v^1 \tag{5.125}$$

which is a null vector

$$(v'^1)^2 + (v'^2)^2 + (v'^3)^2 = (v'^1 + v'^2 + v'^3)^2$$

as it should. The new free motion after the collision is thus:

$$\begin{aligned} \beta^1 &= -v^1\tau, \\ \beta^2 &= (v^2 + 2v^1)\tau + \beta_0^2, \\ \beta^3 &= (v^3 + 2v^1)\tau + \beta_0^3. \end{aligned} \tag{5.126}$$

Using the relationship (5.62) between the velocities and the Kasner exponents, one then finds that the new Kasner exponents are given in terms of the old ones by

$$p_1' = \frac{-p_1}{1 + 2p_1}, \quad p_2' = \frac{p_2 + 2p_1}{1 + 2p_1}, \quad p_1' = \frac{p_3 + 2p_1}{1 + 2p_1},$$

which reproduces exactly the law (1.56) for the change of Kasner exponents derived in Chapter 1. Thus, the reflection corresponds indeed to a change from one Kasner epoch to another.

It is easy to identify the components of the motion parallel to the wall that are unaffected by the reflection, and the orthogonal component that suffers the one-dimensional reflection. From (5.62), one sees that the combinations $\beta^1 + \beta^2$ and $\beta^1 + \beta^3$ remain unchanged under reflection. We thus introduce new coordinates X, Y, Z defined by:

$$\beta^1 = \frac{1}{\sqrt{2}}Y, \tag{5.127}$$

$$\beta^2 = \frac{1}{\sqrt{2}}\left(X - Z - Y\right),$$
$$\beta^3 = \frac{1}{\sqrt{2}}\left(X + Z - Y\right).$$

Equations (5.124) and (5.126) for the incoming and reflected trajectories can be rewritten in the coordinates X, Y, Z. Before the reflection we have:

$$\begin{aligned} Y_{before} &= \sqrt{2} q p_{\bar{1}} \tau, \\ X_{before} &= \frac{1}{\sqrt{2}} q \left(p_{\bar{3}} + p_{\bar{2}} + 2p_{\bar{1}}\right) \tau + \frac{1}{\sqrt{2}} \left(\beta_0^{\bar{3}} + \beta_0^{\bar{2}}\right), \\ Z_{before} &= \frac{1}{\sqrt{2}} q \left(p_{\bar{3}} - p_{\bar{2}}\right) \tau + \frac{1}{\sqrt{2}} \left(\beta_0^{\bar{3}} - \beta_0^{\bar{2}}\right), \end{aligned} \tag{5.128}$$

where q is the normalizing factor relating the Kasner exponents to the velocities. After the reflection, we get:

$$Y_{after} = -Y_{before}, \; X_{after} = X_{before}, \; Z_{after} = Z_{before}, \tag{5.129}$$

showing that X and Z are unaffected.

The new variables X, Y, Z have been chosen to be Minkowskian coordinates. This can directly be checked by relating them to the Lorentzian coordinates $\bar{\beta}^1, \bar{\beta}^2, \bar{\beta}^3$ introduced previously through the relations (5.15). One finds:

$$\begin{aligned} \bar{\beta}^1 &= \frac{1}{\sqrt{3}}\left(2X - Y\right), \\ \bar{\beta}^2 &= \frac{1}{2\sqrt{3}}\left(X + 3Z - 2Y\right), \\ \bar{\beta}^3 &= \frac{1}{2}\left(X - Z - 2Y\right) \end{aligned} \tag{5.130}$$

from which it follows that

$$-\left(d\bar{\beta}^1\right)^2 + \left(d\bar{\beta}^2\right)^2 + \left(d\bar{\beta}^3\right)^2 = -dX^2 + dY^2 + dZ^2, \tag{5.131}$$

i.e., the new coordinates are also Minkowskian and the coordinate change (5.130) is a Lorentz transformation.

Collision in Hyperbolic Space

The collision law can also be described in hyperbolic space by radially projecting the previous derivation on the upper sheet of the unit hyperboloid. By virtue of the properties of the vector model of $\mathbb{H}_2$, one finds that the reflection against a wall is again of mirror type. This is because (i) an hyperplane H_ω in Minkowski space intersects the upper sheet of the unit hyperboloid along a hyperplane of $\mathbb{H}_2$ denoted also H_ω; (ii) the normal to H_ω in Minkowski space is tangent to the upper sheet of the unit hyperboloid and normal to the corresponding hyperplane H_ω of $\mathbb{H}_2$; and (iii) the radial component of the velocity does not contribute to the scalar product with the normal to H_ω. Thus, the reflection of the trajectory in $\mathbb{H}_2$ against a wall is given by the same formula

$$v'^a = v^a - 2\frac{(w_{A'} \cdot v)w^a_{A'}}{(w_{A'} \cdot w_{A'})}$$

as in the three-dimensional Minkowski space, but where everything refers now to the two-dimensional motion in $\mathbb{H}_2$. As an element of the orthochronous Lorentz group, this transformation is an isometry of $\mathbb{H}_2$.

The collision law implies in particular that the angle of incidence is equal to the angle of reflection. It is instructive to check this property explicitly for the bounce against the gravitational wall that we have just analyzed.

In the same way as we introduced the coordinates (ρ, r, ϕ), we can define analogous coordinates (ρ, v, u) (with the same ρ) through:

$$X = \rho \cosh v,\ Y = \rho \sinh v \sin u,\ Z = \rho \sinh v \cos u, \tag{5.132}$$

in which the metric on the Lobachevsky surfaces $\rho = const$ take the standard form:

$$\rho^2 \left(dr^2 + \sinh^2 r d\phi^2\right) = \rho^2 \left(dv^2 + \sinh^2 v du^2\right). \tag{5.133}$$

The coordinates change $(r, \phi) \to (v, u)$ on the Lobachevsky surfaces is an isometry transformation. Its exact form can be found by substituting the expressions (5.37) and (5.132) into (5.130). One obtains:

$$\begin{aligned}
\cosh r &= \frac{1}{\sqrt{3}}\left(2\cosh v - \sinh v \sin u\right), \qquad (5.134)\\
\sinh r \sin\phi &= \frac{1}{2\sqrt{3}}\left(\cosh v - 2\sinh v \sin u + 3\sinh v \cos u\right),\\
\sinh r \cos\phi &= \frac{1}{2}\left(\cosh v - 2\sinh v \sin u - \sinh v \cos u\right).
\end{aligned}$$

Any two of these relations are in fact sufficient, since the third one is a consequence of the two being chosen.

The advantage of the coordinates (v, u) over the coordintes (r, ϕ) is that they make the computations of the angle with the gravitational wall H_1^G – which are geometric invariants – much simpler. Indeed, the equation of the billiard cushion (projection of the wall $\beta^1 = 0$) in the coordinates (v, u) is just $u = 0$.

Using (5.132), it is easy to derive the law of reflection in coordinates (v, u). Before the reflection we have:

$$\tan\left(u_{before}\right) = \frac{m(\tau)}{n(\tau)},\quad \tanh\left(v_{before}\right) = \sqrt{m^2(\tau) + n^2(\tau)}, \tag{5.135}$$

where we used the notations:

$$\begin{aligned}
m(\tau) &= \frac{2qp_1\tau}{q\left(p_3 + p_2 + 2p_1\right)\tau + \beta_0^3 + \beta_0^2}, \qquad (5.136)\\
n(\tau) &= \frac{q\left(p_3 - p_2\right)\tau + \left(\beta_0^3 - \beta_0^2\right)}{q\left(p_3 + p_2 + 2p_1\right)\tau + \beta_0^3 + \beta_0^2}.
\end{aligned}$$

After the reflection, one has:

$$u_{after}(\tau) = -u_{before}(\tau), \; v_{after}(\tau) = v_{before}(\tau). \tag{5.137}$$

The contravariant tangent vector T_c (unnormalized) to the cushion $u = 0$ at the point of reflection ($\tau = 0$) can be chosen to be

$$T_c = (1, 0), \tag{5.138}$$

where the first number in the brackets shows the v-component of the vector and the second number corresponds to its u-component. The contravariant tangent vector T_i (unnormalized) to the incoming trajectory (before the reflection) at the point of reflection follows from (5.135) and (5.136):

$$T_i = \left(\frac{dv_{before}}{d\tau}, \frac{du_{before}}{d\tau} \right)_{\tau=0}. \tag{5.139}$$

The contravariant tangent vector T_r (unnormalized) to the reflected trajectory at the point of reflection, in accordance with the law (5.137), is

$$T_r = \left(\frac{dv_{after}}{d\tau}, \frac{du_{after}}{d\tau} \right)_{\tau=0} = \left(\frac{dv_{before}}{d\tau}, -\frac{du_{before}}{d\tau} \right)_{\tau=0}. \tag{5.140}$$

The metric on the Lobachevsky surfaces $\rho = const$, in accordance to (5.133), is $g_{vv} = \rho^2$, $g_{uu} = \rho^2 \sinh^2 v$, $g_{vu} = 0$. Then

$$\cos(angle\ between\ T_c\ and\ T_i) \tag{5.141}$$

$$= \left(\frac{dv_{before}}{d\tau} \right)_{\tau=0} \left[\left(\frac{dv_{before}}{d\tau} \right)^2 + \sinh^2 v \left(\frac{du_{before}}{d\tau} \right) \right]_{\tau=0}^{-1/2},$$

and

$$\cos(angle\ between\ T_c\ and\ T_r) \tag{5.142}$$

$$= \left(\frac{dv_{after}}{d\tau} \right)_{\tau=0} \left[\left(\frac{dv_{after}}{d\tau} \right)^2 + \sinh^2 v \left(\frac{du_{after}}{d\tau} \right)^2 \right]_{\tau=0}^{-1/2}.$$

From these formulas and (5.137), it follows that the reflection is indeed such that the angle of incidence is equal to the angle of reflection.

5.8 Miscellanea

Kasner Frames Versus Iwasawa Frames

It is interesting to compare the time evolution of the logarithmic scales factors in the Iwasawa frames and in the Kasner frames [51].

In Chapter 1, the time evolution was described in "Kasner frames" with respect to which both the spatial metric and the extrinsic curvature are diagonal.

As also explained in that chapter, the time evolution in Kasner frames can be described as a succession of Kasner flights. So, let us assume that the metric is given by

$$g_{ij}(t) = A_1^2(t) l_i l_j + A_2^2(t) m_i m_j + A_3^2(t) r_i r_j \tag{5.143}$$

during a certain Kasner epoch, and by

$$g_{ij}(t) = A_1^2(t)\, l'_i l'_j + A_2^2(t)\, m'_i m'_j + A_3^2(t)\, r'_i r'_j \tag{5.144}$$

during the subsequent Kasner epoch, with some interpolating behavior during the short collision period. We have not written explicitly the x-dependence to emphasize the main points.

The two successive Kasner frames $\{l_i, m_i, r_i\}$, $\{l'_i, m'_i, r'_i\}$ are independent of time to leading order, while the scale factors $A_a(t)$ have the typical Kasner power law behavior before and after the collision ($A_a(t) \approx b_a t^{p_a}$ before the collision, and $A_a(t) \approx b'_a t^{p'_a}$ after, with the new Kasner exponents related to the old ones according to the collision rule (1.56) against a gravitational wall given in Section 1.7 and just rediscussed in the billiard context). Furthermore, as explained in Section 1.9, the transformation between the two successive Kasner frames can be written as:

$$l'_i = l_i, \quad m'_i = m_i + \sigma_m l_i, \quad r'_i = r_i + \sigma_r l_i. \tag{5.145}$$

As we have also seen in Section 1.9, σ_m and σ_r are quantities which are generically of order unity.

Now comes a seeming paradox. As the system gets closer and closer to the singularity, the quantities σ_m and σ_r that characterize the change of Kasner axes during a collision have no reason to get smaller, so that the transformation (5.145) from the old Kasner axes $\{l, m, n\}$ to the new Kasner axes $\{l', m', n'\}$ remain of order unity, no matter how close one gets to the singularity. This means that the Kasner axes generically do not freeze since there is generically an infinite number of collisions. On the other hand, we have seen that the Iwasawa frames *do freeze* (become approximately time-independent) as we go to the singularity, even if there is an infinite number of collisions.

But there is no contradiction as can be seen by relating the Iwasawa frame variables n_1, n_2 and n_3,

$$n_1 = \frac{g_{12}}{g_{11}}, \quad n_2 = \frac{g_{13}}{g_{11}}, \quad n_3 = \frac{g_{23}g_{11} - g_{12}g_{13}}{g_{11}g_{22} - g_{12}^2}.$$

(see (5.26)) to the Kasner frame variables l_i, m_i, n_i before and after the collision under consideration.

Let us thus determine what (5.145) implies for the change in the Iwasawa frame variables. This can easily be done because away from the "collision region," the metric takes the form (5.143) or (5.144) with $A_2^2 \ll A_1^2$ and $A_3^2 \ll A_2^2$. If we substitute these expressions in the above formulas giving n_a, one finds, using $A_3^2 \ll A_2^2 \ll A_1^2$,

$$n_1 = \frac{l_2}{l_1}, \; n_2 = \frac{l_3}{l_1}, \; n_3 = \frac{l_1 m_3 - l_3 m_1}{l_1 m_2 - l_2 m_1}, \tag{5.146}$$

(before the collision) and

$$n_1' = \frac{l_2'}{l_1'}, \; n_2' = \frac{l_3'}{l_1'}, \; n_3' = \frac{l_1' m_3' - l_3' m_1'}{l_1' m_2' - l_2' m_1'}, \tag{5.147}$$

(after the collision). If we now express in this second formula l_i', m_i' and r_i' in terms of l_i, m_i and r_i according to (5.145), we see that the Iwasawa off-diagonal variables n_1, n_2 and n_3 do not change but take the same values before and after the collision, $n_1' = n_1$, $n_2' = n_2$, $n_3' = n_3$. Thanks to the triangular form of the change (5.145) of the Kasner axes, there is no contradiction between the change of Kasner axes (5.145) and the freezing of the off-diagonal Iwasawa variables. The two descriptions are perfectly compatible.

The fact that two different Kasner frames lead in the BKL limit to the same Iwasawa frame might seem contradictory as it would appear to imply that the transformation from Kasner frames to Iwasawa frames is not invertible (another seeming paradox!). But there is again no contradiction, because the transformation "Kasner" $\leftrightarrow$ "Iwasawa" is a transformation in the phase space involving also the scale factors and the conjugate momenta, and it is in this full space that the transformation "Kasner" $\leftrightarrow$ "Iwasawa" is invertible.

The Billiard of Homogeneous Diagonal Type IX Models

In the case of homogeneous diagonal Bianchi IX models, the motion of the scale factors can also be described as a billiard motion, but because of the diagonal property of the metric, the symmetry walls are absent. Only the gravitational walls are present. These walls are

$$w_1(\beta) = 2\beta^1, \;\; w_2(\beta) = 2\beta^2, \;\; w_3(\beta) = 2\beta^3 \tag{5.148}$$

and define a bigger billiard region, forming an "ideal" triangle on the Lobachevsky surface, with all its vertices at infinity. This ideal triangle is showed on Figure 5.2.

This billiard table is the union of six regions isomorphic to the tables of the billiard corresponding to generic systems, where the symmetry walls are present. The same billiard characterize the asymptotic dynamics of diagonal Bianchi VIII models.

So, in the diagonal models, some curvature walls that are subdominant, i.e., hidden behind symmetry walls in the complete billiard, are not hidden anymore because the symmetry walls are removed by the peculiar conditions characterizing diagonal models, and become relevant. In the exact, complete system, the

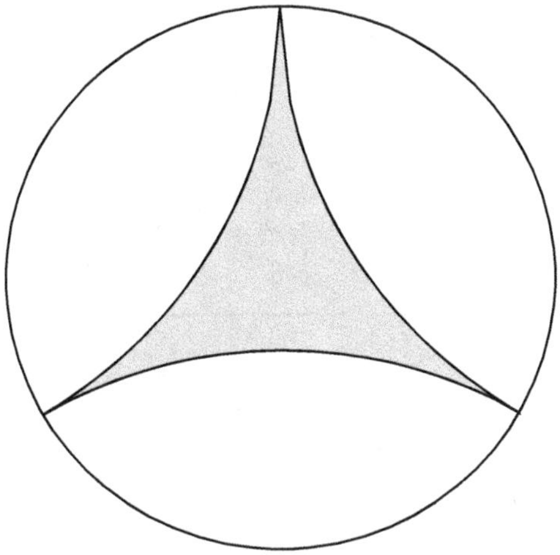

Figure 5.2: The billiard table for diagonal homogeneous cosmological models of Bianchi type IX. It is an ideal equilateral triangle formed by the three curvature (= gravitational) walls. The symmetry walls are absent. Any two different walls are parallel since they meet at infinity. The angles of this triangle are all equal to zero.

curvature walls of course still remain in the system but they cease to be dominant, that is, they have no influence on the asymptotical dynamics any more. Instead the centrifugal walls $\beta^2 - \beta^1 = 0$, $\beta^3 - \beta^2 = 0$ are present and dominant, pressing back the old two behind them.

This ideal triangle billiard table, relevant to homogeneous diagonal Bianchi IX models, is the billiard table described in [35].

Example of a Typical Billiard Motion

One can conclude from our analysis that the motion of the gravitational scale factors is a succesion of reflections in the billiard walls, i.e., defines a word in the reflection group generated by the reflections in the billiard walls.

It is of interest to illustrate a typical such motion. We consider for this purpose the billiard of diagonal models which enables one to visualize better the motion as the billiard table is bigger. We use the Klein representation of hyperbolic space where geodesics are straight lines, but where angles are deformed. Figure 5.3 displays in dotted points the motion corresponding to definite initial conditions. The geometric construction of the reflection of the billiard wall in the second collision is explicitly shown.

P_1, P_2 and P_3 are respectively the three polar points associated with the three sides of the triangles. We denote by r_1, r_2 and r_3 the reflections in the three walls W_1, W_2 and W_3. These are subject to the conditions

$$(r_1)^2 = (r_2)^2 = (r_3)^2 = e \tag{5.149}$$

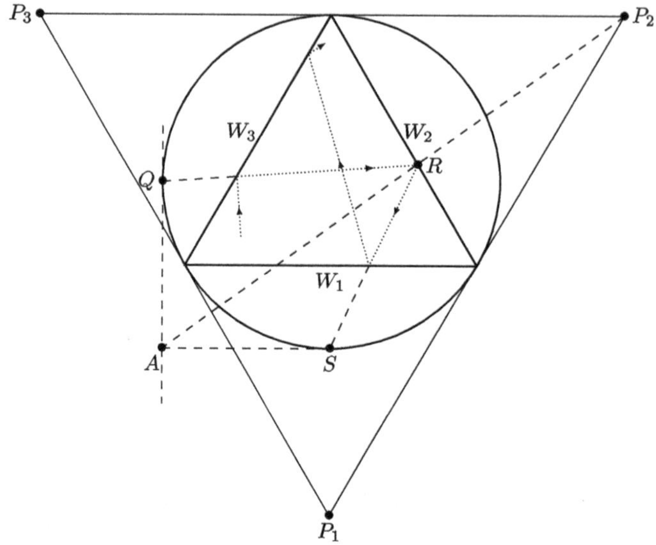

Figure 5.3: The motion is a succesion of reflections in the walls W_1, W_2 and W_3 and defines therefore an element in the group generated by the three reflections r_1, r_2, r_3 in the three walls of the billiard table.

and to no other condition because the sides of the triangle are parallel. The word associated with the particular motion displayed on the picture is (reading from right to left): $\cdots r_3 r_1 r_2 r_3$ and is clearly infinite. A typical trajectory will "fill" the billiard table as $\tau \to \infty$. Note, however, that there are also periodic trajectories, e.g., the trajectory joining successively the "midpoints" of the cushions of the billiard W_3, W_2, W_1, and then again W_3, W_2, W_1 etc. The corresponding infinite word is periodic, $\cdots (r_1 r_2 r_3)(r_1 r_2 r_3)$.

The way to draw a reflection angle equal to the incident angle in the projective Klein model has also been exhibited for the second reflection. The billiard ball arrives along the line QR to hit the wall W_2. The intersection point with the wall is R, while Q is the point where the line meets the boundary circle when continued backwards. The normal to the billiard wall at R is given by the line joining R to the pole P_2 of W_2, and intersects at the point A the tangent to the circle at Q. One draws the second tangent to the circle from A. This tangent intersects the circle at the point S. The angle $\widehat{QRA}$ is equal to the angle $\widehat{ARS}$.

5.9 Chaos and Volume of the Billiard Table

As established in Part I, the infinite number of collisions typical of the BKL behavior exhibits a chaotic feature. In terms of the billiard picture, chaos emerges because the billiard table, as a subset of hyperbolic space, has finite volume [93, 4, 129, 94, 149, 158, 72, 150]. It is thus of fundamental importance to check that the volume (actually, in our two-dimensional case, the area) of the billiard region is indeed finite.

The computation is not entirely trivial because the billiard region, as a subset of $\mathbb{H}_2$, is *non-compact.* Indeed, some of the cushions meet at infinity; in the β-space description, this means that the corresponding hyperplanes intersect on the lightcone. Even though non-compact, the hyperbolic region defined by the billiard walls crucially has, however, finite volume.

To show the finiteness of the volume, it is sufficient to consider the billiard of diagonal Bianchi IX models, since the actual billiard has volume equal to 1/6th of this volume. It is enough to prove that the volume of that part of the table which corresponds to the angle's segment $0 \leqslant \phi \leqslant \pi/2$ (the first quadrant on Figure 5.2) is finite. The finiteness of the whole volume will follow simply from the symmetry of the table. The volume of this part is proportional to the integral:

$$\int_0^{\pi/2} d\phi \int_0^{r_c} \sinh r dr = \int_0^{\pi/2} d\phi \left(\cosh r_c - 1\right) \tag{5.150}$$

where r_c follows from the relation $\coth r_c = \sin\phi + \sqrt{3}\cos\phi$ (see the first equation from (5.34), the right boundary of the part under interest is the curve $\gamma^1 = 0$). Then $\cosh r_c = \left(\sin\phi + \sqrt{3}\cos\phi\right)\left[\left(\sin\phi + \sqrt{3}\cos\phi\right)^2 - 1\right]^{-1/2}$ and the volume we are interested in will be finite if the following limit exists:

$$\lim_{\varepsilon\to 0} \int_0^{\pi/2-\varepsilon} \frac{\sin\phi + \sqrt{3}\cos\phi}{\sqrt{\left(\sin\phi + \sqrt{3}\cos\phi\right)^2 - 1}} d\phi. \tag{5.151}$$

It turns out that the integrand in the last formula has the integrable singularity when $\phi \to \pi/2$. If in this vicinity one represents ϕ as $\phi = \pi/2 - \delta$, $(\delta \to 0)$, then the leading singularity of the integrand will be of the order $\delta^{-1/2}$. Consequently the integral (5.151) exists and the volume of the two-dimensional billiard table is finite.

To summarize, we have shown, exactly as was announced in Chapter 3, that the following crucial features hold: *(i) the billiard table is a region of the two-dimensional space of constant negative curvature* $\mathbb{H}_2$, *(ii) the volume of the billiard is finite, (iii) the two-dimensional trajectories of the billiard ball between reflections against the cushions are geodesics, (iv) the reflections are of the mirror type, i.e., the angles of incidence and reflections are the same.* As we mentioned, it is well known [93, 4, 129, 94, 149, 158, 72, 150] that the motion under these four conditions is chaotic. This holds for both types of billiards, with or without symmetry walls.

For the spatially homogeneous cosmological models that are not of type VIII or IX, some of the billiard walls are missing in such a way that the billiard table has infinite volume. This is why these very special models do not exhibit the generic chaotic behavior from the billiard point of view.

5.10 Coxeter Group for Pure Gravity in Four Dimensions

5.10.1 Billiard Groups

The billiard tables in hyperbolic space described in the previous sections and illustrated in Figures 5.1 and 5.2 have remarkable properties, which imply that the reflection group generated by reflections in its walls – called hereafter the "billiard group" – have also remarkable properties.

First of all, the billiard tables are simplices, i.e., in the two-dimensional case considered here, triangles. Only three walls are relevant to determine them, even though they are more walls in the Lagrangian in the generic case. But the extra walls are subdominant.

Second, the angles between the walls are acute and equal to integer submultiples of π, i.e., of the form $\frac{\pi}{n}$ where n is a positive integer (equal to ∞ when the angle is zero). In the case of the generic billiard of Figure 5.1, the angles are $\frac{\pi}{2}$, $\frac{\pi}{3}$ and $\frac{\pi}{\infty}$. For the billiard of diagonal homogeneous models, the angles are all equal to $\frac{\pi}{\infty}$. So n is an integer equal to 2, 3 or ∞.

Third, the wall forms, i.e., the linear forms $w_A(\beta)$ defining the walls, have all the same length squared, namely 2.

The fact that the angles are integer submultiples of π implies that the groups generated by the reflections in the billiard walls is a Coxeter group. The fact that the angles have the above values implies that the billiard groups are crystallographic, i.e., preserve a lattice. One can in fact identify them with the Weyl groups of definite hyperbolic Kac–Moody algebras.

We shall develop the theory of Coxeter groups and Weyl groups of Lorentzian Kac–Moody algebras in Chapter 7, after we have derived the billiards associated with more general gravitational theories in higher dimensions. Here, we shall only specifically study the groups of reflections associated with the gravitational billiards in four space-time dimensions.

Generic Billiards

We first consider the billiard group of generic solutions and identify it with the modular group $PGL(2, \mathbb{Z})$. This group is defined to be

$$PGL(2, \mathbb{Z}) = PSL(2, \mathbb{Z}) \cup (PSL(2, \mathbb{Z}))^- \tag{5.152}$$

where

$$PSL(2, \mathbb{Z}) = \left\{ \begin{pmatrix} a & b \\ c & d \end{pmatrix} : a, b, c, d \in \mathbb{Z},\ ad - bc = 1,\ \begin{pmatrix} a & b \\ c & d \end{pmatrix} \sim - \begin{pmatrix} a & b \\ c & d \end{pmatrix} \right\} \tag{5.153}$$

and

$$(PSL(2,\mathbb{Z}))^- = \left\{ \begin{pmatrix} a & b \\ c & d \end{pmatrix} : a,b,c,d \in \mathbb{Z},\ ad-bc=-1,\ \begin{pmatrix} a & b \\ c & d \end{pmatrix} \sim - \begin{pmatrix} a & b \\ c & d \end{pmatrix} \right\}. \tag{5.154}$$

The subgroup $PSL(2,\mathbb{Z})$ acts on the upper half plane as linear fractional transformations

$$z \mapsto \frac{az+b}{cz+d} \tag{5.155}$$

while $(PSL(2,\mathbb{Z}))^-$ involves also complex conjugation and acts as

$$z \mapsto \frac{az^*+b}{cz^*+d}. \tag{5.156}$$

While the transformations (5.155) are holomorphic diffeomorphisms of the upper half-plane, the transformations (5.156) are antiholomorphic diffeomorphisms of the upper half-plane.

In order to exhibit the isomorphism of the billiard group with $PGL(2,\mathbb{Z})$, it is convenient to replace the billiard table by an equivalent one obtained from it by an isometry of the hyperbolic space. It is clear that this does not change the billiard group. We shall apply an isometry that moves the origin (where the symmetry walls meet at 60 degrees) to the point of intersection of the gravitational wall with the symmetry wall making with it an angle of 90 degrees. This is because the billiard group contains the rotations around the intersection point by an angle equal to twice the angle made by the reflection axes, and the corresponding matrices do not have integer coefficients when the angle is 60 degrees, making the isomorphism with $PGL(2,\mathbb{Z})$ less evident.

We shall also work in the upper half-plane where the elements of the billiard group are linear or antilinear fractional transformations. By using the well-known theorem that a triangle in the Lobachevsky plane is determined by its angles up to an isometry, we can thus map the billiard table on the region delimited by D_1, D_2 and D_3 in the Poincaré half-space, where

- D_3 is the vertical half-line going from the point $\frac{1}{2} + i\frac{\sqrt{3}}{2}$ upward to infinity,

$$D_3 : z(\lambda) = \frac{1}{2} + i\frac{\sqrt{3}}{2} + i\lambda, \quad \lambda \in \mathbb{R}^+ \equiv [0,\infty); \tag{5.157}$$

- D_2 is the vertical half-line going from the point i upward to infinity,

$$D_2 : z(\lambda) = i - i\lambda, \quad \lambda \in \mathbb{R}^- \equiv (-\infty, 0]; \tag{5.158}$$

- D_1 is the arc of the unit circle centered at the origin joining i to $\frac{1}{2} + i\frac{\sqrt{3}}{2}$,

$$D_1 : z(\lambda) = \sin\lambda + i\cos\lambda, \quad \lambda \in [0, \frac{\pi}{6}]. \tag{5.159}$$

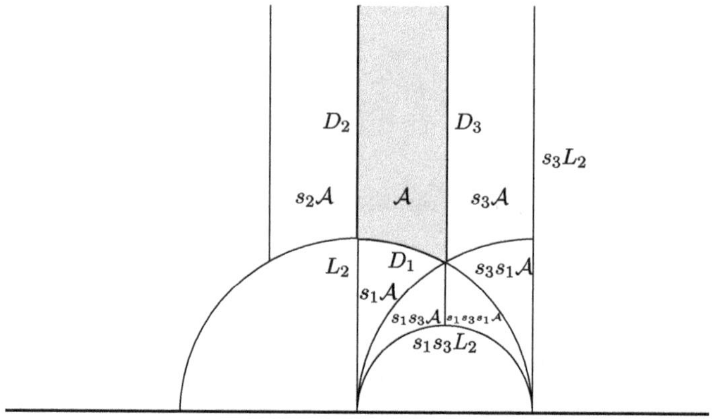

Figure 5.4: The billiard table for four-dimensional pure gravity (one billiard system at each point in space) in the Poincaré disk model. The straight half-line D_2 is the gravitational wall while the arc D_1 and half-line D_3 are the symmetry walls. The billiard table, which lies on the positive side of each wall, is indicated in grey and denoted by A.

The half-line D_2 is the gravitational wall while the arc D_1 and half-line D_3 are the symmetry walls. D_2 and D_1 meet at the origin i. The Poincaré plane description of the billiard is pictured in Figure. 5.4.

We respectively denote by s_1, s_2 and s_3 the reflections in the lines D_1, D_2 and D_3. These are transformations of $PGL(2,\mathbb{Z})$ which are easily computed:

$$s_1 : z \mapsto s_1(z) = \frac{1}{z^*} \tag{5.160}$$

$$s_2 : z \mapsto s_2(z) = -z^* \tag{5.161}$$

$$s_3 : z \mapsto s_1(z) = 1 - z^*. \tag{5.162}$$

The corresponding matrices are:

$$s_1 : \begin{pmatrix} 0 & 1 \\ 1 & 0 \end{pmatrix}, \quad s_2 : \begin{pmatrix} -1 & 0 \\ 0 & 1 \end{pmatrix}, \quad s_3 : \begin{pmatrix} -1 & 1 \\ 0 & 1 \end{pmatrix}. \tag{5.163}$$

The products $s_1 s_2$ and $s_1 s_3$ are rotations by respective angles of $\frac{2\pi}{2}$ and $\frac{2\pi}{3}$,

$$(s_1 s_2)^2 = e, \quad (s_1 s_3)^3 = e \tag{5.164}$$

in agreement with the fact that the corresponding axes meet at $\frac{\pi}{2}$ and $\frac{\pi}{3}$ (the matrix of $s_1 s_2$ (respectively, $s_1 s_3$) has a second power (respectively, third power) equal to $I \sim -I$). By contrast, the product transformation $s_2 s_3$ is a parabolic translation with no condition on $s_2 s_3$

$$z \mapsto z - 1 \tag{5.165}$$

due to the fact that the reflection axes are now parallel.

The reflections s_1, s_2 and s_3 are manifestly elements of $PGL(2,\mathbb{Z})$. It turns out that they completely generate this group. This is because the group generated by s_1, s_2 and s_3 is clearly the same as the group generated by s_2, the inversion $w = s_2 s_1$ and the translation $t = s_3 s_2$. But w and t have matrices

$$w : \begin{pmatrix} 0 & -1 \\ 1 & 0 \end{pmatrix}, \quad t : \begin{pmatrix} 1 & 1 \\ 1 & 0 \end{pmatrix}, \tag{5.166}$$

and these are well known to generate $PSL(2,\mathbb{Z})$. Since any transformation of $PGL(2,\mathbb{Z})$ which is not in $PSL(2,\mathbb{Z})$ can be written as rs_2 with $r \in PSL(2,\mathbb{Z})$, it follows that the group generated by the reflections against the billiard wall is $PGL(2,\mathbb{Z})$.

We thus come to the remarkable conclusion that the billiard group for pure gravity in four dimensions is $PGL(2,\mathbb{Z})$.

We also learn from this analysis that the billiard region is a fundamental domain for the billiard group $PGL(2,\mathbb{Z})$. This is because the region $A \cup s_2 A$ is well known to be a fundamental domain for $PSL(2,\mathbb{Z})$.

Diagonal Bianchi IX Billiards

The billiard associated with homogeneous, diagonal Bianchi models is given by the union $\mathcal{B}$ of the six regions $\mathcal{A}$, $s_1\mathcal{A}$, $s_3\mathcal{A}$, $s_1s_3\mathcal{A}$, $s_3s_1\mathcal{A}$ and $s_1s_3s_1\mathcal{A} = s_3s_1s_3\mathcal{A}$ depicted on Figure 5.4. Its walls are the straight line $L_2 = D_2 \cup s_1 D_2$, $s_3 L_2$ and $s_1 s_3 L_2$. The corresponding billiard group is generated by reflections in these walls

$$s_2 : z \mapsto -z^*, \tag{5.167}$$

$$s_3 s_2 s_3 : z \mapsto 2 - z^*, \tag{5.168}$$

$$s_1 s_3 s_2 s_3 s_1 : z \mapsto \frac{z^*}{2z^* - 1}, \tag{5.169}$$

or in terms of matrices,

$$s_2 : \begin{pmatrix} -1 & 0 \\ 0 & 1 \end{pmatrix}, \tag{5.170}$$

$$s_3 s_2 s_3 : \begin{pmatrix} -1 & 2 \\ 0 & 1 \end{pmatrix} \tag{5.171}$$

$$s_1 s_3 s_2 s_3 s_1 : \begin{pmatrix} 1 & 0 \\ 2 & -1 \end{pmatrix}. \tag{5.172}$$

The transformation $s_3 s_2 s_3$ is a reflection with respect to the line $s_3 L_2$, while the transformation $s_1 s_3 s_2 s_3 s_1$ is a reflection with respect to the line $s_1 s_3 L_2$, i.e., an inversion in the circle centered at $(\frac{1}{2}, 0)$ with radius $\frac{1}{2}$. The products of the wall reflections are all parabolic translations.

The ideal triangle group can be equivalently described as the subgroup $G\Gamma(2)$ of $PGL(2,\mathbb{Z})$ defined by considering only matrices equivalent to the identity

$$\begin{pmatrix} 1 & 0 \\ 0 & 1 \end{pmatrix}$$

modulo 2. Indeed, this subgroup is the kernel of the group homomorphism

$$PGL(2,\mathbb{Z}) \to SL(2,\mathbb{Z}_2) \tag{5.173}$$

obtained by identifying integers that differ by multiples of 2 (note that then $1 \sim -1$). The group $SL(2,\mathbb{Z}_2)$ has six elements, which may be taken to be

$$\begin{pmatrix} 1 & 0 \\ 0 & 1 \end{pmatrix}, \begin{pmatrix} 0 & 1 \\ 1 & 0 \end{pmatrix}, \begin{pmatrix} -1 & 1 \\ 0 & 1 \end{pmatrix}, \begin{pmatrix} 0 & 1 \\ -1 & 1 \end{pmatrix}, \begin{pmatrix} 1 & -1 \\ 1 & 0 \end{pmatrix}, \begin{pmatrix} 1 & 0 \\ 1 & -1 \end{pmatrix}. \tag{5.174}$$

One may equivalently view these six matrices g_i as coset representatives of the classes $g_i G\Gamma(2)$ in $PGL(2,\mathbb{Z})$, which surject to $SL(2,\mathbb{Z}_2)$. This implies that a fundamental domain of $G\Gamma(2)$ is given by

$$\cup_i \, g_i \mathcal{A}. \tag{5.175}$$

But the matrices g_i are precisely the matrices defining $e, s_1, s_3, s_1s_3, s_3s_1, s_1s_3s_1$. Hence a fundamental domain of $G\Gamma(2)$ is given by

$$\mathcal{A} \cup s_1\mathcal{A} \cup s_3\mathcal{A} \cup s_1s_3\mathcal{A} \cup s_3s_1\mathcal{A} \cup s_1s_3s_1\mathcal{A} \tag{5.176}$$

which is, as we have seen above, nothing but the ideal triangle. This proves the assertion.

The billiard table and all its images by the billiard group provides a tiling of hyperbolic space by identical tiles. The tiling is particular notable in the case of diagonal models since one gets a regular tilings with identical ideal triangles having all three vertices at infinity, with all angles between the edges equal to zero.

Coxeter Graphs

One can associate a so-called "Coxeter graph" to Coxeter reflection groups. The Coxeter graph encodes the relations $(s_i s_j)^{m_{ij}} = e$ among the generating reflections. If there is no relation among s_i and s_j, one conventionally declares $m_{ij} = \infty$. The graph contains as many vertices as generating reflections. For each pair of vertices, one draws an edge except when $m_{ij} = 2$, which corresponds to commuting reflections, $s_i s_j = s_j s_i$. Over each edge, one writes explicitly m_{ij}, except when $m_{ij} = 3$ which is the "default value."

In the case of the two reflection groups $PGL(2,\mathbb{Z})$ and $G\Gamma(2)$, there are three vertices and the Coxeter graphs read respectively:

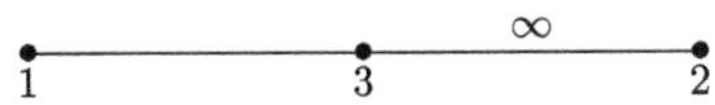

The Coxeter graph for $PGL(2,\mathbb{Z})$

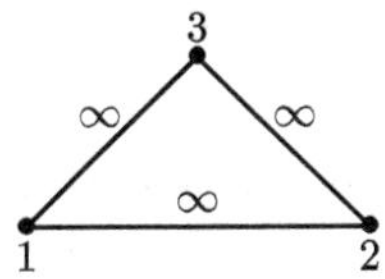

The Coxeter graph for $G\Gamma(2)$

5.10.2 Cartan Matrices

To introduce the concept of "Cartan matrix" and the connection with Cartan subalgebras of Kac–Moody algebras, it is convenient to use the three-dimensional β-space description.

We have seen that the motion near a cosmological singularity can be described as a zigzag future-directed lightlike trajectory in Minkowski space. The successive collisions in the timelike hyperplanes defining the walls are mirror reflections explicitly given by

$$v'^a = v^a - 2\frac{(w_{A'} \cdot v)w^a_{A'}}{(w_{A'} \cdot w_{A'})} \tag{5.177}$$

where the vectors $w_{A'}$ are orthogonal to the respective wall hyperplanes. The three reflections (5.177) are elements of the orthochronous group $O^+(2,1)$ and generate the billiard group.

The three vectors $w_{A'}$ are linearly independent and therefore form a basis of $\mathbb{R}^3$. One can describe the fundamental reflections (5.177) in that basis, i.e., suppressing coordinate indices,

$$w_{B'} \to w'_{B'} = w_{B'} - 2\frac{(w_{A'} \cdot w_{B'})w_{A'}}{(w_{A'} \cdot w_{A'})} \tag{5.178}$$

(no summation over A') or

$$w_{B'} \to w'_{B'} = w_{B'} - A_{A'B'}w_{A'} \tag{5.179}$$

where A is the matrix with matrix elements

$$A_{A'B'} = 2\frac{w_{A'} \cdot w_{B'}}{(w_{A'} \cdot w_{A'})}. \tag{5.180}$$

The matrix A for the generic gravitational billiard is

$$A_{A'B'} = \begin{pmatrix} 2 & -1 & 0 \\ -1 & 2 & -2 \\ 0 & -2 & 2 \end{pmatrix} \tag{5.181}$$

while it is

$$A_{A'B'} = \begin{pmatrix} 2 & -2 & -2 \\ -2 & 2 & -2 \\ -2 & -2 & 2 \end{pmatrix} \tag{5.182}$$

for the billiard of the diagonal Bianchi IX model.

A remarkable feature of these matrices is that their matrix elements are integers. This implies that the lattice $\sum_{A'} \mathbb{Z} w_{A'}$ of vectors that are sums of integer multiples of the $w_{A'}$s is preserved by the fundamental reflections (5.179). The billiard groups $PGL(2, \mathbb{Z})$ and $G\Gamma(2)$ generated by these fundamental reflections are thus evidently *crystallographic*. But there is more: the matrices A are in both cases "Cartan matrices" of Lorentzian type, i.e., with one negative principal value inherited from the signature of the metric G_{ab} (see definition of Cartan matrices in Chapter 7).

Any Cartan matrix can be associated with a Lie algebra of Kac–Moody type. The Cartan matrices (5.182) and (5.182) define infinite-dimensional hyperbolic Kac–Moody algebras of rank 3. The algebra corresponding to (5.182) has a variety of names depending on the authors who introduced it. Besides the name A_1^{++} used here, it also called AE_3 in [51], H_3 in [111] and $A_1^{\wedge\wedge}$ in [48]. The matrix (5.182) corresponds to the hyperbolic Kac–Moody algebra having number 7 in the list of nineteen possible algebras of this type provided in reference [146].

The connection between the Kac–Moody algebras and the reflection groups is that the latter are the *Weyl groups* of the former. All these concepts are explained in Chapter 7 and Appendix D.

The appearance of Lie algebras as one studies the equations near a cosmological singularity has led to the conjecture that the corresponding Lie symmetry group might somehow be hidden in the system and moreover be present in the exact Einstein equations, and not only for their limits in the vicinity to the cosmological singularity. If so, the limiting structure near the singularity should be considered just as an auxiliary instrument by means of which this symmetry is brought to light. We will return to this intriguing conjecture later when it will become clear that the appearance of infinite-dimensional Kac–Moody algebras in connection with the oscillatory character of the cosmological singularity turns out to be a much more general phenomenon than what we observed in this section restricted to pure gravity in four space-time dimensions.

6

General Cosmological Billiards

6.1 Models – Hamiltonian Form of the Action

The extension of the billiard description from four-dimensional vacuum gravity to higher-dimensional gravity consistently coupled to matter fields in $D \equiv d+1$ space-time dimensions is direct. It provides an alternative to the analysis carried out in Chapter 4, on which it sheds some very useful light.

The Einstein-bosonic matter models for which we shall construct the billiard description are composed of the metric $g_{\alpha\beta}$, a "dilaton" field ϕ and a number of p-form fields $A^{(p)}_{\alpha_1\cdots\alpha_p}$ (for $p \geq 0$), with general (second-order) action of the form

$$S[g_{\alpha\beta}, \phi, A^{(p)}] = \int d^D x \sqrt{-{}^{(D)}g} \left[R - \partial_\mu \phi \partial^\mu \phi \right.$$
$$\left. - \frac{1}{2} \sum_p \frac{1}{(p+1)!} e^{\lambda_p \phi} F^{(p)}_{\mu_1\cdots\mu_{p+1}} F^{(p)\,\mu_1\cdots\mu_{p+1}} \right] + \dots \tag{6.1}$$

Greek indices are space-time indices running from 0 to d. The action contains the standard Einstein–Hilbert term for the metric, the Klein–Gordon term for the dilaton (conventionally normalized with weight one with respect to the Ricci scalar) and the Maxwell terms for the p-forms. The determinant of the space-time metric is denoted by ${}^{(D)}g$. The p-form field strengths $F^{(p)} = dA^{(p)}$ are normalized in the standard way as

$$F^{(p)}_{\mu_1\cdots\mu_{p+1}} = (p+1)\partial_{[\mu_1} A^{(p)}_{\mu_2\cdots\mu_{p+1}]} \equiv \partial_{\mu_1} A^{(p)}_{\mu_2\cdots\mu_{p+1}} \pm p \text{ permutations.} \tag{6.2}$$

The dots in the action (6.1) indicate possible modifications of the field strength by additional Yang–Mills or Chapline–Manton-type couplings [27, 33], such as $F_B = dB^{(2)} + A^{(1)} dA^{(1)}$ for a 2-form $B^{(2)}$ and a 1-form $A^{(1)}$. Chern–Simons terms are also possibly included in the action, as it is necessary to cover, for instance, the action for $D = 11$ supergravity [41] and other models related to string theory [142]. The real parameter λ_p is called the coupling strength of the p-form $A^{(p)}$ to the dilaton.

Two comments are in order:

- A scalar field with $\lambda_0 \neq 0$ plays the same role as a p-form and must be counted among them. Such a zero-form is also called an "axion." By contrast, 0-forms with $\lambda_0 = 0$ are other "dilatons." They must be distinguished from the p-forms and added to the set of dilatons. For simplicity, we assume that $\lambda_0 \neq 0$ for all zero-forms so that there is only one dilaton. This is done mostly for notational convenience at this stage but will be relaxed below when we discuss the corresponding billiards. We shall see that dilatons and axions play very distinct roles in the BKL analysis. Dilatons behave like additional logarithmic scale factors and increase the dimension of the hyperbolic billiard. Axions contribute to additional walls for the billiard motion.
- The only (bosonic) fields that couple consistently to gravity without involving higher-order derivatives are scalar fields and p-forms. From this point of view, the field content of the second-order action (6.1) is the most general one. Higher spin fields necessitate a cosmological constant and an infinite number of space-time derivatives [74, 10] and so fall out of the scope of this book.

One may derive the generic BKL behavior either along the lines of Part I, from the appropriately generalized Kasner solution, or through the Hamiltonian formalism. It is this second path that we shall follow, imposing no symmetry conditions on the metric $g_{\alpha\beta}$, the dilaton field(s) ϕ or the p-form fields $A^{(p)}_{\mu_1\cdots\mu_p}$ in order to be able to match general initial data. It will turn out again, however, that the evolution equations near the singularity will be asymptotically "ultralocal" as in the case of vacuum four-dimensional gravity. We stress again that this simplification emerges as a direct consequence of the general dynamics without imposing any extra condition of dimensional reduction type. (*"Dimensional reduction without dimensional reduction."*)

The analysis proceeds exactly along the same lines as in Chapter 5, which paved the way to the derivation of the BKL derivation for general systems. To focus on the features relevant to the billiard picture, we drop first the Chern–Simons terms or couplings of the exterior form gauge fields through a modification of the curvatures $F^{(p)}$, which are thus taken to be abelian, $F^{(p)} = dA^{(p)}$ (if any). Although these interaction terms are quite important at subleading orders, they do not change the billiard analysis. This will be verified after the billiard picture has been derived.

In addition to working in a pseudo-Gaussian slicing (zero shift, $N^i = 0$) with the singularity occurring at $t = 0$, we impose the so-called temporal gauge for the p-forms,

$$A^{(p)}_{0i_2\cdots i_p} = 0. \tag{6.3}$$

As is well known, this choice only fixes the gauge partially, since the freedom of performing time-independent gauge transformations remains.

The Hamiltonian action reads then

$$S\left[g_{ij}, \pi^{ij}, \phi, \pi_\phi, A^{(p)}_{j_1\cdots j_p}, \pi_{(p)}^{j_1\cdots j_p}, \tilde{N}\right] =$$

$$\int dx^0 \int d^d x \left(\pi^{ij}\dot{g}_{ij} + \pi_\phi\dot{\phi} + \frac{1}{p!}\sum_p \pi_{(p)}^{j_1\cdots j_p}\dot{A}^{(p)}_{j_1\cdots j_p} - H\right) \tag{6.4}$$

where $\tilde{N}$ is again the rescaled lapse $\tilde{N} = \frac{N}{\sqrt{g}}$ and where the Hamiltonian density H is equal to

$$H \equiv \tilde{N}\mathcal{H} \tag{6.5}$$

$$\mathcal{H} = \mathcal{K} + \mathcal{M} \tag{6.6}$$

with

$$\mathcal{K} = \pi^{ij}\pi_{ij} - \frac{1}{d-1}\pi^i{}_i\pi^j{}_j + \frac{1}{4}\pi_\phi^2 + \sum_p \frac{e^{-\lambda_p\phi}}{2\,p!}\pi_{(p)}^{j_1\cdots j_p}\pi_{(p)\,j_1\cdots j_p} \tag{6.7}$$

$$\mathcal{M} = -gR + gg^{ij}\partial_i\phi\partial_j\phi + \sum_p \frac{e^{\lambda_p\phi}}{2\,(p+1)!}\,g\,F^{(p)}_{j_1\cdots j_{p+1}}F^{(p)\,j_1\cdots j_{p+1}}. \tag{6.8}$$

Here, R is the spatial curvature scalar of the d-dimensional hypersurfaces of constant x^0. The variables π^{ij}, π_ϕ and $\pi_{(p)}^{j_1\cdots j_p}$ are the momenta respectively conjugate to the spatial metric, the dilaton and the p-forms. The momenta $\pi_{(p)}^{j_1\cdots j_p}$ are equal to the electric field components.

The equations of motion for the Hamiltonian conjugate pairs are obtained by varying the above action with respect to the spatial components of the metric, the dilaton, the spatial components of the p-forms and their conjugate momenta. There are also constraints on the canonical variables,

$$\mathcal{H} \approx 0 \quad (\text{“Hamiltonian constraint”}), \tag{6.9}$$

which follows from extremization with respect to the rescaled lapse, and

$$\mathcal{H}_i \approx 0 \quad (\text{“momentum constraint”}), \tag{6.10}$$

$$\varphi_{(p)}^{j_1\cdots j_{p-1}} \approx 0 \quad (\text{“Gauss law” for each } p\text{-form}) \tag{6.11}$$

which must be imposed by hand since we have set the corresponding Lagrange multipliers equal to zero in our gauge choice. Here,

$$\mathcal{H}_i = -2\pi^j{}_{i|j} + \pi_\phi\partial_i\phi + \sum_p \frac{1}{p!}\pi_{(p)}^{j_1\cdots j_p}F^{(p)}_{ij_1\cdots j_p} \tag{6.12}$$

$$\varphi_{(p)}^{j_1\cdots j_{p-1}} = \pi_{(p)}^{j_1\cdots j_{p-1}j_p}{}_{|j_p} \tag{6.13}$$

where the subscript $|j$ stands again for the spatially covariant derivative.

6.2 Geometry of the Space of Scale Factors

Supermetric

As in four-dimensional pure gravity, the kinetic term $\mathcal{K}$ in the Hamiltonian defines a metric in configuration space. We explore here the corresponding geometry.

We shall see that in the BKL limit, the terms in $\mathcal{K}$ involving the electric fields play a role analogous to the centrifugal terms involving the off-diagonal terms of the metric. That is, they are more naturally included among the contributions to the potential for the logarithmic scale factors. Anticipating on this result established below, we drop the electric contributions to $\mathcal{K}$ (including those from the 0-forms) and examine here only the geometry defined by the remaining metric and dilaton terms.

The ultralocal form of $\mathcal{K}$ enables one to study this geometry independently at each individual spatial point, which reduces the analysis to a finite-dimensional one. The fastest way to access this geometry is to assume that all the fields depend only on time, since spatial gradients are absent from $\mathcal{K}$.

Setting $A^{(p)} = 0$ in (6.1) in accordance with the fact that we are not considering the role of the electric fields at this stage, and assuming that all the other fields depend only on time, one gets the reduced action

$$S[g_{ij}, \phi, \tilde{N}] = \int dx^0 \tilde{N}^{-1} \left[\frac{1}{4} \left(\mathrm{tr}\, (\mathrm{g}^{-1}\dot{\mathrm{g}})^2 - (\mathrm{tr}\, \mathrm{g}^{-1}\dot{\mathrm{g}})^2 \right) + \dot{\phi}^2 \right], \qquad (6.14)$$

up to an integral $\int d^d x$ over the spatial volume which is irrelevant for the present considerations and which is therefore dropped. The symmetric matrix $\mathrm{g} \in SL(d, \mathbb{R})/SO(d) \times \mathbb{R}^+$ stands now for the $d \times d$ matrix (g_{ij}) built out of the spatial components of the metric at each spatial point.

From the action (6.14) we get the configuration space “supermetric”

$$d\sigma^2 = \tfrac{1}{4} \left[\mathrm{tr}\, (\mathrm{g}^{-1} d\mathrm{g})^2 - (\mathrm{tr}\, \mathrm{g}^{-1} d\mathrm{g})^2 \right] + d\phi^2 \qquad (6.15)$$

that generalizes the De Witt supermetric for pure four-dimensional Einstein gravity. There is only one dilaton in (6.15) but it is easy to take into account several dilatons ϕ^i (for $i = 1, \ldots, n$) if more are present. The single term $d\phi^2$ in equation (6.15) should simply be replaced by $\Sigma_{i=1}^n (d\phi^i)^2$. Each dilaton is on the same footing and adds a spacelike direction to the supermetric.

The histories $g_{ij}(x^0)$, $\phi(x^0)$, $\tilde{N}(x^0)$ that extremize the action (6.14) are the null geodesics of the supermetric (6.15), the “zero-mass constraint”

$$\tfrac{1}{4} \left(\mathrm{tr}\, (\mathrm{g}^{-1}\dot{\mathrm{g}})^2 - (\mathrm{tr}\, \mathrm{g}^{-1}\dot{\mathrm{g}})^2 \right) + \dot{\phi}^2 = 0 \qquad (6.16)$$

following from extremization over the rescaled lapse $\tilde{N}$. As before, an affine parameter along those geodesics is obtained by taking $\tilde{N} = 1$, i.e., $N = \sqrt{g}$, yielding $d\tau = -dt/\sqrt{g}$. In terms of the affine parametrization τ, the geodesic equations of motion become:

$$\frac{d}{d\tau}\left(\mathrm{g}^{-1}\frac{d\mathrm{g}}{d\tau}\right) = 0, \quad \frac{d^2}{d\tau^2}\phi = 0. \tag{6.17}$$

Diagonal Metrics

The configuration space is $SL(d,\mathbb{R})/SO(d) \times \mathbb{R}^+$ for the metric times $\mathbb{R}^n$ for the dilatons. One can parametrize the symmetric space $SL(d,\mathbb{R})/SO(d)$ through an Iwasawa decomposition that generalizes the Iwasawa decomposition described in the previous chapter for metrics in three dimensions. This Iwasawa decomposition exhibits the flat subspaces associated with the scale factors and will be given explicitly in the next subsection. We focus first on the scale factors by assuming the metric to be diagonal, so that the off-diagonal terms are absent and the flat subspaces parametrized by the scale factors are clearly put in the limelight.

For diagonal metrics, which we parametrize as,

$$\mathrm{g}_D = \exp\left[\mathrm{diag}(-2\beta)\right] \quad \Longleftrightarrow \quad g^D_{ij} = \exp(-2\beta^i)\delta_{ij}, \tag{6.18}$$

the supermetric (6.15) reduces to an expression that is the direct generalization of the one encountered in Chapter 5,

$$\begin{aligned} d\sigma^2 &= \mathrm{tr}\, d\beta^2 - (\mathrm{tr}\, d\beta)^2 + d\phi^2 \\ &= \sum_{i=1}^{d} (d\beta^i)^2 - (\sum_{i=1}^{d} d\beta^i)^2 + d\phi^2 \equiv G_{\mu\nu} d\beta^\mu d\beta^\nu . \end{aligned} \tag{6.19}$$

In this equation, we have used a unifying notation β^μ, with indices running over $\mu = 1, \cdots, d+1$, to collect together the logarithmic scale factors coming from the spatial metric and the dilatons. The first d coordinates β^i correspond to the logarithmic scale factors of the spatial metric while the $(d+1)$-th coordinate $\beta^{d+1} \equiv \phi$ represents the dilaton. To account for n dilatons, one simply extends the range of indices to $\mu, \nu = 1, \cdots, d+n$. In the sequel, we shall refer to this $(d+1)$-dimensional ($(d+n)$-dimensional) space as the "extended space of (logarithmic) scale factors" or just "the β-space" for short. Note that from the point of view of Kaluza–Klein theory, it is natural to combine the scale factors and the dilaton(s) in a single space, since a dilaton can be identified as the logarithm of a scale factor in one extra spatial dimension.

The supermetric (6.19) induced in β-space does not involve the fields and is therefore again manifestly flat. As in vacuum four-dimensional gravity, it has Lorentzian signature $(- + + \cdots +)$, and, as in that case, a timelike direction in β-space is provided by a rescaling of the metric, i.e., $d\beta^\mu \propto (1, 1, \cdots, 1, 0)$. The dilaton and the volume-preserving deformations of the metric correspond to spacelike displacements.

There is thus only one minus sign in the signature of the supermetric. This minus sign is a characteristic feature of gravity since it comes from the kinetic term of the gravitational action, the dilatons contributing only positively. As we shall see below, it is this profound characteristic feature of gravity which is responsible for the Lorentzian nature of the Kac–Moody algebras which emerge in the analysis of the billiard symmetries [47].

As in four dimensions, the Lorentzian signature of the metric in the space of the scale factors enables one to define the light cone through any point. We define again the time-orientation to be such that future-pointing vectors v^μ have $\sum_i v^i > 0$, so that small volumes (small g) are again associated with large positive values of $\sum_i \beta^i$ while large volumes (large g) are associated with large negative values of $\sum_i \beta^i$. The small volume limit is thus $\sum_i \beta^i \to +\infty$.

Iwasawa Decomposition

We now include the off-diagonal metric components through the Iwasawa parametrization of the metric, generalizing to d dimensions what we did for three dimensions,

$$\mathrm{g} = \mathcal{N}^T \mathcal{A}^2 \mathcal{N} . \tag{6.20}$$

The matrix $\mathcal{N}$ is again upper triangular with 1s on the diagonal: the matrix element $\mathcal{N}^a{}_i$ vanishes for $a > i$, is equal to 1 for $a = i$, and is a nontrivial coordinate only for $a < i$. The matrix $\mathcal{A}$ is diagonal with positive elements. Following the parametrization used in the diagonal case, we set

$$\mathcal{A} = \exp(-\beta) , \quad \beta = \mathrm{diag}\,(\beta^1, \beta^2, \cdots , \beta^d). \tag{6.21}$$

In components, the decomposition (6.20) reads

$$g_{ij} = \sum_{a=1}^{d} e^{-2\beta^a} \mathcal{N}^a{}_i \mathcal{N}^a{}_j . \tag{6.22}$$

The determinant of the matrix $\mathcal{N}$ is equal to 1 since it is (upper) triangular with 1s on the diagonal. Therefore, as in the diagonal case, $g = \det \mathrm{g} = \det \mathcal{A}^2 = \exp\big(-2\Sigma_a \beta^a \big)$.

As in the previous chapter and throughout Part II of this book, the notation β^a will always refer to the logarithmic scale factors with respect to the Iwasawa frame. The Iwasawa decomposition (6.22) of the metric defines a change of variables from the $d(d+1)/2$ metric components g_{ij} to the $d + d(d-1)/2$ variables $(\beta^a, \mathcal{N}^a{}_i)$. Note that again, $(\beta^a, \mathcal{N}^a{}_i)$ are ultralocal functions of g_{ij}: that is, they depend, at each space-time point, only on the value of g_{ij} at that point. So, the Iwasawa decomposition can be used unchanged at each point spatial point in the general inhomogeneous case.

One has

$$g_{ij} dx^i dx^j = \sum_{a=1}^{d} e^{-2\beta^a} \theta^a \otimes \theta^a \tag{6.23}$$

with

$$\theta^a = \mathcal{N}^a{}_i\, dx^i. \tag{6.24}$$

The vectorial frame $\{e_a\}$ dual to the coframe θ^a is given by

$$e_a = \mathcal{N}^i{}_a \frac{\partial}{\partial x^i}. \tag{6.25}$$

where the matrix $\mathcal{N}^i{}_a$ is the upper triangular matrix inverse of $\mathcal{N}^a{}_i$, i.e., $\mathcal{N}^a{}_i \mathcal{N}^i{}_b = \delta^a_b$. The matrix $\mathcal{N}^i{}_a$ has also 1s on the diagonal.

In the Iwasawa parametrization, the "supermetric" (6.15) reads

$$\begin{aligned} d\sigma^2 &= \mathrm{tr}\, d\beta^2 - (\mathrm{tr}\, d\beta)^2 + d\phi^2 \\ &\quad + \frac{1}{2}\mathrm{tr}\left[\mathcal{A}^2 (d\mathcal{N}\mathcal{N}^{-1})\mathcal{A}^{-2}(d\mathcal{N}\mathcal{N}^{-1})^T\right] \end{aligned} \tag{6.26}$$

i.e., in components,

$$\begin{aligned} d\sigma^2 &= \sum_{a=1}^{d} (d\beta^a)^2 - \left(\sum_{a=1}^{d} d\beta^a\right)^2 + d\phi^2 \\ &\quad + \frac{1}{2}\sum_{a<b} e^{2(\beta^b - \beta^a)} \left(d\mathcal{N}^a{}_i \mathcal{N}^i{}_b\right)^2, \end{aligned} \tag{6.27}$$

showing that the off-diagonal entries $\mathcal{N}^a{}_i$ also contribute with positive signs to the signature.

Kasner Solution

The diagonal solutions of the Einstein-dilaton equations that depend only on time are also called Kasner solutions. They are the null geodesics of the supermetric and play the same central role in the billiard description as in the vacuum four-dimensional case, because they also correspond to the free flight motion between two collisions. We therefore examine them here in more detail.

The action for diagonal metrics is

$$S[\beta^\mu, \tilde{N}] = \int dx^0\, \tilde{N}^{-1} G_{\mu\nu} \dot{\beta}^\mu \dot{\beta}^\nu. \tag{6.28}$$

In terms of the affine parameter τ, the equations of motion (6.17) reduce to $\frac{d^2\beta^\mu}{d\tau^2} = 0$ and yield

$$\beta^\mu = v^\mu \tau + \beta_0^\mu, \tag{6.29}$$

where v^μ and β_0^μ are constants of the motion. The constant vector v^μ is the τ-parameter velocity, subject to the "zero-mass" constraint

$$G_{\mu\nu} v^\mu v^\nu = 0. \tag{6.30}$$

The Hamiltonian form of the action (6.28) is

$$S[\beta^\mu, \pi_\mu, \tilde{N}] = \int dx^0 \left[\pi_\mu \dot{\beta}^\mu - \tfrac{1}{4}\tilde{N} G^{\mu\nu}\pi_\mu \pi_\nu\right], \tag{6.31}$$

where $\pi_\mu \equiv (\pi_i, \pi_\phi)$ are the momenta conjugate to β^i and ϕ, respectively, and where $G^{\mu\nu}$ is the inverse of $G_{\mu\nu}$. Explicitly, $G^{\mu\nu}\pi_\mu\pi_\nu$ is given by

$$G^{\mu\nu}\pi_\mu\pi_\nu \equiv \sum_{i=1}^{d} \pi_i^2 - \frac{1}{d-1}\left(\sum_{i=1}^{d}\pi_i\right)^2 + \pi_\phi^2 . \tag{6.32}$$

For the solutions (6.29), one finds

$$\pi_\mu = 2\tilde{N}^{-1} G_{\mu\nu}\dot{\beta}^\nu = 2G_{\mu\nu}\frac{d\beta^\nu}{d\tau} \equiv 2G_{\mu\nu}v^\nu . \tag{6.33}$$

One can express the solutions in terms of the proper time $dt = -\sqrt{g}d\tau$ with $\sqrt{g} = \exp(-\Sigma_i \beta^i)$. Exactly as in four dimensions, one gets that the affine parameter τ along the null geodesics in superspace is related to the proper time t by

$$\tau = -\frac{1}{\sum_i v^i}\ln t + const. \tag{6.34}$$

(where we require $\Sigma_i v^i > 0$ in order for $\tau \to +\infty$ as $t \to 0^+$). This gives the space-time metric

$$ds^2 = -dt^2 + \sum_{i=1}^{d} A_i^2(t)(dx^i)^2 , \quad A_i(t) = b_i t^{p_i} \tag{6.35}$$

$$\phi = -p_\phi \ln t + C_\phi \tag{6.36}$$

which has the characteristic Kasner power-law behavior. Here $b_i \equiv \exp(-\beta_0^i)$ and $C_\phi \equiv \beta_0^{d+1}$ are constants related to the integration constants that arise along the resolution of the equations. The minus sign in front of p_ϕ in (6.36) is purely conventional and included for the sake of uniformity in the formulas below (if there is no dilaton one simply sets $p_\phi = C_\phi = 0$). By rescaling the spatial coordinates if necessary, one can set $b_i = 1$ and obtain the standard form of the Kasner metric, $A_i(t) = t^{p_i}$.

The Kasner exponents $p_\mu = (p_i, p_\phi)$ are related to the τ-velocities v^μ by

$$p_\mu = \frac{v^\mu}{\sum_i v^i}. \tag{6.37}$$

We stress that the sum $\sum_i v^i$ in the denominator is only over the logarithmic scale factors of the metric and does not include the dilaton. One easily verifies that the Kasner exponents $p_\mu = (p_i, p_\phi)$ are subject to the "zero-mass condition"

$$\sum_{i=1}^{d} p_i^2 - \left(\sum_{i=1}^{d} p_i\right)^2 + p_\phi^2 = 0 \tag{6.38}$$

and to the linear constraint

$$\sum_{i=1}^{d} p_i = 1 \tag{6.39}$$

where, again, the sum does not involve the dilaton.

The scale factors associated with negative Kasner exponents blow up as $t \to 0$ while the scale factors associated with positive Kasner exponents contract to zero. In the absence of dilaton, the conditions (6.38) and (6.39) imply that at least one Kasner exponent is negative. Therefore, at least one of the scale factors $A_i(t)$ blows up. This is not true anymore in the presence of a dilaton, since the conditions (6.38) and (6.39) now allow all the Kasner exponents to be simultaneously positive. This phenomenon was already observed in Section 4.5 where the dilaton is just the scalar field of that section. As in four dimensions, there is in all cases an overall contraction of the spatial volume. Indeed, the linear constraint (6.39) implies that the determinant of the spatial metric behaves as

$$g \propto t^2 \tag{6.40}$$

and as a consequence, it goes to zero.

Radial Projection and Hyperbolic Space

Just as in Chapter 5, one can project the motion radially on hyperbolic space $\mathbb{H}_{d+n-1}$ where n is the number of dilatons, which is kept arbitrary in this subsection. This alternative description is very useful for deriving the billiard picture controlling the asymptotic dynamics of general inhomogeneous metrics. The steps are exactly the same as in pure four-dimensional gravity and so we only give here the formulas, referring to Chapter 5 (and the review article [51] on which it is based) for the details. To simplify notations, we set $M \equiv d+n-1$, i.e., the dimension of β-space is $M+1$ and the dimension of the unit hyperboloid on which one projects is M. The unit hyperboloid is a realization of the M-dimensional hyperbolic (Lobachevsky) space $\mathbb{H}_M$.

For large enough values of τ, the β^μ trajectories (6.29) will get inside the future light cone of the origin (shifting, if necessary, the origin in β-space to arrange that $v_\mu \beta_0^\mu < 0$), so that $\beta^\mu \beta_\mu$ is negative,

$$\beta^\mu \beta_\mu = 2 v_\mu \beta_0^\mu \tau + \beta_{0\,\mu} \beta_0^\mu < 0. \tag{6.41}$$

The hyperbolic polar coordinates (ρ, γ^μ) are defined through

$$\beta^\mu = \rho \gamma^\mu, \tag{6.42}$$

where ρ is the radial timelike variable

$$\rho^2 \equiv -\beta^\mu \beta_\mu > 0. \tag{6.43}$$

and where the variables γ^μ, which are constrained by

$$\gamma^\mu \gamma_\mu = -1, \tag{6.44}$$

are coordinates on the future sheet of the unit hyperboloid. In the new coordinates (ρ, γ^μ), the metric in β-space reads $d\sigma^2 = -d\rho^2 + \rho^2 d\Sigma^2$ where $d\Sigma^2$ is the metric on $\mathbb{H}_M$.

We denote by π_ρ and π_γ the momenta respectively conjugate to the radial coordinate and the constrained hyperbolic coordinates γ^μ. The Hamiltonian becomes

$$H_0 = \frac{\tilde{N}}{4}\left[-\pi_\rho^2 + \frac{1}{\rho^2}\pi_\gamma^2\right] \tag{6.45}$$

or equivalently,

$$H_0 = \frac{\tilde{N}}{4\rho^2}\left[-\pi_\lambda^2 + \pi_\gamma^2\right] \tag{6.46}$$

by introducing, as in Chapter 5, the new configuration variable

$$\lambda \equiv \ln \rho \equiv \tfrac{1}{2} \ln\left(-G_{\mu\nu}\beta^\mu \beta^\nu\right) \tag{6.47}$$

and its conjugate momentum $\pi_\lambda = \rho \pi_\rho$.

It is again clear from the form (6.46) of the Hamiltonian that the dynamics is simplest in the gauge

$$\tilde{N} = \rho^2, \tag{6.48}$$

since then the Hamiltonian (6.46) reduces to a free Hamiltonian on the pseudo-Riemannian space with metric $-d\lambda^2 + d\Sigma^2$. With that choice, the momentum π_λ is a constant of the motion. The free motion of the βs is projected onto a geodesic motion on $\mathbb{H}_M$.

The coordinate time T associated with the choice (6.48) of lapse is given by

$$T = -\frac{1}{2v_\mu \beta_0^\mu} \ln \tau + const. \tag{6.49}$$

as one can see by integrating the equation $dT = \frac{d\tau}{\rho^2}$, using the fact that ρ^2 varies linearly with τ (see (6.41)). Recalling that τ itself varies logarithmically with the proper time t, one sees that $T \propto \ln|\ln t|$.

Finally, as in the vacuum four-dimensional case, one can also use the configuration variable (6.47) as an intrinsic time. We find from $\lambda = \ln \rho$,

$$\lambda = \tfrac{1}{2}\ln \tau + const. = \tfrac{1}{2}\ln|\ln t| + const.$$

6.3 Hyperbolic Space in M Dimensions

6.3.1 Subspaces, Hyperplanes, Dihedral Angles

The vector model of hyberbolic space, where $\mathbb{H}_M$ is identified with the upper sheet $x^0 > 0$ of the hyperboloid $-\left(x^0\right)^2 + \left(x^1\right)^2 + \cdots + \left(x^M\right)^2 = -1$ (in

Minkowskian coordinates) of $(M+1)$-dimensional Minkowski space $\mathbb{R}^{M,1}$, is the most convenient for our purposes and was already introduced above.

It is useful to recall a few concepts and definitions. Hereafter, the origin of $\mathbb{R}^{M,1}$ is denoted by O and vectors of $\mathbb{R}^{M,1}$ are written in boldface characters. We recall that a subspace of $\mathbb{R}^{M,1}$ is timelike if the restriction of the Minkowskian metric on it has Lorentzian signature.

A straight line ("geodesic") of π is given by the intersection of a two-dimensional timelike plane of $\mathbb{R}^{M,1}$ containing the origin with the upper sheet of the hyperboloid. Let γ and γ' be two geodesics of $\mathbb{H}_M$ that intersect at the point Q. Their tangent vectors $\mathbf{u}$ and $\mathbf{v}$ are both spacelike and span a two-dimensional spacelike plane of $\mathbb{R}^{M,1}$, contained in the spacelike hyperplane tangent to the hyperboloid at Q. The angle between the two tangent vectors is given by

$$\cos\theta = \frac{\mathbf{u}\cdot\mathbf{v}}{\sqrt{\mathbf{u}\cdot\mathbf{u}}\sqrt{\mathbf{v}\cdot\mathbf{v}}}. \tag{6.50}$$

(One has $|\mathbf{u}\cdot\mathbf{v}| \leq \sqrt{\mathbf{u}\cdot\mathbf{u}}\sqrt{\mathbf{v}\cdot\mathbf{v}}$ because the plane spanned by $\mathbf{u}$ and $\mathbf{v}$ is spacelike and hence governed by the standard rules of Euclidean geometry.) The two timelike 2-planes defining γ and γ' intersect along the line OQ, to which the tangent vectors $\mathbf{u}$ and $\mathbf{v}$ are orthogonal. The dihedral angle between these two planes is by definition the angle between $\mathbf{u}$ and $\mathbf{v}$ and is thus equal to θ.

A lightlike geodesic ℓ not going through the origin O defines a two-dimensional timelike plane containing the origin, obtained by considering all straight lines joining the points of ℓ to O. The intersection of that plane with the upper sheet of the hyperboloid is a straight line of $\mathbb{H}_M$. If ℓ enters the future light cone of O, this is, as we have seen, the radial projection on the upper sheet of the hyperboloid of that portion of ℓ which is within the future light cone at O.

A hyperplane π of $\mathbb{H}_M$ is the intersection of a timelike hyperplane Π of $\mathbb{R}^{M,1}$ containing the origin O with $\mathbb{H}_M$, $\pi = \mathbb{H}_M \cap \Pi$. This intersection is nonempty and defines a $(M-1)$-dimensional surface because Π is timelike. More generally, a k-plane of $\mathbb{H}_M$ is the intersection of a $(k+1)$-dimensional timelike plane of $\mathbb{R}^{M,1}$ containing O with $\mathbb{H}_M$.

Let π be a hyperplane of $\mathbb{H}_M$ and Π the corresponding hyperplane of $\mathbb{R}^{M,1}$ going through the origin. Let $\mathbf{n}$ be the vector orthogonal to Π in $\mathbb{R}^{M,1}$. It is a spacelike vector. The reflection

$$\sigma : \mathbf{v} \to \mathbf{v}' = \mathbf{v} - 2\,(\mathbf{v}\cdot\mathbf{n})\,\mathbf{n} \tag{6.51}$$

is an element of the orthochronous Lorentz group $O(M,1)^+$ that maps the upper sheet of the hyperboloid on itself. Its set of fixed points is given by the hyperplane Π. The transformation σ defines therefore an isometry of $\mathbb{H}_M$, denoted by the same letter σ, which leaves π invariant. This isometry is involutive, $\sigma^2 = e$. The hyperplane π can equivalently be defined as the set of fixed points of the involutive isometry σ. It is totally geodesic (contains all its tangent geodesics).

The hyperplane π of $\mathbb{H}_M$ divide hyperbolic space in two half-spaces, just as the corresponding hyperplane Π of $\mathbb{R}^{M,1}$ divides $\mathbb{R}^{M,1}$ in two half spaces. These half-spaces are exchanged under σ. As in $2+1$ dimensions, we distinguish the two half-spaces by making a choice of "orientation" and declaring one of the two half-spaces to be positive and the other to be negative. We adjust the orientation of the normal such that it points inside the positive region H^+, which can then be described as

$$H^+ = \{\mathbf{x} \in X : \mathbf{x} \cdot \mathbf{n} > 0\} \tag{6.52}$$

where X stands for $\mathbb{R}^{M,1}$ or $\mathbb{H}_M \subset \mathbb{R}^{M,1}$.

Let $P \in \pi$ be a point of $\mathbb{H}_M$ belonging to the hyperplane π. The vector $\mathbf{OP}$ defines with $\mathbf{n}$ a two-dimensional timelike plane, which intersects π along a one-dimensional straight line of $\mathbb{H}_M$, the normal to π at P.

Let π_1 and π_2 be two distinct hyperplanes of $\mathbb{H}_M$ that intersect along the $(M-2)$-dimensional plane δ of $\mathbb{H}_M$, and Π_1 and Π_2 the corresponding hyperplanes of $\mathbb{R}^{M,1}$ going through the origin. Because π_1 and π_2 intersect, the timelike hyperplanes Π_1 and Π_2 intersect along a $(M-1)$-dimensional timelike plane Δ of $\mathbb{R}^{M,1}$ containing the origin, the intersection of which with $\mathbb{H}_M$ is δ.

The $(M-2)$-dimensional plane δ of $\mathbb{H}_M$ (respectively, the $(M-1)$-dimensional plane Δ of $\mathbb{R}^{M,1}$) has codimension one in both π_1 and π_2 (respectively, Π_1 and Π_2). Let $\mathbf{e}_1$ be the vector orthogonal to δ in π_1 and $\mathbf{e}_2$ be the vector orthogonal to δ in π_2. The vector $\mathbf{e}_1$ (respectively, $\mathbf{e}_2$) is also orthogonal to Δ in Π_1 (respectively, in Π_2). The vectors $\mathbf{e}_1$ and $\mathbf{e}_2$ are spacelike and moreover, their linear span is also spacelike since it is contained in the spacelike plane orthogonal to Δ. This means in particular $|\mathbf{e}_1 \cdot \mathbf{e}_2| < 1$ and conversely, if this condition is fulfilled, the 2-plane spanned by $\mathbf{e}_1$ and $\mathbf{e}_2$ is spacelike (recall that $\mathbf{e}_1 \cdot \mathbf{e}_1 = 1 = \mathbf{e}_2 \cdot \mathbf{e}_2$ and that $\mathbf{e}_1$ and $\mathbf{e}_2$ are linearly independent).

The dihedral angle $H_1^+ \cap H_2^+$ between the oriented hyperplanes π_1 and π_2 (or Π_1 and Π_2) is defined to be the angle θ given by

$$\cos\theta = -\mathbf{e}_1 \cdot \mathbf{e}_2, \quad 0 < \theta \leq \pi. \tag{6.53}$$

We have allowed $\theta = \pi$ to include the limiting case where H_1^+ and H_2^+ coincide. The fact that the dihedral angle is between 0 and π reflects the fact that a region such as $H_1^+ \cap H_2^+$ is convex: if $\mathbf{x}, \mathbf{y} \in H_1^+ \cap H_2^+$, then the points on the segment joining $\mathbf{x}$ to $\mathbf{y}$ also belong to $H_1^+ \cap H_2^+$. The region complementary to $H_1^+ \cap H_2^+$ is not convex and has a diedral angle $2\pi - \theta$ greater than π.

It should be pointed out that two arbitrary distinct timelike hyperplanes of $\mathbb{R}^{M,1}$ containing O may not intersect along a $(M-1)$-dimensional timelike plane. The $(M-1)$-dimensional plane of intersection Δ may be null or spacelike, and the 2-plane spanned by the normal vectors $\mathbf{e}_1$ and $\mathbf{e}_2$ is then not spacelike. In that case, the corresponding hyperplanes of hyperbolic space do not intersect since Δ has no point in common with $\mathbb{H}_M$.

If $|\mathbf{e}_1 \cdot \mathbf{e}_2| = 1$, the 2-plane spanned by $\mathbf{e}_1$ and $\mathbf{e}_2$ is null since the metric induced on it has zero determinant and is degenerate. The corresponding hyperplanes π_1 and π_2 of hyperbolic space "intersect at infinity," along the null generator of Δ and are said to be "parallel." If $\mathbf{e}_1 \cdot \mathbf{e}_2 = 1$, the vectors $\mathbf{e}_1$ and $\mathbf{e}_2$ differ by a null vector orthogonal to them and either $H_1^+ \subset H_2^+$ or $H_2^+ \subset H_1^+$. The dihedral angle is equal to π. If $\mathbf{e}_1 \cdot \mathbf{e}_2 = -1$, the vectors $\mathbf{e}_1$ and $-\mathbf{e}_2$ differ by a null vector orthogonal to them and either $H_1^+ \cap H_2^+ = \emptyset$ or $H_1^- \cap H_2^- = \emptyset$. The dihedral angle is equal to 0.

If $|\mathbf{e}_1 \cdot \mathbf{e}_2| > 1$, the 2-plane spanned by $\mathbf{e}_1$ and $\mathbf{e}_2$ is timelike. The corresponding hyperplanes π_1 and π_2 of hyperbolic space are said to be "divergent" since they do not meet, even at infinity. (The corresponding hyperplanes Π_1 and Π_2 of $\mathbb{R}^{n,1}$ meet beyond the line cone.) If $\mathbf{e}_1 \cdot \mathbf{e}_2 > 1$, either $H_1^+ \subset H_2^+$ or $H_2^+ \subset H_1^+$. If $\mathbf{e}_1 \cdot \mathbf{e}_2 < -1$, either $H_1^+ \cap H_2^+ = \emptyset$ or $H_1^- \cap H_2^- = \emptyset$.

6.3.2 Other Models of Hyperbolic Space

Just as for $\mathbb{H}_2$, there exist various useful models of hyperbolic space $\mathbb{H}_M$ besides the vector model. Their construction is identical to that of the corresponding two-dimensional models.

- The Klein projective model is obtained by radially projecting the upper sheet of the unit hyperboloid on the tangent hyperplane H of equation $x^0 = 1$. The range of the coordinates $x^1, \cdots, x^M$ on H fulfills $\sum_{i=1}^M (x^i)^2 < 1$, i.e., the hyperboloid $\mathbb{H}_M$ is mapped on the interior of the unit ball of H. "Infinity" is the boundary unit sphere. The metric reads

$$d\Sigma^2 = \frac{\sum_{i=1}^M (dx^i)^2}{1 - \sum_{i=1}^M (x^i)^2} + \frac{\left(\sum_{i=1}^M x^i dx^i\right)^2}{\left(1 - \sum_{i=1}^M (x^i)^2\right)^2} \tag{6.54}$$

  generalizing (5.45). In this model k-planes (in particular, deodesics and hyperplanes) are represented by k-planes: the model is projective. The model distorts, however, the angles.
- This is not the case for the conformal models. The Poincaré ball model is obtained by projecting through the point $(-1, 0, 0, \cdots, 0)$ the upper sheet $\mathbb{H}_M$ of the hyperboloid on the hyperplane $x^0 = 0$. The range of the Poincaré ball coordinates X^i is again the unit ball $\sum_{i=1}^M (X^i)^2 < 1$ and the metric takes the conformally flat form

$$4\frac{\sum_{i=1}^M (dX^i)^2}{\left(1 - \sum_{i=1}^M (X^i)^2\right)^2}. \tag{6.55}$$

  generalizing (5.49).

- Another conformal model is the Poincaré half-space $u^M > 0$ with metric

$$d\Sigma^2 = \frac{\sum_{i=1}^{M} \left(du^i\right)^2}{\left(u^M\right)^2}, \quad u^M > 0, \tag{6.56}$$

which generalizes (5.53).

6.4 Hamiltonian in Iwasawa Variables and BKL Limit

We return to the general inhomogenous case and investigate the BKL limit as one goes towards a spacelike singularity. The analysis proceeds as in pure four-dimensional gravity and relies on the Iwasawa change of variables at each spatial point, which can be performed because it is ultralocal.

Iwasawa Change of Variables as a Point Canonical Transformation

The Iwasawa change of variables in configuration space can be extended to the momenta as a point canonical transformation

$$\pi^{ij}\dot{g}_{ij} \equiv \sum_a \pi_a \dot{\beta}^a + \sum_a P^i{}_a \dot{\mathcal{N}}^a{}_i. \tag{6.57}$$

Again, the momenta

$$P^i{}_a = \frac{\partial \mathcal{L}}{\partial \dot{\mathcal{N}}^a{}_i} = \sum_b e^{2(\beta^b - \beta^a)} \dot{\mathcal{N}}^a{}_j \mathcal{N}^j{}_b \mathcal{N}^i{}_b \tag{6.58}$$

conjugate to the non-constant off-diagonal Iwasawa components $\mathcal{N}^a{}_i$ are only defined for $a < i$; hence the second sum in (6.57) receives only contributions from $a < i$.

To analyze the BKL limit, it is also useful to express the components $\mathcal{A}_{i_1 \cdots i_p}$ of the p-forms in the Iwasawa basis $\{\theta^a\}$. As in the metric sector, this is a point transformation

$$\mathcal{A}_{a_1 \cdots a_p} = \mathcal{N}^{j_1}{}_{a_1} \cdots \mathcal{N}^{j_p}{}_{a_p} A_{j_1 \cdots j_p}. \tag{6.59}$$

which can be extended to the momenta through

$$\begin{aligned}
&\sum_a P^i{}_a \dot{\mathcal{N}}^a{}_i + \sum_p \sum_{i_1, \ldots, i_p} \frac{1}{p!} \pi^{j_1 \cdots j_p} \dot{A}_{j_1 \cdots j_p} \\
&\quad = \sum_a \mathcal{P}^i{}_a \dot{\mathcal{N}}^a{}_i + \sum_p \sum_{a_1, \ldots, a_p} \frac{1}{p!} \mathcal{E}^{a_1 \cdots a_p} \dot{\mathcal{A}}_{a_1 \cdots a_p}.
\end{aligned} \tag{6.60}$$

The new momenta $\mathcal{P}^i{}_a$ conjugate to $\mathcal{N}^a{}_i$ differ from the old ones $P^i{}_a$ by terms involving $\mathcal{E}$, $\mathcal{N}$ and $\mathcal{A}$ but not the scale factors or their conjugate, because the scale factors do not appear in the change of basis (6.59). One finds in particular from (6.60) that $\mathcal{E}^{a_1 \cdots a_p}$ are just the components of the electric field $\pi^{i_1 \cdots i_p}$ in the basis $\{\theta^a\}$,

$$\mathcal{E}^{a_1 \cdots a_p} \equiv \mathcal{N}^{a_1}{}_{j_1} \mathcal{N}^{a_2}{}_{j_2} \cdots \mathcal{N}^{a_p}{}_{j_p} \pi^{j_1 \cdots j_p}. \tag{6.61}$$

Split of the Hamiltonian

As for pure four-dimensional gravity, we next express the Hamiltonian density $\mathcal{H}$ in terms of the new variables and split it in two parts: (i) the kinetic term $\mathcal{H}_0$ for the local scale factors β^μ (including the dilatons); and (ii) the "potential density" $\mathcal{V}$, which contains everything else. Note, in particular, that we treat the off-diagonal metric components and the p-forms on the same footing. This is quite natural from the point of view of Kaluza–Klein reductions, where the off-diagonal components of the metric in one dimension higher become a one-form. This splitting of the Hamiltonian therefore reads

$$\mathcal{H} = \mathcal{H}_0 + \mathcal{V} \tag{6.62}$$

where the kinetic term of the β variables was already worked out above,

$$\mathcal{H}_0 = \frac{1}{4} G^{\mu\nu} \pi_\mu \pi_\nu . \tag{6.63}$$

The total (weight 2) potential density,

$$\mathcal{V} = \mathcal{V}_S + \mathcal{V}_G + \sum_p \mathcal{V}_{(p)} + \mathcal{V}_\phi \tag{6.64}$$

is the sum of many terms:

- a centrifugal (or "symmetry") contribution equal to the kinetic energy of the off-diagonal components

$$\mathcal{V}_S = \frac{1}{2} \sum_{a<b} e^{-2(\beta^b - \beta^a)} \left(P^j{}_b \mathcal{N}^a{}_j \right)^2 \tag{6.65}$$

  (where $P^j{}_b$ is the function of $\mathcal{P}^j{}_b$, $\mathcal{E}$, $\mathcal{N}$ and $\mathcal{A}$ but *not* of the scale factors defined through (6.60));
- a "gravitational" (or "curvature") potential

$$\mathcal{V}_G = -gR \tag{6.66}$$

  (both terms (i) and (ii) appear already in four-dimensional pure gravity);
- new terms originating from the p-forms,

$$\mathcal{V}_{(p)} = \mathcal{V}^{el}_{(p)} + \mathcal{V}^{magn}_{(p)} \tag{6.67}$$

  which are given by a sum of an "electric" and a "magnetic" contribution for each p,

$$\mathcal{V}^{el}_{(p)} = \frac{1}{2\,p!} \sum_{a_1,a_2,\cdots,a_p} e^{-2e_{a_1\cdots a_p}(\beta)} (\mathcal{E}^{a_1\cdots a_p})^2 \tag{6.68}$$

$$\mathcal{V}^{magn}_{(p)} = \frac{1}{2\,(p+1)!} \sum_{a_1,a_2,\cdots,a_{p+1}} e^{-2m_{a_1\cdots a_{p+1}}(\beta)} (\mathcal{F}_{a_1\cdots a_{p+1}})^2 \tag{6.69}$$

where $\mathcal{F}_{a_1\cdots a_{p+1}}$ are the components of the magnetic field $F_{m_1\cdots m_{p+1}}$ in the basis $\{\theta^a\}$,

$$\mathcal{F}_{a_1\cdots a_{p+1}} = \mathcal{N}^{j_1}{}_{a_1}\cdots\mathcal{N}^{j_{p+1}}{}_{a_{p+1}}F_{j_1\cdots j_{p+1}}. \tag{6.70}$$

The components of the exterior derivative $\mathcal{F}$ of $\mathcal{A}$ in the non-holonomic frame $\{\theta^a\}$ involves the structure coefficients in that frame,

$$\mathcal{F}_{a_1\cdots a_{p+1}} = \partial_{[a_1}\mathcal{A}_{a_2\cdots a_{p+1}]} + C\mathcal{A}\text{-terms}$$

where $\partial_a \equiv \mathcal{N}^i{}_a\partial_i$ is the frame derivative, but not the scale factors. The structure coefficients C are defined through

$$d\theta^a = -\frac{1}{2}C^a{}_{bc}\theta^b\wedge\theta^c. \tag{6.71}$$

We have also set

$$e_{a_1\cdots a_p}(\beta) = \beta^{a_1} + \cdots + \beta^{a_p} - \frac{\lambda_p}{2}\phi \tag{6.72}$$

where the indices a_js are all distinct because $\mathcal{E}^{a_1\cdots a_p}$ is completely antisymmetric. Similarly,

$$m_{a_1\cdots a_{p+1}}(\beta) = \sum_{b\notin\{a_1,a_2,\cdots a_{p+1}\}}\beta^b + \frac{\lambda_p}{2}\phi \tag{6.73}$$

where again all a_js are distinct. It is sometimes useful to express $m_{a_1\cdots a_{p+1}}(\beta)$ in dual notations as $\tilde{m}_{a_{p+2}\cdots a_d}$, where $\{a_{p+2}, a_{p+3}, \cdots, a_d\}$ is the set complementary to $\{a_1, a_2, \cdots a_{p+1}\}$. For instance, one rewrites

$$\tilde{m}_{1\,2\cdots\,d-p-1} = \beta^1 + \cdots + \beta^{d-p-1} + \frac{\lambda_p}{2}\phi = m_{d-p\cdots\,d}\,. \tag{6.74}$$

- Finally, there is a contribution to the potential coming from the spatial gradients of the dilaton:

$$\mathcal{V}_\phi = g g^{ij}\partial_i\phi\partial_j\phi = \sum_a e^{-2\mu_a(\beta)}\mathcal{N}^i{}_a\mathcal{N}^j{}_a\partial_i\phi\partial_j\phi\,, \tag{6.75}$$

with

$$\mu_a(\beta) = \sum_{c\neq a}\beta^c\,. \tag{6.76}$$

The gravitational term can be massaged exactly as in the four-dimensional case, to give

$$\mathcal{V}_G \equiv -gR = \frac{1}{4}{\sum_{a,b,c}}' e^{-2\alpha_{abc}(\beta)}(C^a{}_{bc})^2 - \sum_a e^{-2\mu_a(\beta)}F_a, \tag{6.77}$$

where the prime on $\sum$ indicates that the sum is to be performed only over unequal indices, i.e., $a\neq b, b\neq c, c\neq a$, and where the linear forms $\alpha_{abc}(\beta)$ and $\mu_a(\beta)$ are given by

$$\alpha_{abc}(\beta) = 2\beta^a + \sum_{e\neq a,b,c}\beta^e \qquad (a\neq b,\ b\neq c,\ c\neq a) \tag{6.78}$$

and (6.76). The exponents α_{abc} defined by (6.78) are symmetric under the exchange of b with c. In (6.77), F_a is again some complicated function of its arguments whose explicit form will not be needed. What is important is that F_a depends polynomially on the derivatives of the logarithmic scale factors: it is a polynomial of degree 2 in the derivatives $\partial\beta$ and of degree 1 in $\partial^2\beta$, just as in three spatial dimensions.

BKL Limit

The BKL limit $t \to 0^+$ (or equivalently, $\sqrt{g} \to 0$, $\rho \to +\infty$ or $T \to \infty$) can then be easily taken. The analysis proceeds as before without difficulty because the new terms in the potential are manifestly positive. The limit can be described either in hyperbolic space or in β-space, the two equivalent points of view having their own advantages. All "off-diagonal" degrees of freedom, including the p-form variables, freeze in the limit and the potential become a sum of sharp-wall potentials $\sum \Theta(-2w_A)$ where w_A stands symbolically for all the relevant linear wall forms – i.e., the linear functions of the scale factors that appear in the exponentials in the various potential terms. The argument is just the same as the argument given in four dimensions and need not be repeated here (see Chapter 5 and [51]). Thus, the billiard picture again emerges at each spatial point.

In the hyperbolic space description, the billiard ball moves in the region of hyperbolic space $\mathbb{H}_M$ ($M = d + n - 1$) bounded by the relevant billiard walls (or "cushions"). These walls are determined by the energy of the fields that are asymptotically frozen through the corresponding wall forms. The ball moves on geodesics between two collisions with the walls.

In the β-space description, the motion is a future-oriented lightlike straight line within the future light cone, interrupted by specular reflections off the walls. These (relevant) walls are timelike hyperplanes and the motion takes place within a "polywedge." The reflection law is given by (5.123), with the Latin index a replaced by the Greek index μ.

To determine which wall forms are relevant necessitates a closer examination of the various contributions to the potential, a question to which we now turn. More information on the consistency of the billiard limit can be found in [51], which addresses the issue for general systems described by the action (6.1).

6.5 Walls

6.5.1 Centrifugal (or Symmetry) Walls

The off-diagonal metric components give rise to the so-called "symmetry walls," which are familiar from our previous analysis in Chapter 5. In the BKL limit, the centrifugal potential (6.65) $\mathcal{V}_S$ becomes

$$\mathcal{V}_S = \sum_{a<b} \Theta(-2w^S_{(ab)}(\beta)) \tag{6.79}$$

with

$$w^S_{(ab)}(\beta) \equiv w^S_{(ab)\mu}\beta^\mu \equiv \beta^b - \beta^a \quad (a < b). \tag{6.80}$$

The hyperplanes $w^S_{(ab)}(\beta) = 0$ defining the symmetry walls are again timelike because

$$G^{\mu\nu} w^S_{(ab)\mu} w^S_{(ab)\nu} = +2. \tag{6.81}$$

It follows that the symmetry walls $w^S_{(ab)}(\beta) = 0$ intersect the hyperboloid $G_{\mu\nu}\beta^\mu\beta^\nu = -1$, $\sum_a \beta^a \geq 0$.

As in four dimensions (Chapter 5), we have referred to the walls $w^S_{(ab)}(\beta) = 0$ as the "symmetry walls" to take into account their action on the Kasner exponents. When applying the general collision law (5.123) derived previously to the case of the collision on the wall (6.80), one easily finds, just as in four dimensions, that its effect on the components of the velocity vector v^μ is simply to permute the components v^a and v^b, while leaving unchanged the other components $\mu \neq a, b$.

It is easy to see that some symmetry walls are "behind" the others, and bring no new independent information. This is because all the inequalities enforced by the symmetry walls can in fact be derived from a smaller subset, the subset of "dominant symmetry walls" (see Chapter 5). Indeed, the $d(d-1)/2$ inequalities $w^S_{(ab)}(\beta) \geq 0$ contain the subset of $d-1$ independent inequalities

$$\beta^2 - \beta^1 \geq 0,\ \beta^3 - \beta^2 \geq 0,\ \cdots, \beta^d - \beta^{d-1} \geq 0 \tag{6.82}$$

and, in turn is fully implied by it. For instance, $\beta^3 - \beta^1 = (\beta^3 - \beta^2) + (\beta^2 - \beta^1)$ and therefore $\beta^3 - \beta^1 \geq 0$ if $\beta^3 - \beta^2 \geq 0$ and $\beta^2 - \beta^1 \geq 0$.

Independently of the p-form content, the symmetry walls, related to the kinetic term of the off-diagonal metric components, are always present for generic solutions, i.e., for solutions such that the prefactors of the symmetry exponentials do not vanish. When we come to the Kac–Moody interpretation, the dominant linear forms entering (6.82) are identified with the simple roots of the Lie algebra $SL(d, \mathbb{R})$.

Note that, in principle, the final set of dominant walls can only be decided once one knows the complete list of all the dynamically relevant walls. This phenomenon was seen in vacuum four-dimensional gravity where all the gravitational walls are dominant for diagonal Bianchi models of type IX, but only one of them is dominant if the symmetry walls are included. In all the models examined here, the set of dominant symmetry walls (6.82) will, however, always be part of the final minimal set of dominant walls defining the complete billiard table. For that reason, they play a central role.

6.5.2 Curvature (Gravitational) Walls

We now analyze the gravitational potential (6.77). The discussion proceeds very much along the lines of Chapter 5. The gravitational potential $\mathcal{V}_G$ is composed of two types of terms, the α-type and the μ-type.

As in four space-time dimensions, the α-type terms clearly come with positive prefactors, proportional to the square of a structure function $C^a{}_{bc}$. Furthermore, the hyperplanes $\alpha_{abc}(\beta) = 0$ are timelike, like the symmetry wall hyperplanes, since

$$G^{\mu\nu}(\alpha_{abc})_\mu(\alpha_{abc})_\nu = +2\,. \tag{6.83}$$

By contrast, the μ-type contributions to $\mathcal{V}_G$ do not have a definite sign. However, the linear forms $\mu_a(\beta)$ are all lightlike, i.e., $G^{\mu\nu}(\mu_a)_\mu(\mu_a)_\nu = 0$, as in four space-time dimensions. Therefore, each hyperplane $\mu_a(\beta) = 0$ is tangent to the light cone along some null generator and the light cone is entirely on one side of the hyperplane $\mu_a(\beta) = 0$. By considering the point $\beta^1 = \beta^2 = \cdots = \beta^d = 1$ which is inside the future light cone and which makes all the μ_as positive, the light cone is seen to be on the positive side of each μ-hyperplane. Hence $\mu_a(\beta) > 0$ inside the future light cone for each a and $\Theta[-2\mu_a(\beta)] = 0$. This means that one can actually forget about the potentially troublesome μ-type terms in $\mathcal{V}_G$.

The gravitational potential can therefore be replaced in the BKL limit by the manifestly positive sum

$$\lim_{\rho\to\infty} \mathcal{V}_G = \sum_{a,b,c}{}' \Theta[-2\alpha_{abc}(\beta)] \tag{6.84}$$

of α-type sharp wall potentials. These walls are called the “curvature” or “gravitational” walls.

It follows that the gravitational contribution $\mathcal{V}_G$ to the potential brings in the additional constraints

$$\alpha_{abc}(\beta) \geq 0 \quad (D > 3) \tag{6.85}$$

besides the symmetry inequalities (6.82). Note that the inequalities $\alpha_{abc}(\beta) \geq 0$ evidently imply $\mu_a(\beta) \geq 0$ because μ_a is a linear combination with positive coefficients of the α_{abc}s. Indeed, we can write $\mu_c = (\alpha_{abc} + \alpha_{bca})/2$. This is another manner to see that the μ-walls can be neglected in the BKL limit.

The restriction $D > 3$ arises in (6.85) because in $D = 3$ space-time dimensions, the gravitational walls $\alpha_{abc}(\beta) = 0$ are absent, for the simple technical reason that one cannot find three distinct spatial indices. In this case all gravitational walls are of subdominant type μ_a and thus, in the BKL limit, the gravitational potential is effectively absent,

$$\mathcal{V}_G \simeq \sum_a \left(\pm\, \Theta[-2\mu_a(\beta)]\right) \simeq 0 \quad (D = 3). \tag{6.86}$$

It is quite remarkable that even though it contains contributions of both signs, the potential $\mathcal{V}_G$ (i.e., minus the spatial curvature times g) becomes in the BKL limit, a positive sum of sharp wall potentials. What happens is that the negative contributions to $\mathcal{V}_G$ are subdominant in the limit.

There exists another subleading contribution to the potential $\mathcal{V}$, which is of the same type as the subleading terms in the gravitational potential. It is the dilaton contribution (6.75) which involves the same lightlike walls μ_a. Consequently, at least to leading order, we can neglect $\mathcal{V}_\phi$ in the BKL limit.

To be included also among the gravitational contributions is the cosmological term

$$\mathcal{V}_\Lambda = \Lambda g = \Lambda \exp\Big[-2\sum_a \beta^a \Big] \tag{6.87}$$

(if any). The corresponding wall is however spacelike and, as argued in Chapter 5, can be neglected if the system collapses to a spacelike singularity.

6.5.3 p-Form Walls

The previous walls did not involve the dilaton but only the metric variables. By contrast, the p-form wall may involve the dilaton(s). We consider for definiteness a single type of p-form and omit the decoration "(p)" on the corresponding p-form variables to make the notation lighter. The BKL limit of the electric walls give

$$\mathcal{V}^{el}_{(p)} \simeq \sum_{a_1<a_2<\cdots<a_p} \Theta[-2e_{a_1\cdots a_p}(\beta)] \tag{6.88}$$

while the BKL limit of the magnetic walls yields

$$\mathcal{V}^{magn}_{(p)} \simeq \sum_{a_1<\cdots<a_{d-p-1}} \Theta[-2b_{a_1\cdots a_{d-p-1}}(\beta)]. \tag{6.89}$$

Of course, for these walls to be present, their prefactors should not vanish, which will be the case for generic electric and magnetic fields.

The "billiard ball" representing the scale factors is therefore constrained by the further inequalities

$$e_{a_1\cdots a_p}(\beta) \geq 0, \quad \tilde{m}_{a_1\cdots a_{d-p-1}}(\beta) \geq 0. \tag{6.90}$$

The hyperplanes $e_{a_1\cdots a_p}(\beta) = 0$ and $\tilde{m}_{a_1\cdots a_{d-p-1}}(\beta) = 0$ are called "electric" and "magnetic" walls, respectively. Both walls are timelike because their gradients are spacelike, with squared norm

$$\frac{p(d-p-1)}{d-1} + \left(\frac{\lambda_p}{2}\right)^2 > 0 \tag{6.91}$$

Thus, these walls have a nontrivial intersection with the future sheet of the unit hyperboloid/hyperbolic space and are to be considered. The norm in (6.91) is invariant under the exchange of p with $d-p-1$ showing that electric and magnetic walls for the same p-form have the same norm. This is a sign of electric–magnetic duality.

It is interesting to point out that, contrary to the squared norms of the symmetry and gravitational wall, which are always equal to the integer 2 independently of the space-time dimension, the squared norms of the electric and magnetic walls need not be an integer. It is only for peculiar values of the integers d, p and the continuous parameter λ_p that they will be. We shall come back to this important point in Chapter 7.

6.5.4 Constraints

In addition to the evolution equations that lead to the above billiard picture at each spatial point, one must impose the constraint equations. These constraints, which involve the spatial gradients, need only be imposed at one given time, since they are preserved by the time evolution. The constraints just restrict the initial data.

These restrictions on the initial data do not bring dangerous conditions on the coefficients of the walls in the sense that these may generically take nonzero values. This important property can be explicitly verified for each type of walls. For instance, in the case of the p-form walls, one easily checks that it is consistent with Gauss' law to have non-vanishing electric and magnetic energy densities, which are the coefficients of the electric and magnetic walls. The electric and magnetic walls are thus generically present even when the constraints are fulfilled. Our analysis would otherwise be invalidated. In some non-generic contexts, however, the constraints could force some of the wall coefficients to be zero; the corresponding walls would thus be absent.

It is easy to see that the number of arbitrary physical functions involved in the solution of the asymptotic BKL equations of motion is the same as in the general solution of the complete Einstein-matter equations. The counting is a direct generalization of the counting performed in Chapter 1, and proceeds similarly. Further discussion of the constraints in the BKL context may be found in [3, 52].

6.6 Chapter 4 Revisited

The general billiard picture established in this chapter provides an alternative derivation of the results obtained in Chapter 4, on which it sheds new light. The BKL behavior is universal and applies to gravity in higher dimensions as well as to the scalar and electromagnetic couplings studied in Chapter 4. Each model comes with its own walls, but the extremely simple billiard description is always asymptotically valid.

Now, as we pointed out in Chapters 3 and 5, the geodesic motion in a subregion of hyperbolic space is known to be chaotic or non-chaotic depending on whether the billiard table has finite or infinite volume [93, 4, 129, 94, 149, 158, 72, 150]. If the volume is finite, the evolution corresponding to generic initial data exhibits a never-ending, infinite number of collisions with the walls with strong chaotic features. This is the oscillatory behavior. Of course, one might have periodic trajectories for fine-tuned initial data, as for pure gravity in four space-time dimensions. If the billiard volume is infinite, the evolution is non-chaotic. After a finite time, the system settles for generic initial data in a Kasner-like motion that lasts all the way to the singularity and which is directed towards those portions of infinity in $\mathbb{H}_M$ that are contained in the billiard table. There are only a finite number of collisions with the walls, which take place before the system settles in the final Kasner regime.

As we shall show in Chapter 7, the billiard table has finite volume for pure gravity as long as the spatial dimension d is smaller than or equal to 9, so that pure gravity exhibits the chaotic, never-ending oscillatory behavior up to space-time dimension 10. For $d \geq 10$ ($D \geq 11$), the billiard table has infinite volume and the behavior of the gravitational field ceases to be chaotic. This is in perfect agreement with what we found in Chapter 4. Note, in particular, that the curvature wall forms α_{ijk} coincide with the exponents that control the curvature behavior in the BKL regime introduced in Chapter 4.

Furthermore, the inclusion of a single scalar field (with no other field) increases the dimension of the billiard, but does not bring in new walls, so that it leads to a billiard table with infinite volume and no chaotic behavior. The inclusion of the electromagnetic field (1-form) works in the opposite direction because it does not change the dimension of the billiard table but brings in new walls, so that it can only reinforce chaos when chaos is present. All these features will be worked out in detail in Chapter 7 and confirm our previous findings.

Other results established in Chapter 4 also find a natural interpretation in terms of the billiard picture. For instance, the competition in four space-time dimensions between the energy–momentum tensor of the electromagnetic field and the spatial curvature of the gravitational field in governing Kasner epochs and transitions between them appears here as a mere consequence of the fact that the electromagnetic walls and the gravitational walls are identical for $d = 3$ (the wall forms come with different normalizations but define the same hyperplanes). Their relative importance will depend on the relative weight of the prefactors in the given epochs, but they will both lead to the same collision law, as we found previously in Chapter 4.

As also discussed in Chapter 4, there is no difference between a Yang–Mills gauge field and a collection of abelian gauge fields in the asymptotic BKL dynamics. Indeed, the energy density takes the same form in both cases, as the sum of the electric energy and magnetic energy densities. What enters the electric energy density are the momenta π^i_a (where $a = 1, \cdots, N$ and N is the number of vector fields, i.e., the dimension of the internal Lie algebra in the Yang–Mills case). What enters the magnetic energy density are the spatial components of the field strengths. The field strengths are the abelian field strengths for a collection of free vector fields, and the non-abelian ones in the general Yang–Mills case. The non-abelian corrections do not depend on the logarithmic scale factors, however, and so, the asymptotic dynamics is the same in both cases. From the point of view of the billiard, each electric and magnetic 1-form wall is just repeated N times, but the walls are the same.

The same is true for Chapline–Manton couplings or Chern–Simons terms for p-form gauge fields. This is because the energy density of the p-forms provides the same scale-factors-dependent exponentials in the absence or in the presence of couplings. It is true that the wall coefficients themselves are different functions of the p-form canonical variables, but this difference is irrelevant in the sharp wall limit, where the coefficients can be replaced by one when they are different from zero.

6.7 Miscellanea

We conclude this chapter with a few comments.

1. There exist theories for which the asymptotic BKL billiard is defined by the symmetry walls and the electric walls of the p-form present in the models. All the other walls are subdominant, i.e., behind the symmetry and electric walls. Notable examples are the bosonic sector of maximal supergravity, where the billiard is defined by the symmetry walls and the 3-form electric walls [45], or the pure Einstein–Maxwell system in space-time dimensions $D \geq 5$ [46, 118] (see Chapter 7). Curvature and magnetic walls are then subdominant, i.e., spatial gradients become literally negligible as one approaches the singularity.

   In those cases, the complete billiard dynamics is already visible in the simplest homogeneous models, namely, the "Bianchi I" models where the fields depend only on time. Indeed, the simplifying assumption that the fields depend only on time preserves the symmetry and electric walls, which do not involve spatial gradients, and eliminates only subdominant walls. This is true whether the billiard table has finite volume (chaos) or infinite volume (no chaos).

   Assuming that the fields do not depend on some spatial coordinate is what one does upon dimensional reduction. Thus, if the curvature and magnetic walls can be neglected, the asymptotic equations of motion at each point are exactly the same as the equations of motion obtained by performing a torus dimensional reduction to $1+0$ dimensions. For this reason, the confusion arises sometimes in the context of maximal supergravity (or M-theory) that the BKL analysis emerges as a result of dimensional reduction. This is incorrect. The remarkable fact that one gets a set of effective ODEs with respect to time is not a consequence of any dimensional reduction, but is a dynamical result following from the equations of motion analyzed in a general setting. No homogeneity assumption was ever made above. The effective torus dimensional reduction follows from the dynamics and is not imposed by hand. One gets an independent billiard dynamics at each spatial point, and the BKL analysis produces a general solution adjustable to generic inhomogeneous initial data, i.e., involving a complete number of arbitrary functions of space.

It is interesting to compare the above Hamiltian with the Hamiltonian one would have obtained in a space-time of Euclidean signature. One finds

$$\mathcal{H} = \mathcal{K}' + \varepsilon \mathcal{V}' \tag{6.92}$$

where $\varepsilon = \pm 1$ according to whether the space-time signature is Lorentzian ($\varepsilon = 1$) or Euclidean ($\varepsilon = -1$). Here,

$$\mathcal{K}' = \mathcal{H}_0 + \mathcal{V}_S + \mathcal{V}_{(p)}^{el} \tag{6.93}$$

contains all the kinetic terms, and

$$\mathcal{V}' = \mathcal{V}_G + \mathcal{V}_{(p)}^{magn} + \mathcal{V}_\phi \tag{6.94}$$

contains the terms with spatial derivatives. We see that for the above theories, the asymptotic BKL limit is completely governed by $\mathcal{K}'$, i.e., by the limit $\varepsilon = 0$, since $\mathcal{V}'$ contains only spatial gradients and can thus be dropped (and spatial gradients are only there). The limit $\varepsilon = 0$, sometimes called the "zero signature limit" [151], lies halfway between space-times of Minkowskian and Lorentzian signature. It amounts to considering a vanishing velocity of light. The underlying geometry is built on the Carroll contraction of the Lorentz group [85]. In the paper [97], that same limit was termed "strong coupling limit" because the Hamiltonian density $\mathcal{H}$ can be rewritten as

$$\mathcal{H} = G_N \mathcal{K}' + G_N^{-1} \mathcal{V}' \tag{6.95}$$

with appropriate rescalings. Taking the large G_N limit (with rescaling of the lapse to absorb the factor of G_N in front of $\mathcal{K}'$) is equivalent to setting ε equal to zero. The ultrarelativistic "Carrollian" limit described by the Hamiltonian density $\mathcal{K}'$ has recently attracted renewed interest [57, 126, 2, 80].

2. **Historical Note**

Hamiltonian methods for investigating the behavior of the gravitational field near the cosmological singularity were developed in [131]. The hyperbolic billiard description of the (3+1)-dimensional homogeneous Bianchi IX system was first worked out by Chitre [35] and Misner [132], who showed that the dynamics of diagonal models could be rephrased as a succession of free flight motions interrupted by reflections in the walls given by the sides of the ideal triangle in hyperbolic space. The picture was subsequently generalized to inhomogeneous metrics in [115, 100]. The extension to higher dimensions with perfect fluid sources was considered in [117], without symmetry walls.

Exterior p-form sources were investigated in [101, 102, 103] for special classes of metric and p-form configurations. The uniform approach presented here and based on a systematic use of the Iwasawa decomposition of the spatial metric, which simplifies enormously the discussion and brings in the symmetry walls, has been developed in [45, 46, 47, 48, 51].

7

Hyperbolic Coxeter Groups

7.1 Introduction

We saw in the previous chapter that in spite of the complicated structure of the gravitational field equations, the asymptotic dynamics near a cosmological singularity can be described in surprisingly simple terms.

Indeed, the asymptotic dynamics are governed by the scale factors – including the dilaton(s) if any – while the other variables (off-diagonal metric components, p-form fields) tend to become mere "spectators" which get asymptotically frozen.

The motion of the scale factors at each spatial point takes place, in the γ-space picture, in a subset of hyperbolic space $\mathbb{H}_M$ delimited by hyperplanes which can be of four types: symmetry (centrifugal) walls, curvature (gravitational) walls, p-form electric walls, and p-form magnetic walls. Between two collisions, the motion is a geodesic in hyperbolic space.

Equivalently, in the β-space picture, the motion is a future-directed zigzag null line in the Minkowski space $\mathbb{R}^{M,1}$ of the scale factors, interrupted by collisions with the symmetry, curvature, electric or magnetic hyperplanes, which are all timelike.

Given the action (6.1) with definite space-time dimension, menu of fields and dilaton couplings, one can determine the dimension of the billiard and display the wall forms by the methods of Chapter 6. The dimension M of the billiard is given by the number of independent scale factors, i.e., the number of spatial dimensions d minus one (to take into account the Hamiltonian constraints) plus the number n of dilatons ($M = d + n - 1$). The walls come in the four varieties described before: symmetry walls, curvature walls, electric walls and magnetic walls, the last two involving the dilaton couplings (if dilatons are present).

As we already experienced with pure gravity in four space-time dimensions, we expect that not all the wall forms are relevant, since only the dominant ones define the billiard, the other ones being associated with walls that are behind the dominant walls. To compute the billiards, one needs therefore to determine which walls H_i are dominant. Once these are determined, the billiard domain is simply the intersection $\cap_i H_i^+$ of the half spaces on the positive side to each dominant wall.

It turns out that the billiard regions associated with the known physically interesting theories have remarkable features and that the groups generated by the reflections in the billiard walls enjoy also remarkable properties that generalize those of the billiard group of four-dimensional pure gravity exhibited in Chapter 5. This chapter is devoted to exploring these questions by first laying down the necessary concepts to address these issues.

7.2 Convex Polyhedra in Hyperbolic Space

We shall use in this chapter two models for hyperbolic space $\mathbb{H}_M$. The first is the vector model where a hyperplane H_i in hyperbolic space is the intersection of a timelike hyperplane $\tilde{H}_i$ in the space $\mathbb{R}^{M,1}$ with the upper sheet of the hyperboloid. H_i has dimension $M-1$ while $\tilde{H}_i$ has dimension M. The second is the Klein model where hyperbolic space is projectively realized as the interior K_M of the unit ball B_M in the space $\mathbb{R}^M$ with equation $x^0 = 1$, which is tangent to the hyperboloid at $x^0 = 1$, $x^1 = 0 = \cdots = x^M$. The hyperplane H_i is then the intersection with the interior of the unit ball K_M of an hyperplane $\hat{H}_i$ of $\mathbb{R}^M$ that cuts it. In the Klein model, one sometimes call the hyperplane $\mathbb{R}^M$ defined by $x^0 = 1$ on which $\mathbb{H}_M$ is realized as the interior of the unit ball, the "ambient space" (of the Klein model).

Let H_i be a hyperplane of $\mathbb{H}_M$. We assume that an orientation has been chosen for H_i, and we denote by H_i^+ the positive half-space bounded by H_i. The hyperplane H_i is included in H_i^+, i.e., if $\mathbf{e}_i$ is the unit vector orthogonal to the corresponding hyperplane $\tilde{H}_i$ in $\mathbb{R}^{M,1}$ and pointing towards the positive region, $\tilde{H}_i^+$ is defined as

$$\tilde{H}_i^+ = \{\mathbf{x} \in \mathbb{R}^{M,1} : \mathbf{x} \cdot \mathbf{e}_i \geq 0\} \tag{7.1}$$

with $\geq$ (equality included) and not the strict inequality $>$ while H_i^+ is defined as the intersection of $\tilde{H}_i^+$ with the unit hyperboloid. Similar considerations hold in the Klein model.

Two hyperplanes H_i and H_j may or may not intersect in $\mathbb{H}_M$. However, the corresponding timelike hyperplanes $\tilde{H}_i$ and $\tilde{H}_j$ always intersect in $\mathbb{R}^{M,1}$ along a subspace that might be timelike (in which case H_i and H_j intersect in $\mathbb{H}_M$), null (in which case H_i and H_j are parallel in $\mathbb{H}_M$) or spacelike (in which case H_i and H_j are divergent in $\mathbb{H}_M$). By abuse of language, we shall say that H_i and H_j meet at infinity or beyond infinity in the latter two cases.

A convex polyhedron P in $\mathbb{H}_M$ is the intersection of a set of positive half-spaces with nonempty interior,

$$P = \cup_{i=1}^N H_i^+ \tag{7.2}$$

where H_i $(i = 1, \cdots, N)$ is a collection of oriented hyperplanes. We assume that all hyperplanes are relevant, i.e., that if one removes any of them from (7.2), one gets a different region of hyperbolic space. If not, we just drop the irrelevant

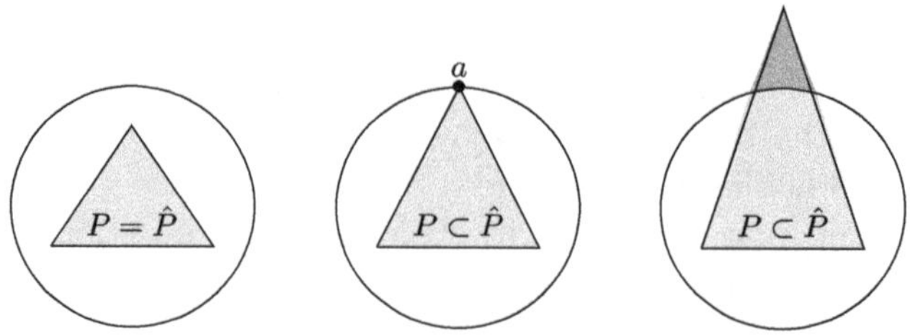

Figure 7.1: Three polygons are shown on the Klein model K_2 of hyperbolic space. In the first case, P is entirely contained in the interior of the disk and coincides with the polygon $\hat{P}$ defined by the same straight lines in the ambient space $\mathbb{R}^2$. In the second and third cases, P is a proper subset of $\hat{P}$. In the second case, the point at infinity a on the boundary circle belongs to $\hat{P}$ but not to P. In the third case, the tip of the triangle beyond the circle at infinity, shown in dark grey, and the arc of circle at infinity within the triangle are in $\hat{P}$ but not in P. Note that the polygon contains in that latter case entire lines and half-spaces bounded by them, e.g., the line joining the two points of intersection of the triangle with the boundary circle, as well as the half-space above it are in the polygon.

hyperplanes. This implies that no $\mathbf{e}_i$ can be written as linear combination with nonnegative coefficients of the others.

The faces of the polyhedron are the intersections $P \cap H_i$. In two dimensions, a polyhedron is called a polygon.

The polyhedron P is convex, which means that if two points belong to P, then the segment that joins these points lies entirely within P. This means in particular that there is no dihedral angle greater than π.

In the Klein model K_M, one can consider the convex polyhedron $\hat{P}$ in the ambient Euclidean space $\mathbb{R}^M$ bounded by the same hyperplanes. The polyhedron P is the intersection of $\hat{P}$ with the interior K_M of the unit ball. It may coincide with $\hat{P}$ or be a proper subset of it (see Figure 7.1).

Gram Matrix

To the polyhedron P one can associate a symmetric matrix formed by the scalar products of the unit vectors $\mathbf{e}_i$,

$$(\mathbf{e}_i \cdot \mathbf{e}_j) \tag{7.3}$$

called a *Gram matrix*. This matrix has 1s on its diagonal. Furthermore, its rank is at most $M + 1$ since one cannot find more than $M + 1$ linearly independent vectors in $\mathbb{R}^{M,1}$.

A polyhedron is non-degenerate when the rank is exactly $M+1$ and we assume from now on that this is the case. This implies $N \geq M + 1$ and that among the N vectors $\mathbf{e}_i$ one can find a subset of $M+1$ vectors that are linearly independent and form a basis. The Gram matrix restricted to any such subset has signature $(M, 1)$.

The case $N = M + 1$ corresponds to a simplex. The case $N > M + 1$ defines a polyhedron that is not a simplex.

Acute-Angled Polyhedra

If $\mathbf{e}_i$ and $\mathbf{e}_j$ span a two-dimensional spacelike plane, then $|\mathbf{e}_i \cdot \mathbf{e}_j| < 1$ and the dihedral angle θ_{ij} between the faces H_i and H_j is given by

$$\cos\theta_{ij} = -\mathbf{e}_i \cdot \mathbf{e}_j. \tag{7.4}$$

It is therefore acute ($\theta_{ij} \leq \frac{\pi}{2}$) if and only if $\mathbf{e}_i \cdot \mathbf{e}_j \leq 0$.

If $\mathbf{e}_i$ and $\mathbf{e}_j$ span a two-dimensional lightlike ($|\mathbf{e}_i \cdot \mathbf{e}_j| = 1$) or timelike ($|\mathbf{e}_i \cdot \mathbf{e}_j| > 1$) plane, our non-redundancy hypothesis implies that the scalar product is in fact negative, since otherwise one of the half-spaces H_i, H_j is contained in the other (see Section 6.3).

One defines an acute-angled polyhedron as one for which the off-diagonal entries of the Gram matrix are non-positive,

$$\mathbf{e}_i \cdot \mathbf{e}_j \leq 0, \quad i \neq j \tag{7.5}$$

(the diagonal ones being equal to 1). So, a polyhedron is acute-angled if and only if two bounding hyperplanes make an acute angle whenever they intersect.

In the null case $\mathbf{e}_i \cdot \mathbf{e}_j = -1$, the hyperplanes H_i and H_j are parallel and meet at infinity with a dihedral angle equal to 0. In the hyperbolic case $\mathbf{e}_i \cdot \mathbf{e}_j < -1$, the hyperplanes H_i and H_j are divergent and meet "beyond infinity."

One can show ([1], page 106) that for acute-angled polyhedra, if the faces $P \cap H_i$ and $P \cap H_j$ are not adjacent, then the hyperplanes H_i and H_j do not intersect. In fact, the acute-angled condition need only be checked for adjacent faces.

Coxeter Polyhedra

A Coxeter polyhedron is a convex polyhedron that has the property that for all i, j such that the hyperplanes H_i, H_j intersect, the dihedral angle θ_{ij} is a submultiple of π,

$$\theta_{ij} = \frac{\pi}{m_{ij}}, \quad m_{ij} \in \mathbb{N}, \quad m_{ij} \geq 2.$$

A Coxeter polyhedron is clearly acute-angled.

When the hyperplanes H_i and H_j do not meet, we set $\theta_{ij} = 0$ and $m_{ij} = \infty$.

Volumes of Polyhedra

The volume of a non-degenerate polyhedron P is finite if the polyhedron $\hat{P}$ is contained in K_M since the polyhedron defines then a compact subspace of hyperbolic space. This corresponds to the first case of Figure 7.1. If the polyhedron

$\hat{P}$ is contained in the closure $\bar{K}_M$ of the unit ball, P has finite volume [1] but is non-compact if it has vertices at infinity, like in the second case of Figure 7.1. Finally, P has infinite volume if $\hat{P}$ is not contained in $\bar{K}_M$, like in the third case of Figure 7.1.

These properties generalize what we found in two-dimensional hyperbolic space.

7.3 Coxeter Groups: General Considerations

Given a polyhedron in hyperbolic space, one can consider the group of reflections generated by the reflections s_i in the faces of the polyhedron. These reflections obey

$$s_i^2 = 1 \tag{7.6}$$

since they are just mirror reflections.

Consider two mirror hyperplanes H_i and H_j and the product $s_i s_j$. Let $\tilde{H}_i$ and $\tilde{H}_j$ be the corresponding hyperplanes in the Minkowski space $\mathbb{R}^{M,1}$ of the vector model. The intersection $\tilde{H}_i \cap \tilde{H}_j$ can be timelike, null or spacelike.

We thus have three cases:

1. If the intersection $\tilde{H}_i \cap \tilde{H}_j$ is timelike, the hyperplanes H_i and H_j meet in hyperbolic space and the product $s_i s_j$ is an isometry of elliptic type.
2. If the intersection $\tilde{H}_i \cap \tilde{H}_j$ is lightlike, the hyperplanes H_i and H_j are parallel in hyperbolic space and the product $s_i s_j$ is a parabolic translation.
3. If the intersection $\tilde{H}_i \cap \tilde{H}_j$ is spacelike, the hyperplanes H_i and H_j are divergent in hyperbolic space and the product $s_i s_j$ is a hyperbolic translation.

In the first case, the transformation is a rotation by an angle $2\theta_{ij}$ about $H_i \cap H_j$, which is left pointwise invariant. In the case of a Coxeter polyhedron, $\theta_{ij} = \frac{\pi}{m_{ij}}$ and so the rotation angle is an integer multiple of 2π,

$$2\theta_{ij} = \frac{2\pi}{m_{ij}}.$$

This implies that

$$(s_i s_j)^{m_{ij}} = 1. \tag{7.7}$$

In the second and third cases, there is no power of $s_i s_j$ that gives the identity. It is nevertheless convenient to consider that (7.7) holds, but with $m_{ij} = \infty$.

Definition

It is useful to abstract the above properties and introduce the concept of the "Coxeter group."

A Coxeter group $\mathfrak{C}$ is a group generated by a finite number of elements s_i $(i = 1, \cdots, n)$ subject to relations that take the form

$$s_i^2 = 1 \tag{7.8}$$

and

$$(s_i s_j)^{m_{ij}} = 1, \tag{7.9}$$

where the integers m_{ij} associated with the pairs (i, j) fulfill

$$\begin{aligned} m_{ij} &= m_{ji}, \\ m_{ij} &\geq 2 \qquad (i \neq j). \end{aligned} \tag{7.10}$$

Note that equation (7.8) is a particular case of equation (7.9) with $m_{ii} = 1$. If there is no power of $s_i s_j$ that gives the identity, we conventionally set $m_{ij} = \infty$, as we just explained. The generators s_i are called "reflections" because of equation (7.8), even though a geometric interpretation of the group is not given a priori.

The number n of generators is called the rank of the Coxeter group. The Coxeter group is completely specified by the integers m_{ij}. It is useful to draw the set $\{m_{ij}\}$ pictorially in a diagram Γ, called a "Coxeter graph," or "Coxeter scheme."

The Coxeter graph is constructed according to the same rules as in Chapter 5. With each reflection s_i, one associates a node. Thus there are n nodes in the diagram. If $m_{ij} > 2$, one draws a line between the node i and the node j and writes m_{ij} over the line, except if m_{ij} is equal to 3, in which case one writes nothing. The default value is thus "3." When there is no line between i and j $(i \neq j)$, the exponent m_{ij} is equal to 2.

As we have seen in Chapter 5, the group $PGL(2, \mathbb{Z})$ of the vacuum Einstein theory in four space-time dimensions is a Coxeter group with Coxeter graph

The Coxeter graph for $PGL(2, \mathbb{Z})$

while the Coxeter group relevant for homogeneous cosmological models of type IX has Coxeter graph

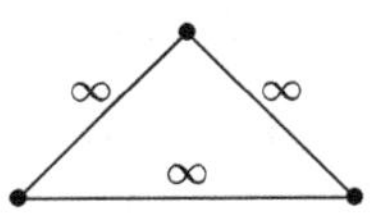

The Coxeter graph for $G\Gamma(2)$

The group generated by the reflections in the faces of a Coxeter polyhedron is a Coxeter group. It is indeed clear that the reflections in the faces of a Coxeter polyhedron obey the relations (7.8) and (7.9) defining a Coxeter group. The only question that arises, then, is whether there are not additional, independent relations, fulfilled by these reflections. That this is not so is discussed in [143], Chapter 7. For a group of reflections, the Gram matrix of the polyhedron determines its Coxeter graph. The converse is true only if no m_{ij} is infinite. If $m_{ij} = \infty$, one knows that the corresponding hyperplanes do not intersect, i.e., that $e_i \cdot e_j \leq -1$, but one does not have sufficient information to determine the exact value of that scalar product.

Note that if $m_{ij} = 2$, the generators s_i and s_j commute, $s_i s_j = s_j s_i$. Thus, a Coxeter group $\mathfrak{C}$ is the direct product of the Coxeter subgroups associated with the connected components of its Coxeter graph. For that reason, we can restrict the analysis to Coxeter groups associated with connected (also called irreducible) Coxeter graphs.

A Coxeter group is defined abstractly in terms of generators and relations. However, one can show that a Coxeter group always admits a geometric realization as a reflection group. In our case, the group actually appears in the first place as a reflection group generated by the reflections in the faces of a given Coxeter polyhedron (the billiard table), so the question of a geometric realization does not really arise.

The theory of Coxeter groups is extremely rich and we refer to the books [29, 95, 143, 154] and to the review [90] for more information. We shall give here only a glimpse at the theory.

Fundamental Domain

In order to describe the action of the Coxeter group, it is useful to introduce the concept of the *fundamental domain*. A fundamental domain is by definition such that any orbit of the group intersects it once and only once.

One can show that the Coxeter polyhedron defining the Coxeter group through reflections in its faces is a fundamental domain. In particular, the billiard tables of pure gravity in four dimensions in the generic case and in the homogeneous context, are fundamental domains for the groups $PGL(2, \mathbb{Z})$ and $G\Gamma(2)$, respectively. The acute-angled condition is quite important here, for otherwise one would not get a fundamental domain. This is illustrated in the third example of the next section.

7.4 Coxeter Groups: Examples

The Coxeter groups appearing in gravity are groups of reflections in hyperbolic space. Simpler, and more familiar examples of Coxeter groups, are reflection groups on the sphere or in Euclidean space. We give here examples of each.

The dihedral group $I_2(3) \equiv A_2$

Consider the dihedral group $I_2(3)$ of symmetries of the equilateral triangle in the Euclidean plane shown in Figure 7.2. Its order is equal to 6.

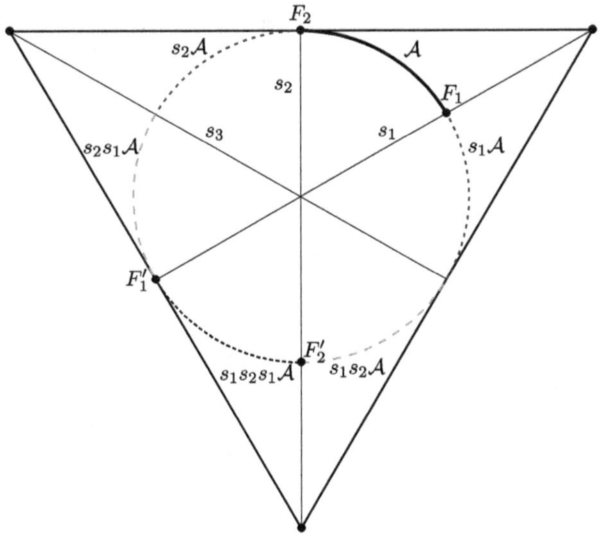

Figure 7.2: The equilateral triangle has a symmetry group of order 6, which is a Coxeter group of rank 2 generated by the two reflections s_1 and s_2. This group acts on the 1-sphere S_1 (circle). The reflections s_1 and s_2 are reflexions in the zero-dimensional faces of the "polyhedron" $\mathcal{A}$ in black, which is a fundamental domain. The images of $\mathcal{A}$ under the group are the six regions $\mathcal{A}$, $s_1\mathcal{A}$, $s_2\mathcal{A}$ etc. represented in the figure. The group acts as a permutation subgroup of these six regions.

The elements of the group are easily listed. These are the identity, three reflections s_1, s_2 and s_3 about the three medians, the rotation R_1 of $2\pi/3$ about the origin and the rotation R_2 of $4\pi/3$ about the origin,

$$I_2(3) = \{I, s_1, s_2, s_3, R_1, R_2\}. \tag{7.11}$$

All the elements of the dihedral group $I_2(3)$ can be written as products of the two reflections s_1 and s_2:

$$\begin{aligned} &I = s_1^0, \qquad && s_1 = s_1, \qquad && s_2 = s_2, \\ &R_1 = s_1 s_2, \qquad && R_2 = s_2 s_1, \qquad && s_3 = s_1 s_2 s_1 \end{aligned}$$

Hence, the dihedral group $I_2(3)$ is generated by s_1 and s_2, which are subject to the relations,

$$s_1^2 = 1, \qquad s_2^2 = 1, \qquad (s_1 s_2)^3 = 1 \tag{7.12}$$

(the mirror lines of reflexion of s_1 and s_2 make an angle of $\pi/3$). There is no other independent relation between the generators s_1 and s_2 because any product of them can be reduced, using the Coxeter relations, to one of the six products written above, and these are independent. Note that $s_1 s_2 s_1 = s_2 s_1 s_2$ because of the Coxeter relations. This shows that $I_2(3)$ is a Coxeter group of rank 2.

The dihedral group $I_2(3)$ is also denoted A_2 because it is the Weyl group of the simple Lie algebra A_2 (see Appendix D). It is isomorphic to the permutation group S_3 of three objects. Its Coxeter graph is

The Coxeter graph for $I_2(3) \equiv A_2$

The dihedral group $I_2(3)$ can be viewed as acting on the circle, i.e., on the 1-sphere S_1. In fact, one can describe $I_2(3)$ as a reflection group on S_1 generated by the reflections in the faces of a Coxeter polyhedron. The situation is a bit degenerate because of the low dimension, but completely fits with the above construction of Coxeter groups.

The reflections s_1 and s_2 can indeed be viewed as reflections in the zero-dimensional faces F_1 and F_2 of the one-dimensional polyhedron $\mathcal{A}$ drawn as a thick black circular arc in Figure 7.2. The antipodal points F_1 and F_1' define a hyperplane H_1 on the circle. We orient it so that the regions $\mathcal{A}$, $s_2\mathcal{A}$ and $s_2 s_1 \mathcal{A}$ stand on its positive side. Similarly, we orient the hyperplane H_2 consisting of the two antipodal points F_2 and F_2' so that the regions $\mathcal{A}$, $s_1\mathcal{A}$ and $s_1 s_2 \mathcal{A}$ stand on its positive side. The polyhedron $\mathcal{A}$ is given by $H_1^+ \cap H_2^+$. The orbit of any point on the circle intersects $\mathcal{A}$ once and only once, which is a fundamental domain.

One can show that acute-angled polyhedra on the sphere S_n are necessary simplices (see, e.g., [154]). This means that the number of faces is equal to $n+1$ and that the vectors e_i defining the bounding hyperplanes are linearly independent and form a basis of the ambient Euclidean space. The Gram matrix is non-degenerate and positive definite. The corresponding Coxeter scheme (or graph) is called "elliptic." A Coxeter group is finite if and only if its Coxeter scheme is elliptic.

Elliptic Coxeter schemes – and thus finite Coxeter groups – have been completely classified and are given in Table 7.1 (see [29, 95, 143, 154, 90]).

Another Example: C_2^+

The next example is an infinite Coxeter group of "affine type," which acts on the two-dimensional Euclidean space.

Consider the group of isometries of the Euclidean plane generated by reflections in the faces of the triangle $\mathcal{A}$ bounded by the following three straight lines (hyperplanes in two dimensions): (i) the x-axis (s_1), (ii) the straight line joining the points $(1,0)$ and $(0,1)$ (s_2), and (iii) the y-axis (s_3). We orient these lines in such a way that H_1^+ is $y \geq 0$, H_2^+ is $1 - x - y \geq 0$ and H_3^+ is $x \geq 0$.

The Coxeter exponents are finite and equal to 4 ($m_{12} = m_{21} = m_{23} = m_{32} = 4$) and 2 ($m_{13} = m_{31} = 2$). The Coxeter graph is

The Coxeter graph for C_2^+

Table 7.1: *Finite Coxeter groups.*

Name	Coxeter graph
A_n	
$B_n \equiv C_n$	
D_n	
$I_2(m)$	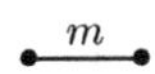
F_4	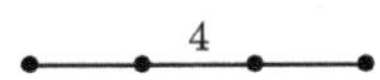
E_6	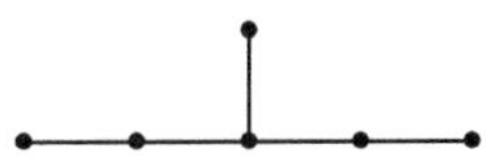
E_7	
E_8	
H_3	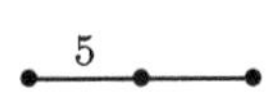
H_4	

The Coxeter group is the symmetry group of the regular paving of the plane by squares of length 2. This group contains translations and is therefore infinite. Indeed, the product $s_2s_1s_2$ is a reflection in the line parallel to the y-axis going through $(1,0)$ and thus the product $t = s_2s_1s_2s_3$ is a translation by $+2$ in the x-direction. All powers of t are distinct.

The region $\mathcal{A}$ is a fundamental domain since any orbit intersects it once and only once. All the triangles on the picture can be mapped on $\mathcal{A}$. We have drawn in grey on Figure 7.3 a few images of $\mathcal{A}$ under the action of the group. This Coxeter group is denoted C_2^+ or B_2^+ (see Appendix D). It is a group of reflections in Euclidean space. Such groups are called *affine* Coxeter groups and the corresponding Coxeter graphs (or schemes) are called *parabolic*.

The fundamental domain of C_2^+ is a simplex. This is not an accident. Indeed one can show, just as for the sphere, that acute-angled polyhedra in Euclidean space are simplices (see again, e.g., [154]). Because the fundamental domain is a simplex, the $n+1$ vectors orthogonal to its $n+1$ bounding hyperplanes form

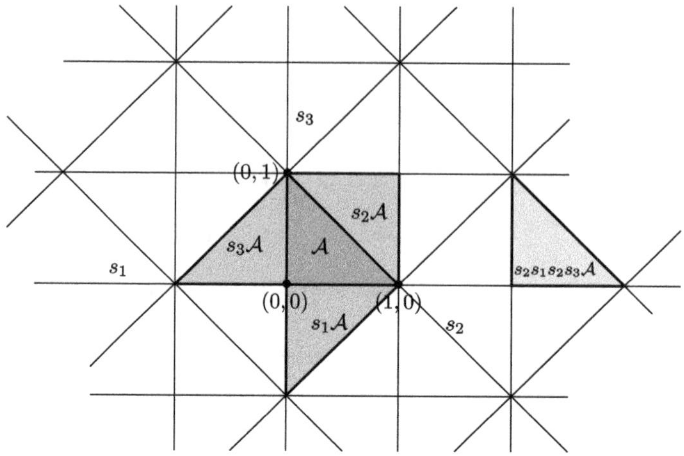

Figure 7.3: The reflection group C_2^+ generated by the reflections s_1, s_2 and s_3 in the lines H_1 ($y = 0$), H_2 ($1 - x - y = 0$) and H_3 ($x = 0$). The domain $\mathcal{A}$ on the positive side of each line is a fundamental domain. The group contains translations. For instance, $t \equiv s_2s_1s_2s_3$ is a translation by $+2$ along the x-axis. The group is therefore infinite since all powers of t are distinct.

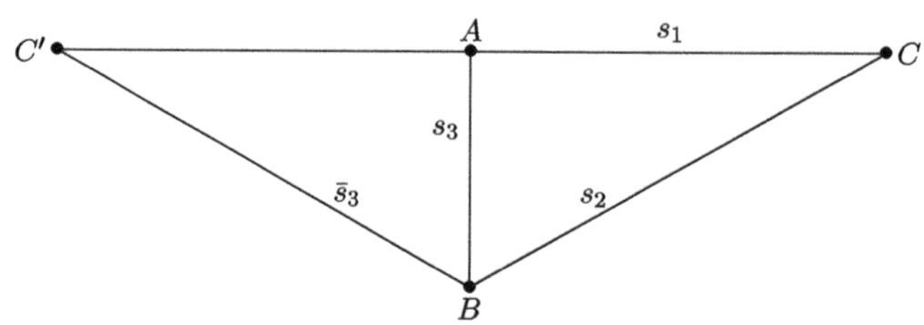

Figure 7.4: The polyhedron $\mathcal{P}$ defining the Coxeter group through reflections in its faces must be acute-angled to be a fundamental domain (see text).

an overcomplete set of vectors of rank n in n-dimensional Euclidean space. The Gram matrix is positive semi-definite, with only one null vector.

All affine Coxeter groups have been determined [29, 95, 143, 154, 90]. They are listed in Table 7.2.

Why Acute-Angled?

We have insisted above on the "acute-angled" condition. To illustrate its importance, we consider the affine group of reflections G_2^+, which is generated by reflections in the sides of the triangle ABC in Figure 7.4, with respective angles $\frac{\pi}{2}, \frac{\pi}{3}, \frac{\pi}{6}$.

The reflections obey

$$(s_1s_2)^6 = 1, \quad (s_1s_3)^2 = 1, \quad (s_2s_3)^3 = 1.$$

Table 7.2: *Affine Coxeter groups.*

Name	Coxeter graph
A_1^+	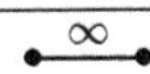
A_n^+ $(n > 1)$	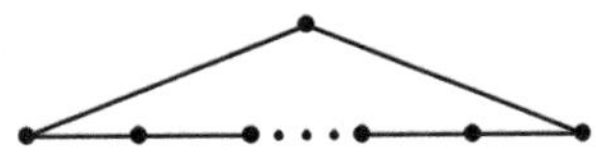
B_n^+	
C_n^+	
D_n^+	
G_2^+	
F_4^+	
E_6^+	
E_7^+	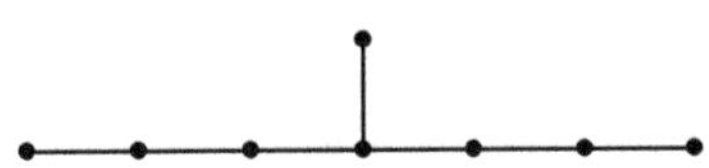
E_8^+	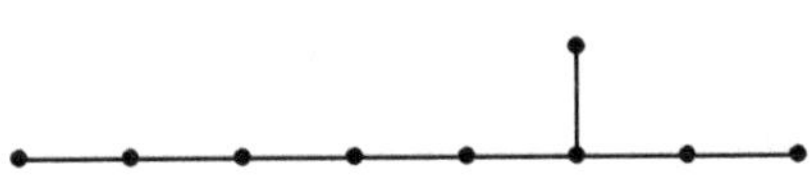

By acting with s_1, s_2 and s_3 on the triangle ABC, one generates a tiling of the plane by an infinite number of right triangles congruent to ABC. We have only represented the image $s_3ABC = ABC'$ by the reflection s_3. The triangle ABC is a fundamental domain for the action of the Coxeter group generated by s_1, s_2 and s_3.

Now, consider the triangle $C'BC = ABC \cup ABC'$. The reflections s_1, s_2 and $\bar{s}_3$ in its faces generate exactly the same Coxeter group because one can express s_3 in terms of s_2 and $\bar{s}_3$, and vice versa, one can express $\bar{s}_3$ in terms of s_2 and s_3,

$$s_3 = s_2\bar{s}_3s_2, \quad \bar{s}_3 = s_2s_3s_2.$$

Furthermore, the generating reflections obey the Coxeter relations

$$(s_1 s_2)^6 = 1, \quad (s_1 \bar{s}_3)^6 = 1, \quad (s_2 \bar{s}_3)^3 = 1.$$

However, the triangle $C'BC$ is not a fundamental domain, even though the reflections in its sides also define the Coxeter group G_2^+. The orbits have generically two intersection points with the triangle $C'BC$. The triangle $C'BC$ is not acute-angled and has angles $\frac{\pi}{6}, \frac{2\pi}{3}, \frac{\pi}{6}$. There is a reflection axis bisecting the obtuse angle and cutting $C'BC$ in two fundamental domains.

7.5 Coxeter Groups and Weyl Groups

7.5.1 Lorentzian Coxeter Groups

Reflection groups acting on hyperbolic space are called Lorentzian. These are discrete subgroups of the orthochronous Lorentz group $O^+(k,1)$. One reserves the terminology *hyperbolic* only to those reflection groups defined by Coxeter polyhedra with finite volume.

Hyperbolic reflection groups are very special structures. One can show that there exist no Coxeter polyhedra of finite volume in hyperbolic space H_n for $n \geq 996$. If one imposes the even stronger condition of compactness (no vertices at infinity), one can show that there are no bounded Coxeter polyhedra in hyperbolic space H_n for $n \geq 30$ [154].

7.5.2 Simplex Coxeter Groups

While Coxeter polyhedra in the sphere S_n or in the Euclidean plane E_n are necessarily simplices, this is not the case in hyperbolic space H_n. There exist Coxeter polyhedra that are not simplices. Coxeter groups defined by simplices are called "simplex Coxeter groups." It turns out that these are the only ones that we shall encounter.

For Coxeter simplices in H_n, the $n+1$ vectors e_i defining the bounding hyperplanes are linearly independent in the ambient vector space $\mathbb{R}^{n,1}$. Consequently, the Gram matrix is invertible and has Lorentzian signature.

Finite volume Coxeter simplices exist only in H_n with $n \leq 9$ (10 bounding walls), and there are only three of them of maximal rank 10. The stronger condition that the simplex should be bounded can only be fulfilled in H_n with $n \leq 4$ [154].

We list in Table 7.3 the three hyperbolic simplex Coxeter graphs of rank 10. As we shall see, all three of them are realized in gravitational theories.

7.5.3 Hyperbolicity Criterion

It can be shown (see, e.g., [95, 90]) that hyperbolic Coxeter groups are described by Lorentzian Coxeter graphs that have the property that any proper subgraph is

Table 7.3: *Hyperbolic simplex Coxeter groups of rank* 10.

Name	Coxeter graph
$B_8^{++} \equiv BE_{10}$	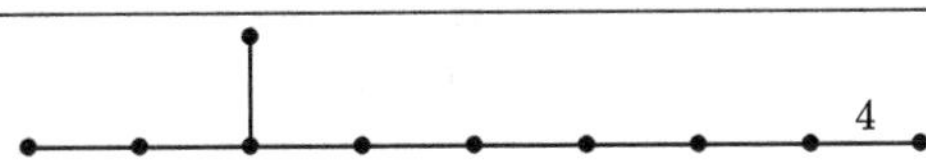
$D_8^{++} \equiv DE_{10}$	
$E_8^{++} \equiv E_{10}$	

either of finite or affine type. [Note that by removing a node, one might get a non-irreducible diagram even if the original diagram is connected. Each irreducible component must be of finite or affine type.]

This provides an easy criterion to verify hyperbolicity, which we shall explicitly use below in the context of pure gravity in any space-time dimension. We can verify it in the case of $PGL(2,\mathbb{Z})$, described by the Coxeter graph

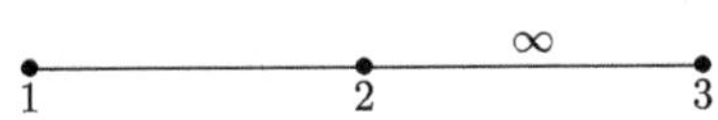

which we know is hyperbolic. The Gram matrix is

$$\frac{1}{2}\begin{pmatrix} 2 & -1 & 0 \\ -1 & 2 & -2 \\ 0 & -2 & 2 \end{pmatrix}$$

and is clearly of Lorentzian signature. If we remove the node number 1, we get the graph,

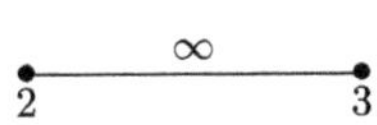

which is of affine type (A_1^+). If we remove the node number 2, we get the graph,

which is of finite type ($A_1 \oplus A_1$). And finally, if we remove the node 3, we get the graph

which is also of finite type (A_2), confirming hyperbolicity.

7.5.4 Crystallographic Coxeter Groups

We consider for definiteness Lorentzian Coxeter groups and work in the vector model. By definition, a Lorentzian Coxeter group is crystallographic if it stabilizes a lattice in the ambient Minkowski space $\mathbb{R}^{M,1}$. This lattice need not be the lattice generated by the unit vectors e_is orthonormal to the bounding hyperplanes $\tilde{H}_i$, because of normalization questions (see below).

As shown, e.g., in [95], a Coxeter group is crystallographic if and only if two conditions are satisfied: (i) the integers m_{ij} $(i \neq j)$ are restricted to the set $\{2, 3, 4, 6, \infty\}$, and (ii) for any closed circuit in the Coxeter graph of $\mathfrak{C}$, the number of edges labeled 4 or 6 is even. To impose that the integers m_{ij} $(i \neq j)$ are restricted to the set $\{2, 3, 4, 6, \infty\}$ is equivalent to requiring that the scalar products $e_i \cdot e_j$ $(i \neq j)$ are restricted to the set $\{0, -\frac{1}{2}, -\frac{\sqrt{2}}{2}, -\frac{\sqrt{3}}{2}, -1\}$. (When $m_{ij} = \infty$, we assume that $\tilde{H}_i$ and $\tilde{H}_j$ are parallel, i.e., $e_i \cdot e_j = -1$, as this is the case met in practice in supergravity models (when the value $m_{ij} = \infty$ arises). If the hyperplanes diverge $(e_i \cdot e_j < -1)$, one must require that $(e_i \cdot e_j)^2$ be an integer or a half-integer, but this situation, which can easily be covered, is not considered below.)

Given a crystallographic Coxeter group, it is easy to exhibit a lattice L stabilized by it. We can construct a basis for that lattice as follows. The basis vectors α_i of the lattice are multiples of the unit vectors e_i, $\alpha_i = c_i e_i$ for some scalars c_i which we determine by applying the following rules:

- $m_{ij} = 3 \Rightarrow c_i = c_j$;
- $m_{ij} = 4 \Rightarrow c_i = \sqrt{2} c_j$ or $c_j = \sqrt{2} c_i$;
- $m_{ij} = 6 \Rightarrow c_i = \sqrt{3} c_j$ or $c_j = \sqrt{3} c_i$;
- $m_{ij} = \infty \Rightarrow c_i = c_j$

where the choices of unequal c_ks must be adjusted around closed loops by consistency requirements, something that is guaranteed to be possible thanks to the second condition in the definition of a crystallographic Coxeter group. This can be seen by starting from an arbitrary node, say e_1, for which one takes $c_1 = 1$. One then proceeds to the next nodes in the (connected) Coxeter graph by applying the above rules. If there is no closed circuit, there is no consistency problem since there is only one way to proceed from e_1 to any given node. If there are closed circuits, one must make sure that one comes back to the same vector after one turn around any circuit. This can be arranged if the number of steps where one multiplies or divides by $\sqrt{2}$ (respectively, $\sqrt{3}$) is even.

One easily verifies that the coefficients A_{ij} appearing in the expression of the image of α_j by the reflection s_i in the hyperplane $\tilde{H}_i$,

$$s_i(\alpha_j) = \alpha_j - A_{ij}\alpha_i,$$

$$A_{ij} = 2\frac{\alpha_i \cdot \alpha_j}{\alpha_i \cdot \alpha_i},$$

are all integers. Hence the lattice $L \equiv \sum_i \mathbb{Z}\alpha_i$ is indeed stabilized.

The integers A_{ij} are called Cartan integers and related to the m_{ij}s as follows:

- $m_{ij} = 2 \Leftrightarrow A_{ij} = 0$;
- $m_{ij} = 3 \Leftrightarrow A_{ij} = -1$;
- $m_{ij} = 4 \Leftrightarrow A_{ij} = -1$ or $A_{ij} = -2$ with $A_{ij}A_{ji} = 2$;
- $m_{ij} = 6 \Leftrightarrow A_{ij} = -1$ or $A_{ij} = -3$ with $A_{ij}A_{ji} = 3$;
- $m_{ij} = \infty \Leftrightarrow A_{ij} = -2$.

If $c_i < c_j$, one has $A_{ij} > A_{ji}$ (if $A_{ij} \neq 0$). We should also point out that when $m_{ij} = \infty$, the choice $2c_i = c_j$ yielding $A_{ij} = -4$ and $A_{ji} = -1$ is also possible provided the number of edges labeled ∞ is even for any closed circuit in the Coxeter graph of $\mathfrak{C}$.

Our construction shows that the lattice L is not unique. If there are only two different lengths for the lattice vectors α_i, it is convenient to normalize the lengths so that the longest lattice vectors have length squared equal to two.

If the Coxeter graph is irreducible, as we assume, then the Cartan matrix with entries A_{ij} is *indecomposable.* A matrix A_{ij} is called *decomposable* if, after reordering of its indices, it decomposes as a nontrivial direct sum, i.e., if one can split the indices i, j in two sets J and Λ such that $A_{ij} = 0$ whenever $i \in J, j \in \Lambda$ or $i \in \Lambda, j \in J$. The indecomposability of the Cartan matrix follows from the fact that if it were decomposable, the corresponding Coxeter graph would be disconnected as no line would join a point in the set Λ to a point in the set J.

The rank 10 hyperbolic Coxeter groups are all crystallographic. The lattices preserved by E_{10} and DE_{10} are unique up to an overall rescaling because the nontrivial m_{ij} $(i \neq j)$ are all equal to 3 and there is no ambiguity in the ratios c_i/c_j, which are all equal to 1 (first rule above). The Coxeter group BE_{10} preserves two (dual) lattices corresponding to the two possible choices $c_i = \sqrt{2}c_j$ or $c_j = \sqrt{2}c_i$ permitted for the last two nodes, which have $m_{ij} = 4$.

7.5.5 Cartan Matrices and Dynkin Diagrams

The Cartan matrix $(A_{ij}) = \left(2\frac{\alpha_i \cdot \alpha_j}{\alpha_i \cdot \alpha_i}\right)$ formed by the Cartan integers encodes more information than the Coxeter scheme from which it originates, corresponding to the fact that the same Coxeter group, corresponding to a definite Coxeter scheme, can preserve different lattices, corresponding to different Cartan matrices.

Cartan matrices are central to the development of the theory of Kac–Moody algebras. Furthermore, it turns out that the billiard dynamics near a cosmological singularity provides more than just the Coxeter reflection group in the billiard walls, but a definite Cartan matrix. For this reason, it is useful to abstract the concept of "Cartan matrix."

An $n \times n$ matrix A is called a "generalized Cartan matrix" or just "Cartan matrix" if it satisfies the following conditions:*

$$A_{ii} = 2 \qquad \forall i = 1, \cdots, n, \tag{7.13}$$

$$A_{ij} \in \mathbb{Z}_- \qquad (i \neq j), \tag{7.14}$$

$$A_{ij} = 0 \quad \Rightarrow \quad A_{ji} = 0, \tag{7.15}$$

where $\mathbb{Z}_-$ denotes the non-positive integers.

We assume the Cartan matrix to be indecomposable.

The Cartan matrices originating from the billiard dynamics fulfill two additional conditions. (i) The first one is that $\det A \neq 0$; this is because the Gram matrix $(e_I \cdot e_j)$ has also a non-vanishing determinant. (ii) The second one is that A is symmetrizable, i.e., that there exists an invertible diagonal matrix D with positive elements ϵ_i and a symmetric matrix S such that

$$A = DS. \tag{7.16}$$

The matrix S is called a symmetrization of A. That A is symmetrizable just follows from its construction from the Gram matrix.

Note that the symmetrization S of A is unique up to an overall positive factor because A is indecomposable. To prove this, choose the first (diagonal) element $\epsilon_1 > 0$ of D arbitrarily. Since A is indecomposable, there exists a nonempty set J_1 of indices j such that $A_{1j} \neq 0$. One has $A_{1j} = \epsilon_1 S_{1j}$ and $A_{j1} = \epsilon_j S_{j1}$. This fixes the ϵ_js > 0 in terms of ϵ_1 since $S_{1j} = S_{j1}$. If not all the elements ϵ_j are determined at this first step, we pursue the same construction with the elements $A_{jk} = \epsilon_j S_{jk}$ and $A_{kj} = \epsilon_k S_{kj} = \epsilon_k S_{kj}$ with $j \in J_1$ and, more generally, at step p, with $j \in J_1 \cap J_2 \cdots \cap J_p$. As the matrix A is assumed to be indecomposable, all the elements ϵ_i of D and S_{ij} of S can be obtained, depending only on the choice of ϵ_1. One gets no contradicting values for the ϵ_js because the matrix A is assumed to be symmetrizable.

In the symmetrizable case, one can characterize the Cartan matrix according to the signature of (any of) its symmetrization(s). For gravitational billiards, S has Lorentzian signature – the signature of the generalized De Witt supermetric.

One can encode the Cartan matrix in terms of a Dynkin diagram, which is obtained as follows:

1. For each $i = 1, \cdots, n$, one associates a node in the diagram.
2. One draws a line between the node i and the node j if $A_{ij} \neq 0$; if $A_{ij} = 0$ $(= A_{ji})$, one draws no line between i and j.
3. One writes the pair (A_{ij}, A_{ji}) over the line joining i to j. When the products $A_{ij} \cdot A_{ji}$ are all ≤ 4 (which is the only situation we shall meet in practice), this third rule can be replaced by the following rules:

* We are employing the convention of Kac [111] for the Cartan matrix. There exists an alternative definition of Kac–Moody algebras in the literature, in which the transposed matrix A^T is used instead.

Table 7.4: BE_{10} *and* $A_{15}^{(2)+}$.

Name	Coxeter graph
$B_8^{++} \equiv BE_{10}$	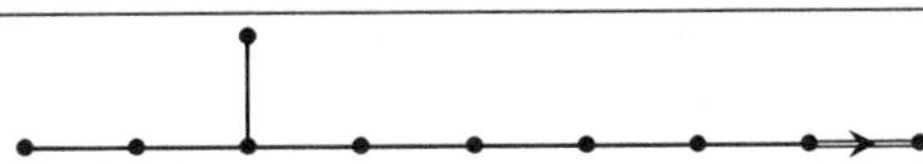
$A_{15}^{(2)+}$	

(a) one draws a number of lines between i and j equal to $\max(|A_{ij}|, |A_{ji}|)$;
(b) one draws an arrow from j to i if $|A_{ij}| > |A_{ji}|$.

For instance, there are two Dynkin diagrams associated with the Coxeter group BE_{10}, denoted by B_8^{++} and $A_{15}^{(2)+}$. The first choice makes all α_is of length squared two, except the last one, which is shorter and of length squared one. The other choice is opposite and makes all α_is "short" and of length squared one, with the last α_i "long" and of length squared two. Both Dynkin diagrams are given in Table 7.4.

The lattices of B_8^{++} and of $A_{15}^{(2)+}$ are both preserved by the same Coxeter group of reflections in $\mathbb{R}^{9,1}$.

7.5.6 Weyl Groups of Kac–Moody Algebras

Simplex, crystallographic Coxeter groups have an additional important property: they can be identified with the Weyl groups of Kac–Moody algebras, traditionally denoted in the same way.

Kac–Moody algebras are infinite-dimensional Lie algebras that generalize familiar finite-dimensional simple Lie algebras [111]. The infinite-dimensional generalization preserves the key property of possessing a triangular decomposition such that the finite-dimensional concepts of Cartan subalgebra, raising and lowering operators, roots, positive roots, Weyl group, etc. can all be introduced.

A Kac–Moody algebra is defined in terms of generators and relations involving a Cartan matrix A. The Cartan matrix completely determines the Kac–Moody algebra. The relation between a simplex, crystallographic Coxeter group and the Kac–Moody algebra of which it is the Weyl group is that they share the same Cartan matrix. Actually, since different Cartan matrices might correspond to the same reflection group, there might be more than one Kac–Moody algebra associated with a given Coxeter group; different Kac–Moody algebras might have the same Weyl group. This phenomenon is in fact already present in the finite-dimensional case (B_n and C_n).

The definition of Kac–Moody algebras and some of their properties are given in Appendix D. Let us simply mention here that in the Lie algebra context, the Coxeter reflections are called "Weyl reflections," the fundamental domain given by the Coxeter simplex is called "the fundamental Weyl chamber" and the vector α_i orthogonal to the Coxeter wall are identified with the simple roots. The Cartan matrix and the Dynkin diagram that encodes it provide the information about the relative lengths and angles made by the simple roots.

7.6 Coxeter Groups Associated with Gravitational Theories

We have seen that the dynamics of gravitational theories near a spacelike singularity can be phrased in terms of a billiard motion in a region of hyperbolic space which is called the billiard table.

What makes the connection with Kac–Moody algebras possible for the theories of physical interest (pure gravity, supergravities) is that the billiard table has the remarkable properties:

- it is a Coxeter polyhedron;
- the polyhedron is in fact a simplex;
- the Coxeter group generated by reflections in its faces is crystallographic.

Thus, this Coxeter group is equal to the Weyl group of some Kac–Moody algebra. For the most interesting theories, a fourth property holds:

- the Coxeter group is hyperbolic.

Consequently, the corresponding Kac–Moody algebra is also hyperbolic (see Appendix D) and not just Lorentzian.

These remarkable properties are not automatic. It is the purpose of this subsection to indicate how they arise.

Crucial in the analysis is the fact established in the previous chapter that the billiards for all gravitational theories are defined by convex polyhedra lying on the positive side of hyperplanes in $\mathbb{R}^{M,1}$. These hyperplanes are the zero loci of linear forms that come with a definite normalization dictated by the Lagrangian. The fact that it is $\mathbb{R}^{M,1}$ that is relevant for the computations – leading to the emergence of hyperbolic space $\mathbb{H}^M$ – comes from the De Witt supermetric in the space of the scale factors (including the dilatons, if any), which indeed has a Lorentzian signature. Dilations of the spatial metric define timelike directions, while deformations of the metric that do not change its volume, or matter fields deformations, correspond to spacelike directions. Gravity is thus essential to get the Lorentzian signature and hence, hyperbolic space.

Given the billiard table, the De Witt supermetric enables one to compute the matrix

$$A_{ij} = 2\frac{(\alpha_i, \alpha_j)}{(\alpha_i, \alpha_j)}$$

of the scalar products of the wall forms $\alpha_i(\beta)$. If the matrix elements A_{ij} are all integers, the reflection group is crystallographic.

We shall now examine the reasons that make each of the above properties emerge.

7.6.1 Pure Gravity in $D = d + 1$ Space-Time Dimensions

Since gravity is essential, and is always part of the theory, while the menu of p-forms is theory-dependent, we start with pure gravity in $D = d + 1$ dimensions. This already reveals key features.

In $D = d + 1$ dimensions, there are d scale factors β^a in the Minkowski space $\mathbb{R}^{d-1,1}$ and the motion takes place in hyperbolic space $\mathbb{H}_{d-1}$. As we have seen in the previous chapter, the billiard table of pure gravity is the region on the positive side of the following walls:

- Symmetry walls:

$$w_{ab} \equiv \beta^b - \beta^a \geq 0, \quad b > a. \tag{7.17}$$

- Curvature walls:

$$\alpha_{a|b|c} \equiv 2\beta^a + \sum_{e \neq a,b,c} \beta^e \geq 0 \quad (a \neq b, a \neq c, b \neq c). \tag{7.18}$$

The scalar products are computed with the De Witt supermetric. For handy reference, we recall the formula: if $x(\beta) = \sum_{a=1}^{d} x_a \beta^a$ and $y(\beta) = \sum_{a=1}^{d} y_a \beta^a$, then

$$(x, y) = \sum_{a=1}^{d} x_a y_a - \frac{1}{d-1} \left(\sum_{a=1}^{d} x_a \right) \left(\sum_{a=1}^{d} x_a \right). \tag{7.19}$$

The inequalities (7.17) and (7.18) are not all independent. As we have also seen in the previous chapter, the following symmetry walls are dominant,

$$\alpha_i \equiv \beta^{i+1} - \beta^i \geq 0, \quad i = 1, \cdots, d-1. \tag{7.20}$$

In the region defined by (7.20), there is only one dominant curvature wall, namely

$$\alpha_d \equiv \alpha_{1|d-1|d} = 2\beta^1 + \beta^2 + \cdots \beta^{d-2} \geq 0. \tag{7.21}$$

All other curvature inequalities are consequences of (7.20) and (7.21). This is because α_d is the smallest of the sums (7.18) once the β^a are ordered as in (7.20).

In fact, all curvature walls can be obtained from α_d through permutations of the indices. This is a general rule: the walls of a given type (curvature, p-electric, p-magnetic) can be obtained from any one of them of the same type through permutations. This is the imprint left by diffeomorphism invariance in the space of the walls. As the permutations are the transformations generated by the reflections in the symmetry walls, we see that these walls play a distinguished role.

From this observation, we can anticipate what will be the situation when matter fields are included. The symmetry walls (7.20) will always be present among the dominant walls. Among the other walls, only one per given type, the "smallest," will then have to be considered. This smallest wall may or may not be dominant. It must be kept as relevant billiard wall only in the former case, and can be dropped in the latter case.

In pure gravity, there is only one extra type, the curvature walls, and among them, the smallest (7.21) is the only dominant one. Out of the $\frac{d(d-1)}{2} + \frac{d(d-1)(d-2)}{2} = \frac{d(d-1)^2}{2}$ walls, only $(d-1)+1 = d$ are dominant. These d walls define a simplex in hyperbolic space $\mathbb{H}_{d-1}$.

Cartan Matrix and Dynkin Diagram

The angles made by the faces of the simplex can be computed from the matrix $A_{ij} = 2\frac{(\alpha_i,\alpha_j)}{(\alpha_i,\alpha_i)}$ (where the scalar products are computed with the De Witt supermetric), which is equal to $A_{ij} = (\alpha_i, \alpha_j)$ since the symmetry and curvature walls have squared length equal to 2. One gets explicitly

$$\begin{pmatrix} 2 & -1 & 0 & \cdots & 0 & 0 & -1 \\ -1 & 2 & -1 & \cdots & 0 & 0 & 0 \\ 0 & -1 & 2 & \cdots & 0 & 0 & 0 \\ \vdots & \vdots & \vdots & \vdots & \vdots & \vdots & \vdots \\ 0 & 0 & 0 & \cdots & 2 & -1 & -1 \\ 0 & 0 & 0 & \cdots & -1 & 2 & 0 \\ -1 & 0 & 0 & \cdots & -1 & 0 & 2 \end{pmatrix} \tag{7.22}$$

which is the Cartan matrix of the Kac–Moody algebra A^{++}_{d-2}, with Dynkin diagram

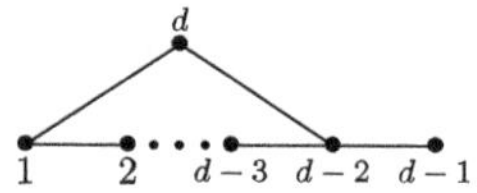

The Dynkin diagram of A^{++}_{d-2}

In the case $d = 3$, the nodes 1 and $d-2$ coincide, and the Dynkin diagram collapses to that of A^{++}_1,

The Dynkin diagram of A^{++}_1

with Cartan matrix

$$\begin{pmatrix} 2 & -1 & -2 \\ -1 & 2 & 0 \\ -2 & 0 & 2 \end{pmatrix}. \tag{7.23}$$

In both cases, one gets a Coxeter simplex since the angles between the faces are acute and given by $\frac{\pi}{2}$ and $\frac{\pi}{3}$ $(d > 3)$ or $\frac{\pi}{2}$, $\frac{\pi}{3}$ and $\frac{\pi}{\infty}$ $(d = 3)$. Since all wall forms have equal length, the Coxeter graph coincides with the Dynkin diagram for $d > 3$, and for that reason we shall sometimes use the terminology "Dynkin–Coxeter diagrams." For $d = 3$, the Coxeter graph is the Coxeter graph of $PGL(2,\mathbb{Z})$ drawn above and studied extensively in Section 5.10.

The importance of the symmetry walls for getting a Coxeter simplex cannot be overstressed. That these are needed to get a simplex is not obvious in four space-time dimensions where the number of curvature walls is equal to three, so that the polyhedron formed by the curvature walls happens to be a simplex – the billiard table of diagonal models, which is an ideal triangle in $\mathbb{H}_2$. However, this is not true anymore in higher space-time dimensions. For instance, in five space-time dimensions, there are twelve curvature walls given by $2\beta^a + \beta^b$ $(a, b = 1, 2, 3, 4,\ a \neq b)$. All these walls are relevant and so do not define a simplex. Furthermore, some dihedral angles are even obtuse and equal to $\frac{2\pi}{3}$.

Given their importance, one calls the Dynkin subdiagram formed by the symmetry walls the "gravity line," as these are brought in by gravity (and ultimately, diffeomorphism invariance). It is

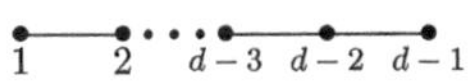

The Dynkin diagram of A_{d-1}

The gravity line is the Dynkin diagram of the simple Lie group $SL(d)$, corresponding to the fact that the Weyl group of $SL(d)$ is the permutation group of d objects.

Hyperbolicity

We now turn to hyperbolicity and determine, using the hyperbolicity criterion of Subsection 7.5.3, that the Coxeter groups A^{++}_{d-2} are hyperbolic provided $d \leq 9$. For $d \geq 10$, they are not hyperbolic any more.

Indeed, the Dynkin–Coxeter diagram of A_8^{++} is

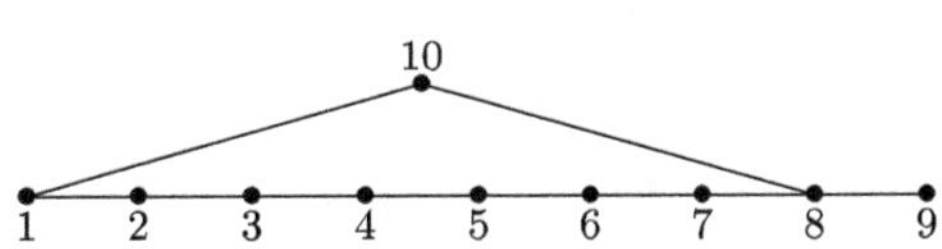

The Dynkin–Coxeter diagram of A_8^{++}

If one removes node 4, one gets the Coxeter–Dynkin diagram

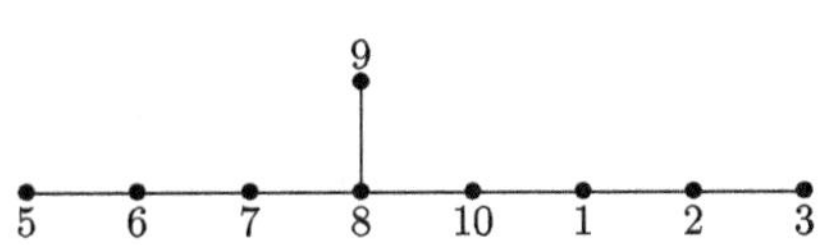

which is not in the list of Coxeter diagrams of affine or finite type given above. It is in fact E_7^{++} in the classification given below. The same is actually true if one removes node 3. Hence, A_8^{++} is not hyperbolic. A similar argument shows that A_k^{++} is not hyperbolic for $k \geq 8$.

For A_7^{++}, however, the situation is different. The Dynkin–Coxeter graph is

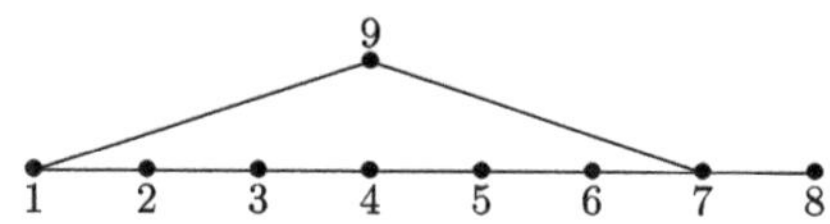

The Dynkin–Coxeter diagram of A_7^{++}

This defines a hyperbolic Coxeter group. Indeed,

- If one removes the node 1, one gets D_8, which is finite;
- If one removes the node 2, one gets E_8, which is finite;
- If one removes the node 3, one gets E_7^+, which is affine;
- If one removes the node 4, one gets E_8, which is finite;
- If one removes the node 5, one gets D_8, which is finite;
- If one removes the node 6, one gets A_8, which is finite;
- If one removes the node 7, one gets $A_7 \oplus A_1$, which is finite;
- If one removes the node 8, one gets A_7^+, which is affine;
- If one removes the node 9, one gets A_8, which is finite.

Therefore, the Coxeter group A_7^{++} is hyperbolic, as are all the A_k^{++} with $k \leq 7$.

In terms of the billiard motion, hyperbolicity means finite volume for the billiard table, while non-hyperbolicity means an infinite volume for the billiard table. In the first case, the never-ending chaotic BKL behavior prevails. In the second case, chaos is absent and the system settles for ever (as one goes towards the singularity) in an unperturbed Kasner regime. We thus see that the phenomenon uncovered in [63, 62] on the disappearance of the BKL behavior for pure gravity in space-time dimensions $D \geq 11$ ($d \geq 10$) has a simple algebraic interpretation in terms of Coxeter groups and Kac–Moody algebras [48].

7.6.2 Including p-Forms (No Scalar)

Adding matter fields to gravity generically destroys the Coxeter properties that we have exhibited. However, in the case of (extended) supergravity models, the space-time dimension, the menu of p-forms and the dilaton couplings conspire to always yield a simplex, crystallographic, Coxeter group.

To emphasize that this is quite exceptional, we consider first the explicit case of a single 3-form coupled to gravity in $D = d+1$ space-time dimensions. This case is of course motivated by M-theory and maximal supergravity, the bosonic sector of which is precisely described in $D = 11$ dimensions by the gravity-3-form system.

Its Lagrangian is given in [41]. We leave the dimension unspecified here, however, in order to show how special and spectacular $D = 11$ is from the Coxeter group perspective. We only assume the space-time dimension to be at least equal to 6 since, below 6, a 3-form is equivalent to a scalar in $D = 5$ (more precisely, a dilaton, since there is no other scalar) or carries no local degree of freedom at all ($D \leq 4$).

In addition to the symmetry walls (7.17) and the curvature walls (7.18), one has electric walls

$$e_{abc} \equiv \beta^a + \beta^b + \beta^c, \quad (a \neq b, a \neq c, b \neq c) \geq 0 \tag{7.24}$$

and magnetic walls

$$m_{a_1 a_2 \cdots a_{d-4}} \equiv \beta^{a_1} + \beta^{a_2} + \cdots + \beta^{a_{d-4}} \geq 0, \quad (a_i \neq a_j \text{ for } i \neq j). \tag{7.25}$$

Because of the symmetry walls, we need only consider e_{123} among the electric walls and $m_{123 \cdots d-4}$ among the magnetic walls. The billiard is defined by the dominant symmetry walls (7.20) and those walls that are dominant among the electric wall e_{123}, the magnetic wall $m_{123 \cdots d-4}$ and the gravitational wall $\alpha_{1|d-1|d}$.

It is easy to see that the curvature wall is subdominant, because

$$\alpha_{1|d-1|d} = e_{123} + m_{145 \cdots d-2}$$

and so the inequalities $e_{123} \geq 0$ and $m_{123 \cdots d-4} \geq 0$ together with (7.20) imply $\alpha_{1|d-1|d} \geq 0$. The subdominance of the curvature walls holds because there are nontrivial magnetic walls ($d > 4$) and would still be true in the presence of a dilaton coupled to the 3-form, because the dilaton term comes with opposite signs in electric and magnetic walls and so, cancel in the sum $e_{123} + m_{145 \cdots d-2}$.

More generally, one shows by exactly the same argument that *the curvature walls are always subdominant in the presence of p-forms, in the dimensions where these are not dual to scalars.*

Let us come back to our coupled gravity + 3-form system. Except when $d = 7$ in which case electric and magnetic walls coincide, either the electric walls are dominant ($d > 7$) or the magnetic walls are dominant ($d = 5$ or $d = 6$). This is because one can write the walls with more exponents as linear combinations with positive coefficients of the walls with less exponents, e.g.,

- $d = 5$:

$$e_{123} = m_1 + m_2 + m_3$$

- $d = 6$:

$$e_{123} = \frac{1}{2}(m_{12} + m_{23} + m_{13})$$

- $d = 8$:

$$m_{1234} = \frac{1}{3}(e_{123} + e_{124} + e_{134} + e_{234})$$

etc. Thus, in space-time dimensions 6 and 7, the billiard is defined by the dominant symmetry walls (7.20) and the magnetic wall $m_1 \geq 0$ ($D = 6$) or $m_{12} \geq 0$ ($D = 7$). In space-time dimensions $D \geq 9$, the billiard is defined by the dominant symmetry walls (7.20) and the electric wall $e_{123} \geq 0$. And in space-time dimension 8, where electric-magnetic duality is a symmetry, the billiard is defined by the dominant symmetry walls (7.20) and the wall $\beta^1 + \beta^2 + \beta^3 \geq 0$, which is at the same time electric ($e_{123} \geq 0$) and magnetic ($m_{123} \geq 0$) ($D = 8$).

This implies that the billiard table is a simplex.

Chaos Is Present . . . But the Billiard Table Is Almost Never a Coxeter Simplex!

It is proved in [46], using the equivalent formulation in terms of Kasner exponents, that the billiard table has finite volume. Thus, the billiard of the coupled gravity + 3-form system is a finite volume simplex in $D \geq 6$.

More generally, for gravity coupled to a single p-form ($p \geq 1$), the billiard table is a finite-volume simplex in the space-time dimensions strictly above the dimension $p+2$ where the p-form is equivalent to a dilaton. This is demonstrated explicitly in [44] to which we refer the reader for the details. It follows from this analysis that p-forms (without additional scalar fields) bring in chaos. (The restriction that there is no scalar field comes because these can eliminate chaos, as we have seen in the previous chapters.)

One can include more than one species of p-forms (but no scalar, even in disguise) and reach the same conclusions. Indeed, among the walls brought in by the p-forms, there will always be only one dominant wall in addition to the dominant symmetry walls: the electric or magnetic wall with the smallest number of β^as (which will be always equal to at least one).

When is the billiard table a Coxeter simplex? Since finite volume Coxeter simplices exist only in hyperbolic space of dimension ≤ 9, the answer to this question is "almost never" because in space-time dimension D, the relevant hyperbolic space is $\mathbb{H}_{D-2}$, and $D-2$ exceeds 9 whenever the space-time dimension exceeds 11. Thus, the coupled gravity + p-form system (without scalar fields) can be described by a Coxeter simplex only in space-time dimension ≤ 11.

The Special Role of $D = 11$ in the 3-Form Case

In fact, for a 3-form, one gets a Coxeter simplex only when the space-time dimension is equal to 11. And furthermore, the Coxeter group is then crystallographic and isomorphic to the Weyl group of E_{10}. In the other space-time dimensions ≤ 11, the simplex is not a Coxeter one. To repeat the words of [44]: "From this point of view, $D = 11$ is thus quite special for the gravity-3-form system, irrespective of any supersymmetry consideration. If one did not know about supergravity, but had some independent reason for considering a 3-form, one could discover from this independent line of insight the peculiar role of $D = 11$." We recall that the gravity-3-form system is the bosonic sector of eleven-dimensional supergravity, and that the dimension $D = 11$ is derived in that context from supersymmetry considerations.

The verification of these statements on the peculiar role played by $D = 11$ is a straightforward exercise, which is left to the reader – the explicit derivation is given in [44]. We shall only prove here that it is indeed the Weyl group of E_{10} that emerges in $D = 11$.

The billiard walls are then the dominant symmetry walls (7.20) and the electric wall $\alpha_{10} \equiv e_{123}$ which has squared length equal to two. This electric wall is orthogonal to all dominant symmetry walls, except $\alpha_3 = \beta^4 - \beta^3$, with which it has a scalar product equal to -1. This yields the Dynkin–Coxeter diagram of E_{10},

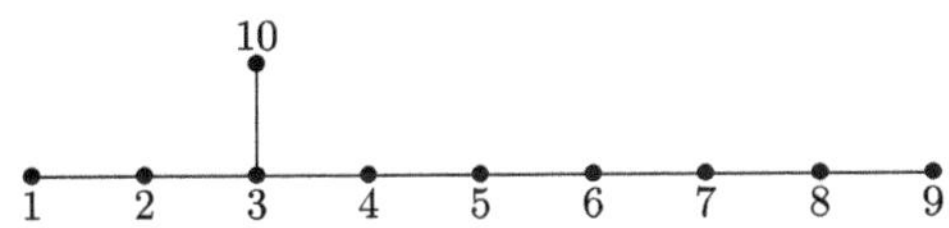

with Cartan matrix

$$\begin{pmatrix} 2 & -1 & 0 & 0 & 0 & 0 & 0 & 0 & 0 & 0 \\ -1 & 2 & -1 & 0 & 0 & 0 & 0 & 0 & 0 & 0 \\ 0 & -1 & 2 & -1 & 0 & 0 & 0 & 0 & 0 & -1 \\ 0 & 0 & -1 & 2 & -1 & 0 & 0 & 0 & 0 & 0 \\ 0 & 0 & 0 & -1 & 2 & -1 & 0 & 0 & 0 & 0 \\ 0 & 0 & 0 & 0 & -1 & 2 & -1 & 0 & 0 & 0 \\ 0 & 0 & 0 & 0 & 0 & -1 & 2 & -1 & 0 & 0 \\ 0 & 0 & 0 & 0 & 0 & 0 & -1 & 2 & -1 & 0 \\ 0 & 0 & 0 & 0 & 0 & 0 & 0 & -1 & 2 & 0 \\ 0 & 0 & -1 & 0 & 0 & 0 & 0 & 0 & 0 & 2 \end{pmatrix} \tag{7.26}$$

7.6.3 Introducing Scalars

As we have seen, the scalars can eliminate chaos and render the billiard table of infinite volume. This depends on the values of the various dilaton couplings. Since these couplings are continuous parameters, the angles between the walls depend also on continuous parameters, and for a generic system involving dilatons, the billiard table will not be a Coxeter polyhedron.

All situations can arise, as one can see by dimensionally reducing the previous dilaton-free models. If one dimensionally reduces pure gravity from dimension above the critical dimension 11 down to a lower dimension, one gets a coupled system involving gravity, dilatons, 0-forms and 1-forms that define a Coxeter simplex of infinite volume. By slightly changing the dilaton couplings, one destroys the Coxeter property but keeps the infinite volume. Similarly, if one dimensionally reduces the gravity-3-form system from 11 down to a lower dimension, one gets a system with dilatons and p-forms corresponding to the same simplex Coxeter group E_{10}, but if one starts from $D \neq 11$, one gets a finite-volume billiard that does not enjoy the Coxeter property. (Dimensional reduction of billiard systems is studied at length in [44].)

We shall consider here two models that illustrate the main points and that give rise to the other two hyperbolic Coxeter groups of rank 10. These models are pure supergravity in $D = 10$ dimensions, and the same model coupled to the super-Maxwell multiplet. Both models are related to string theory and involve one dilaton ϕ. The relevant hyperbolic space is thus again $\mathbb{H}_9$ embedded in $\mathbb{R}^{9,1}$ since the dilaton counts as a scale factor. The scalar products are computed with the modified De Witt supermetric that takes into account the dilaton contribution. We recall that these scalar products are computed as follows: if $x(\beta,\phi) = \sum_{a=1}^{9} x_a\beta^a + \lambda\phi$ and $y(\beta,\phi) = \sum_{a=1}^{9} y_a\beta^a + \mu\phi$, then

$$(x, y) = \sum_{a=1}^{9} x_a y_a - \frac{1}{8}\left(\sum_{a=1}^{9} x_a\right)\left(\sum_{a=1}^{9} x_a\right) + \lambda\mu. \tag{7.27}$$

These models, with their Lagrangians, are, for instance, given in [142].

The Coxeter Group $DE_{10} \equiv D_8^{++}$

Pure supergravity contains a 2-form in addition to the dilaton and the metric. Therefore, the billiard table is determined by the symmetry walls and the curvature walls of gravity, as well as the electric walls

$$e_{ab} = \beta^a + \beta^b + \frac{\sqrt{2}}{2}\phi \tag{7.28}$$

(where the dilaton couplings are read from [142]) and the magnetic walls

$$m_{a_1\cdots a_6} = \beta^{a_1} + \beta^{a_2} + \beta^{a_3} + \beta^{a_4} + \beta^{a_5} + \beta^{a_6} - \frac{\sqrt{2}}{2}\phi \tag{7.29}$$

brought in by the 2-form. The dominant walls are the dominant symmetry walls (7.20), as always, as well as the smallest walls of each type (electric and magnetic), i.e.,

$$\alpha_9 \equiv e_{12} = \beta^1 + \beta^2 + \frac{\sqrt{2}}{2}\phi \tag{7.30}$$

and

$$\alpha_{10} \equiv m_{123456} = \beta^1 + \beta^2 + \beta^3 + \beta^4 + \beta^5 + \beta^6 - \frac{\sqrt{2}}{2}\phi. \tag{7.31}$$

These walls are independent because of the dilaton terms. They define a simplex.

A straightforward computation reveals that the Cartan matrix defined by these walls is given by

$$\begin{pmatrix} 2 & -1 & 0 & 0 & 0 & 0 & 0 & 0 & 0 & 0 \\ -1 & 2 & -1 & 0 & 0 & 0 & 0 & 0 & -1 & 0 \\ 0 & -1 & 2 & -1 & 0 & 0 & 0 & 0 & 0 & 0 \\ 0 & 0 & -1 & 2 & -1 & 0 & 0 & 0 & 0 & 0 \\ 0 & 0 & 0 & -1 & 2 & -1 & 0 & 0 & 0 & 0 \\ 0 & 0 & 0 & 0 & -1 & 2 & -1 & 0 & 0 & -1 \\ 0 & 0 & 0 & 0 & 0 & -1 & 2 & -1 & 0 & 0 \\ 0 & 0 & 0 & 0 & 0 & 0 & -1 & 2 & 0 & 0 \\ 0 & -1 & 0 & 0 & 0 & 0 & 0 & 0 & 2 & 0 \\ 0 & 0 & 0 & 0 & 0 & -1 & 0 & 0 & 0 & 2 \end{pmatrix} \tag{7.32}$$

with the Dynkin–Coxeter diagram of $DE_{10} \equiv D_8^{++}$,

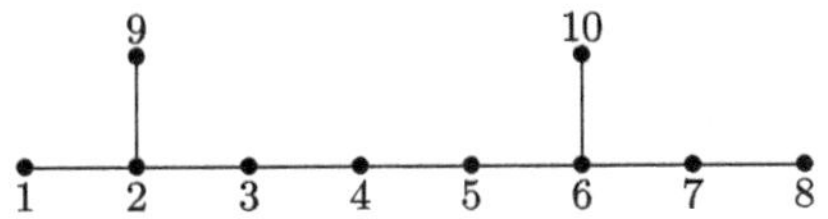

The Coxeter Group $BE_{10} \equiv B_8^{++}$

Supergravity coupled to a super-Maxwell multiplet contains the same 2-form and a 1-form in addition to the dilaton and the metric. Therefore, the billiard table is determined by the symmetry walls and the curvature walls of gravity, the electric walls

$$e_{ab} = \beta^a + \beta^b + \frac{\sqrt{2}}{2}\phi \tag{7.33}$$

and the magnetic walls

$$m_{a_1 \cdots a_6} = \beta^{a_1} + \beta^{a_2} + \beta^{a_3} + \beta^{a_4} + \beta^{a_5} + \beta^{a_6} - \frac{\sqrt{2}}{2}\phi \tag{7.34}$$

of the 2-form, and the electric walls

$$\tilde{e}_a = \beta^a + \frac{\sqrt{2}}{4}\phi \tag{7.35}$$

and the magnetic walls

$$\tilde{m}_{a_1 \cdots a_7} = \beta^{a_1} + \beta^{a_2} + \beta^{a_3} + \beta^{a_4} + \beta^{a_5} + \beta^{a_6} + \beta^{a_7} - \frac{\sqrt{2}}{4}\phi \tag{7.36}$$

brought in by the 1-form.

The dominant walls are the dominant symmetry walls (7.20), as always, as well as the smallest electric wall of the 1-form and the smallest magnetic wall of the 2-form, i.e.,

$$\alpha_9 \equiv \tilde{e}_1 = \beta^1 + \frac{\sqrt{2}}{4}\phi \tag{7.37}$$

and

$$\alpha_{10} \equiv m_{123456} = \beta^1 + \beta^2 + \beta^3 + \beta^4 + \beta^5 + \beta^6 - \frac{\sqrt{2}}{2}\phi. \tag{7.38}$$

These walls define a simplex.

A straightforward computation shows that the Cartan matrix is now

$$\begin{pmatrix} 2 & -1 & 0 & 0 & 0 & 0 & 0 & 0 & -1 & 0 \\ -1 & 2 & -1 & 0 & 0 & 0 & 0 & 0 & 0 & 0 \\ 0 & -1 & 2 & -1 & 0 & 0 & 0 & 0 & 0 & 0 \\ 0 & 0 & -1 & 2 & -1 & 0 & 0 & 0 & 0 & 0 \\ 0 & 0 & 0 & -1 & 2 & -1 & 0 & 0 & 0 & 0 \\ 0 & 0 & 0 & 0 & -1 & 2 & -1 & 0 & 0 & -1 \\ 0 & 0 & 0 & 0 & 0 & -1 & 2 & -1 & 0 & 0 \\ 0 & 0 & 0 & 0 & 0 & 0 & -1 & 2 & 0 & 0 \\ -2 & 0 & 0 & 0 & 0 & 0 & 0 & 0 & 2 & 0 \\ 0 & 0 & 0 & 0 & 0 & -1 & 0 & 0 & 0 & 2 \end{pmatrix} \tag{7.39}$$

with the Dynkin–Coxeter diagram of $BE_{10} \equiv B_8^{++}$,

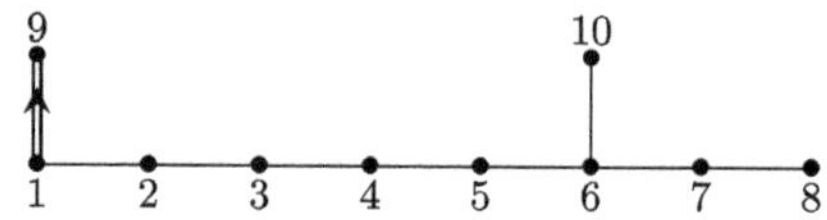

We thus see that all maximum rank-10 simplex hyperbolic Coxeter groups are realized in supergravity models related to string theories. In terms of the Kac–Moody algebras, we get three out of four rank-10 hyperbolic algebras. The missing one is the dual to BE_{10}, which is denoted as we have seen by $A_{15}^{(2)+}$ (with the arrow pointing in the reversed direction). It has the same Weyl group. It is not known whether there is a field-theoretical model that realizes it at the time of writing.

Among the rank-10 hyperbolic algebras, $E_{(10)}$ possesses notable properties. These are reviewed in Appendix D and make maximal supergravity somewhat privileged.

The billiard tables of other supergravity models have been investigated in [44, 88], to which we refer for more information.

7.7 The Kac–Moody Symmetry Conjecture

With the identification of the billiard group with the Weyl group of a Lorentzian Kac–Moody algebra, the motion of the scale factors near the spacelike singularity can be viewed as taking place in the fundamental Weyl chamber of the (Cartan subalgebra of the) Kac–Moody algebra.

We have emphasized in this section that the emergence of Weyl groups is by no means automatic, since it requires a fine tuning of the space-time dimension, of the menu of p-forms and of the dilaton couplings. While the billiard description is always valid for generic choices of these parameters, it is only for specific choices that one gets the Weyl group of a Lorentzian Kac–Moody algebra. It is in that sense that this emergence is remarkable.

This is the reason why it is expected that the Coxeter groups might signal a bigger symmetry structure. The explicit appearance of infinite crystallographic Coxeter groups in the billiard limit suggests indeed that gravitational theories might be invariant under the huge symmetry described by the corresponding Lorentzian Kac–Moody algebras.

While the precise form of the conjecture has not been formulated yet, so that its demonstration is still a task for the future, attempts to reformulate the gravitational Lagrangians in a way that makes the conjectured symmetry manifest has yielded very intriguing and encouraging results. However, to the extent that the aim of this book is to describe well-established properties, which are facts, in contrast to features that are still conjectural and need further development, we shall not report these fascinating results here. We refer to [49, 51] for more information.

The existence of "hidden symmetries" underlying gravitational theories goes back to the pioneering work by Ehlers [69] and Geroch [78, 79] (see [30, 137] for more information). The subject received a remarkable boost through the discovery of the E_7 duality symmetry in maximal supergravity [41, 39]. That this result could be the signal of the existence of a more profound symmetry structure was then discussed in [106, 107, 108] and triggered further interest in [136, 110]. The emergence of E_{10} through the cosmological billiards was uncovered in [47] and how to make the symmetry manifest was explored in [49]. A possible role of the non-hyperbolic algebra E_{11} was advocated in [109, 155].

Appendices

Note on Appendices

In the appendices, the following conventions are adopted: conventions of Part I hold in Appendices A, B and C. Conventions of Part II hold in Appendix D.

Appendix A

Various Technical Derivations

A.1 Perturbations to Kasner-Like Asymptotics

To find the linear perturbation to the Kasner-like regime we write

$$g_{\alpha\beta} = g^{(0)}_{\alpha\beta} + h_{\alpha\beta} \,, \tag{A.1}$$

where $g^{(0)}_{\alpha\beta}$ is the metric tensor (1.20) of the leading approximation and $h_{\alpha\beta}$, together with all its derivatives, are considered as small quantities. Indices in $h_{\alpha\beta}$ are raised and lowered with the unperturbed metric $g^{(0)}_{\alpha\beta}$. To calculate $h_{\alpha\beta}$, we write the exact vacuum gravitational equations (1.17) in the form

$$\dot{\kappa}_{\alpha\beta} + \frac{1}{2}\kappa\kappa_{\alpha\beta} - \kappa_{\alpha\gamma}\kappa^{\gamma}_{\beta} = -2P_{\alpha\beta} \tag{A.2}$$

and demand that the correction to the main part of these equations, that is to its left-hand side, is generated by the right-hand side calculated with the metric $g^{(0)}_{\alpha\beta}$. In this way we obtain the following equations for the corrections:

$$\ddot{h}_{\bar{\alpha}\bar{\alpha}} + \frac{1-4p_{\bar{\alpha}}}{t}\dot{h}_{\bar{\alpha}\bar{\alpha}} + \frac{4p^2_{\bar{\alpha}}}{t^2}h_{\bar{\alpha}\bar{\alpha}} + p_{\bar{\alpha}}t^{2p_{\bar{\alpha}}-1}\dot{h} = -2P^{(0)}_{\bar{\alpha}\bar{\alpha}}, \; (\textit{no sum over } \bar{\alpha}), \tag{A.3}$$

and

$$\ddot{h}_{\bar{\alpha}\bar{\beta}} + \frac{1-2p_{\bar{\alpha}}-2p_{\bar{\beta}}}{t}\dot{h}_{\bar{\alpha}\bar{\beta}} + \frac{4p_{\bar{\alpha}}p_{\bar{\beta}}}{t^2}h_{\bar{\alpha}\bar{\beta}} = -2P^{(0)}_{\bar{\alpha}\bar{\beta}}, \; \bar{\alpha} \neq \bar{\beta}, (\textit{no sum over } \bar{\alpha}, \bar{\beta}). \tag{A.4}$$

In these equations we already performed the projections of the three-dimensional tensors into the triad $l^{\bar{\alpha}}_{\alpha}(x^{\mu})$. The quantity h in (A.3) is the contraction $h = h^{\bar{\mu}}_{\bar{\mu}} = \sum_{\bar{\mu}} t^{-2p_{\bar{\mu}}} h_{\bar{\mu}\bar{\mu}}$ and $P^{(0)}_{\bar{\alpha}\bar{\beta}} = l^{\alpha}_{\bar{\alpha}} l^{\beta}_{\bar{\beta}} P^{(0)}_{\alpha\beta}$, where $P^{(0)}_{\alpha\beta}$ is the Ricci tensor corresponding to $g^{(0)}_{\alpha\beta}$.

Finding the general solution of the system (A.3)–(A.4) does not represent any difficulty, but we need only that particular part of the solution which is generated by $P^{(0)}_{\bar{\alpha}\bar{\beta}}$. The general solution of the homogeneous equations (A.3)–(A.4), as we already explained in the main text, just reproduces the structure of the metric

$g^{(0)}_{\alpha\beta}$. So, the only relevant solution of the equations (A.4) for the non-diagonal $h_{\bar{\alpha}\bar{\beta}}$ is:

$$h_{\bar{\alpha}\bar{\beta}} = \frac{1}{(p_{\bar{\beta}} - p_{\bar{\alpha}})}\left[t^{2p_{\bar{\alpha}}}\int t^{1-2p_{\bar{\alpha}}}P^{(0)}_{\bar{\alpha}\bar{\beta}}dt - t^{2p_{\bar{\beta}}}\int t^{1-2p_{\bar{\beta}}}P^{(0)}_{\bar{\alpha}\bar{\beta}}dt\right], \tag{A.5}$$
$$(\bar{\alpha} \neq \bar{\beta} \textit{ and no sum over } \bar{\alpha}, \bar{\beta}).$$

The solution for the diagonal components $h_{\bar{\alpha}\bar{\alpha}}$ is a little bit more sophisticated. To derive it, we first sum all three equations (A.3), obtaining thereby a simply solvable equation for the contraction h. Then we insert this result for h back into (A.3), whereupon the problem of integration for $h_{\bar{\alpha}\bar{\alpha}}$ becomes simple. At the end we should check the consistency of the procedure (the resulting contraction $h = \sum_{\bar{\mu}} t^{-2p_{\bar{\mu}}} h_{\bar{\mu}\bar{\mu}}$ must coincide with its value calculated previously). The final solution for $h_{\bar{\alpha}\bar{\alpha}}$ can be written in the form:

$$h_{\bar{\alpha}\bar{\alpha}} = 2t^{2p_{\bar{\alpha}}}\left(\int \frac{p_{\bar{\alpha}}Q - Q_{\bar{\alpha}}}{t}dt - \frac{p_{\bar{\alpha}}}{t}\int Qdt\right), \ (\textit{no sum over } \bar{\alpha}), \tag{A.6}$$

where Q and $Q_{\bar{\alpha}}$ are defined as

$$Q_{\bar{\alpha}} = \int t^{1-2p_{\bar{\alpha}}}P^{(0)}_{\bar{\alpha}\bar{\alpha}}dt\ , \quad Q = \sum_{\bar{\beta}} Q_{\bar{\beta}}\ , \ (\textit{no sum over } \bar{\alpha}). \tag{A.7}$$

In the last three formulas, all integrations mean the operation of taking only the antiderivative of the integrand, i.e., indefinite integrals without additive time-independent terms.

It is essential that we need to evaluate the corrections only in the vicinity of the singularity, that is in the limit $t \to 0$. The fact that all functions in the integrands of the expressions (A.5)–(A.7) effectively are of power law character with respect to time t (see Section 1.6 and corresponding footnote there) make such an evaluation straightforward. Indeed, it follows from (A.6) and (A.7) that for each value of $\bar{\alpha}$ the quantity $t^{-2p_{\bar{\alpha}}}h_{\bar{\alpha}\bar{\alpha}}$ in the vicinity to the singularity behaves as a linear superposition of the three terms $t^{2-2p_{\bar{\alpha}}}P^{(0)}_{\bar{\alpha}\bar{\alpha}}$ $(\bar{\alpha} = \bar{1}, \bar{2}, \bar{3})$. Schematically:

$$t^{-2p_{\bar{\alpha}}}h_{\bar{\alpha}\bar{\alpha}} \sim \sum_{\bar{\mu}} t^{2-2p_{\bar{\mu}}}P^{(0)}_{\bar{\mu}\bar{\mu}}\ , \tag{A.8}$$

and the non-diagonal components of the corrections, as can be seen from (A.5) behave as

$$h_{\bar{\alpha}\bar{\beta}} \sim t^2 P^{(0)}_{\bar{\alpha}\bar{\beta}},\ (\bar{\alpha} \neq \bar{\beta}). \tag{A.9}$$

The conditions $h_{\alpha\beta} \ll g^{(0)}_{\alpha\beta}$ of smallness of the corrections in comparison with the leading approximation, once projected into the triad $l^{\alpha}_{\bar{\alpha}}$, gives $h_{\bar{\alpha}\bar{\beta}} \ll \eta_{\bar{\alpha}\bar{\beta}}$. For the diagonal components, these inequalities are:

$$t^{-2p_{\bar{\alpha}}}h_{\bar{\alpha}\bar{\alpha}} \ll 1,\ (\bar{\alpha} = \bar{1}, \bar{2}, \bar{3}). \tag{A.10}$$

For the non-diagonal $h_{\bar{\alpha}\bar{\beta}}$, the conditions of smallness take the following form (since $\eta_{\bar{\alpha}\bar{\beta}}$ is diagonal):

$$h_{\bar{\alpha}\bar{\beta}} \ll \sqrt{\eta_{\bar{\alpha}\bar{\alpha}}\eta_{\bar{\beta}\bar{\beta}}} = t^{p_{\bar{\alpha}}+p_{\bar{\beta}}}, \ \bar{\alpha} \neq \bar{\beta}, \ \ (no\ sum\ over\ \bar{\alpha}, \bar{\beta}). \tag{A.11}$$

Consequently, the result is that all the corrections will be negligibly small in the limit $t \to 0$ if for any value of $\bar{\alpha}$ and $\bar{\beta}$ (equal to each other or not) the following inequalities are satisfied:

$$P^{(0)}_{\bar{\alpha}\bar{\beta}} \ll t^{p_{\bar{\alpha}}+p_{\bar{\beta}}-2}, \ t \to 0. \tag{A.12}$$

That is, we obtained the conditions (1.36) written in the main text.

A.2 Frame Components of the Ricci Tensor

In this section we give the general expression for the Ricci tensor projected into a frame for N-dimensional space. The capital Latin indices $A, B, \ldots$ take the values 1,2, ...,N and the corresponding frame indices $\bar{A}, \bar{B}, \ldots$ run over the same values $\bar{1}$,$\bar{2}$, ...$\bar{\text{N}}$. The metric is:

$$-ds^2_{(N)} = G_{AB}dx^A dx^B, \ \ G_{AB} = \gamma_{\bar{A}\bar{B}} L^{\bar{A}}_A L^{\bar{B}}_B \ , \ L^{\bar{A}}_A = L^{\bar{A}}_A(x^1, x^2, \ldots, x^N) \ , \tag{A.13}$$

where $\gamma_{\bar{A}\bar{B}}$ is any constant (i.e., not depending on any coordinate x^A) symmetric $N \times N$ matrix. The fastest way to calculate the Ricci tensor in terms of its projections into the frame $L^{\bar{A}}_A$ (see, e.g., ([68]) is based on the standard formula for the commutation of the second covariant derivatives of a vector applied to the vectors of the frame $L^{\bar{A}}_A$, that is $L^{\bar{A}}_{B;D;C} - L^{\bar{A}}_{B;C;D} = L^{\bar{A}}_F G^{FE} R_{EBDC}$, where for the projections of the Ricci tensor we have: $R_{\bar{A}\bar{B}} = L^B_{\bar{A}} L^C_{\bar{B}} R_{BC} = L^B_{\bar{A}} L^C_{\bar{B}} G^{ED} R_{EBDC} = L^C_{\bar{A}} L^B_{\bar{B}} L^D_{\bar{E}} (L^{\bar{E}}_{B;D;C} - L^{\bar{E}}_{B;C;D})$. Here, as usual, $L^A_{\bar{A}}$ are vectors inverse to the original frame vectors $L^{\bar{A}}_A$, that is $L^A_{\bar{A}} L^{\bar{B}}_A = \delta^{\bar{B}}_{\bar{A}}$ and $L^{\bar{A}}_A L^B_{\bar{A}} = \delta^B_A$. The covariant derivatives of the frame vectors can be expressed in terms of the Ricci coefficients $\rho^{\bar{A}}_{\bar{B}\bar{C}}$ which are defined by the relations:

$$L^{\bar{A}}_{A;B} = \rho^{\bar{A}}_{\bar{B}\bar{C}} L^{\bar{B}}_A L^{\bar{C}}_B. \tag{A.14}$$

If we substitute this representation for the covariant derivatives of the frame vectors into the formula for $R_{\bar{A}\bar{B}}$, we obtain:

$$R_{\bar{A}\bar{B}} = -\rho^{\bar{C}}_{\bar{A}\bar{B},D} L^D_{\bar{C}} + \rho^{\bar{C}}_{\bar{A}\bar{C},D} L^D_{\bar{B}} + \rho^{\bar{C}}_{\bar{D}\bar{C}} \rho^{\bar{D}}_{\bar{A}\bar{B}} - \rho^{\bar{C}}_{\bar{A}\bar{D}} \rho^{\bar{D}}_{\bar{B}\bar{C}}. \tag{A.15}$$

This result is not yet suitable for practical calculations because the Ricci coefficients are rather complicated, they contain covariant derivatives. To get rid of this inconvenience, one introduces the coefficients $\sigma^{\bar{A}}_{\bar{B}\bar{C}}$:

$$\sigma^{\bar{A}}_{\bar{B}\bar{C}} = (L^{\bar{A}}_{B,C} - L^{\bar{A}}_{C,B}) L^B_{\bar{B}} L^C_{\bar{C}} \ , \tag{A.16}$$

which contain only simple derivatives. At the same time the Ricci coefficients are expressible linearly in terms of the σ-coefficients:

$$\rho^{\bar{A}}_{\bar{B}\bar{C}} = \frac{1}{2}(\sigma^{\bar{A}}_{\bar{B}\bar{C}} - \gamma^{\bar{A}\bar{E}}\gamma_{\bar{B}\bar{D}}\sigma^{\bar{D}}_{\bar{E}\bar{C}} + \gamma^{\bar{A}\bar{E}}\gamma_{\bar{C}\bar{D}}\sigma^{\bar{D}}_{\bar{B}\bar{E}}), \tag{A.17}$$

where $\gamma^{\bar{A}\bar{B}}$ is inverse to $\gamma_{\bar{A}\bar{B}}$. Now we insert this expression for $\rho^{\bar{A}}_{\bar{B}\bar{C}}$ into the previous expression for $R_{\bar{A}\bar{B}}$ and obtain the following final result:

$$\begin{aligned} R_{\bar{A}\bar{B}} = -\frac{1}{2}[&(\gamma_{\bar{A}\bar{E}}\sigma^{\bar{E}}_{\bar{B}\bar{F},D} + \gamma_{\bar{B}\bar{E}}\sigma^{\bar{E}}_{\bar{A}\bar{F},D})L^{D}_{\bar{C}}\gamma^{\bar{F}\bar{C}} + \sigma^{\bar{C}}_{\bar{C}\bar{A},D}L^{D}_{\bar{B}} \\ &+ \sigma^{\bar{C}}_{\bar{C}\bar{B},D}L^{D}_{\bar{A}} + \gamma_{\bar{F}\bar{E}}\gamma^{\bar{C}\bar{G}}\sigma^{\bar{F}}_{\bar{C}\bar{A}}\sigma^{\bar{E}}_{\bar{G}\bar{B}} + \sigma^{\bar{D}}_{\bar{C}\bar{A}}\sigma^{\bar{C}}_{\bar{D}\bar{B}} \\ &+ \gamma^{\bar{D}\bar{G}}\sigma^{\bar{C}}_{\bar{C}\bar{D}}(\gamma_{\bar{A}\bar{E}}\sigma^{\bar{E}}_{\bar{B}\bar{G}} + \gamma_{\bar{B}\bar{E}}\sigma^{\bar{E}}_{\bar{A}\bar{G}}) - \frac{1}{2}\gamma_{\bar{A}\bar{E}}\gamma_{\bar{B}\bar{G}}\gamma^{\bar{F}\bar{C}}\gamma^{\bar{H}\bar{D}}\sigma^{\bar{E}}_{\bar{H}\bar{C}}\sigma^{\bar{G}}_{\bar{D}\bar{F}}]. \end{aligned} \tag{A.18}$$

This expression is valid for any constant symmetric matrix $\gamma_{\bar{A}\bar{B}}$, but for the case when $\gamma_{\bar{A}\bar{B}}$ represents the diagonal Euclidean frame metric, i.e., when $\gamma_{\bar{A}\bar{B}} = diag(1, 1, \ldots, 1)$ this formula can be written in a simpler way. In each term in (A.18) the up and down positions of the indices follow the rigorous conventional rules. However, due to the specifically simple structure of the Euclidian frame metric we can disregard these rules in favor of simplest ones that do overload the expressions by many trivial and excessive symbols $\gamma_{\bar{A}\bar{B}}$ and $\gamma^{\bar{A}\bar{B}}$, namely, we can write the expression (A.18) for such a particular case as

$$\begin{aligned} R_{\bar{A}\bar{B}} = -\frac{1}{2}[&\sigma^{\bar{B}}_{\bar{A}\bar{C},D}L^{D}_{\bar{C}} + \sigma^{\bar{A}}_{\bar{B}\bar{C},D}L^{D}_{\bar{C}} + \sigma^{\bar{C}}_{\bar{C}\bar{A},D}L^{D}_{\bar{B}} + \sigma^{\bar{C}}_{\bar{C}\bar{B},D}L^{D}_{\bar{A}} \\ &+ \sigma^{\bar{D}}_{\bar{C}\bar{A}}\sigma^{\bar{C}}_{\bar{D}\bar{B}} + \sigma^{\bar{C}}_{\bar{D}\bar{A}}\sigma^{\bar{C}}_{\bar{D}\bar{B}} + \sigma^{\bar{C}}_{\bar{C}\bar{D}}\sigma^{\bar{B}}_{\bar{A}\bar{D}} + \sigma^{\bar{C}}_{\bar{C}\bar{D}}\sigma^{\bar{A}}_{\bar{B}\bar{D}} - \frac{1}{2}\sigma^{\bar{A}}_{\bar{C}\bar{D}}\sigma^{\bar{B}}_{\bar{C}\bar{D}}], \end{aligned} \tag{A.19}$$

where summation over the repeated indices is still implied independently of their (up or down) positions.

A.3 Exact Solution for Transition Between Two Kasner Epochs

To analyze the system (1.52)–(1.53), it is convenient to pass to another time variable τ which we define (separately at each 3-space point x^α) through the relation:

$$d\tau = -(abc)^{-1}dt. \tag{A.20}$$

(In the case of Kasner-like dynamics (when $abc \sim t$) the singularity $t \to 0$ corresponds to $\tau \to \infty$). For the functions a, b, c, we introduce the notations:

$$a = \exp(-\beta^{\bar{1}}),\ b = \exp(-\beta^{\bar{2}}),\ c = \exp(-\beta^{\bar{3}}). \tag{A.21}$$

Now the system (1.52)–(1.53) takes the form:

$$\frac{\partial^2\beta^{\bar{1}}}{\partial\tau^2} = \frac{\lambda^2}{2}e^{-4\beta^{\bar{1}}},\ \frac{\partial^2\beta^{\bar{2}}}{\partial\tau^2} = -\frac{\lambda^2}{2}e^{-4\beta^{\bar{1}}},\ \frac{\partial^2\beta^{\bar{3}}}{\partial\tau^2} = -\frac{\lambda^2}{2}e^{-4\beta^{\bar{1}}}, \tag{A.22}$$

$$\sum_{\bar{\alpha}}(\frac{\partial\beta^{\bar{\alpha}}}{\partial\tau})^2 - \left(\sum_{\bar{\alpha}}\frac{\partial\beta^{\bar{\alpha}}}{\partial\tau}\right)^2 + \frac{\lambda^2}{2}e^{-4\beta^{\bar{1}}} = 0. \tag{A.23}$$

The exact general solution of these equations is:

$$
\begin{aligned}
a^2 &= \frac{2q\left|p_{\bar 1}\right|}{|\lambda| \cosh\left[2qp_{\bar 1}(\tau-\tau_0) - \ln\frac{|\lambda| a_0^2}{4q\left|p_{\bar 1}\right|}\right]}\,, \\
b^2 &= \frac{q\,|\lambda|}{2c_0^2\left|p_{\bar 1}\right|} e^{-2q(p_{\bar 2}+p_{\bar 1})(\tau-\tau_0)} \cosh\left[2qp_{\bar 1}(\tau-\tau_0) - \ln\frac{|\lambda|\, a_0^2}{4q\left|p_{\bar 1}\right|}\right]\,, \\
c^2 &= \frac{q\,|\lambda|}{2b_0^2\left|p_{\bar 1}\right|} e^{-2q(p_{\bar 3}+p_{\bar 1})(\tau-\tau_0)} \cosh\left[2qp_{\bar 1}(\tau-\tau_0) - \ln\frac{|\lambda|\, a_0^2}{4q\left|p_{\bar 1}\right|}\right]\,, \\
q &= a_0 b_0 c_0\,,
\end{aligned}
\tag{A.24}
$$

where a_0, b_0, c_0 and τ_0 are arbitrary three-dimensional functional parameters (with the condition $q > 0$) and $p_{\bar 1}, p_{\bar 2}, p_{\bar 3}$ are Kasner exponents satisfying the relations (1.21)–(1.22). From (A.24) and (A.21) it follows that the first derivatives of the functions $\beta^{\bar 1}, \beta^{\bar 2}, \beta^{\bar 3}$ with respect to the time τ are:

$$
\begin{aligned}
\frac{d\beta^{\bar 1}}{d\tau} &= qp_{\bar 1} \tanh\left[2qp_{\bar 1}(\tau-\tau_0) - \ln\frac{|\lambda|\, a_0^2}{4q\left|p_{\bar 1}\right|}\right]\,, \\
\frac{d\beta^{\bar 2}}{d\tau} &= q\left\{p_{\bar 2} + p_{\bar 1} - p_{\bar 1} \tanh\left[2qp_{\bar 1}(\tau-\tau_0) - \ln\frac{|\lambda|\, a_0^2}{4q\left|p_{\bar 1}\right|}\right]\right\}\,, \\
\frac{d\beta^{\bar 3}}{d\tau} &= q\left\{p_{\bar 3} + p_{\bar 1} - p_{\bar 1} \tanh\left[2qp_{\bar 1}(\tau-\tau_0) - \ln\frac{|\lambda|\, a_0^2}{4q\left|p_{\bar 1}\right|}\right]\right\}\,.
\end{aligned}
\tag{A.25}
$$

If we choose the order $p_{\bar 1} < p_{\bar 2} < p_{\bar 3}$ (i.e., $p_{\bar 1} < 0$ and $p_{\bar 3} > p_{\bar 2} > 0$) for the Kasner exponents then it easy to calculate the change of asymptotics of the scale factors a^2, b^2, c^2 between the initial ($\tau \to -\infty$, $t \to \infty$) and the final ($\tau \to \infty$, $t \to 0$) Kasner epochs. For the initial epoch, we have:

$$
\begin{aligned}
&\tau \to -\infty,\ t = e^{-q\tau} \to \infty, \\
a^2 &= a_0^2 e^{-2qp_{\bar 1}\tau} = a_0^2 t^{2p_{\bar 1}}, \\
b^2 &= b_0^2 e^{-2qp_{\bar 2}\tau} = b_0^2 t^{2p_{\bar 2}}, \\
c^2 &= c_0^2 e^{-2qp_{\bar 3}\tau} = c_0^2 t^{2p_{\bar 3}},
\end{aligned}
\tag{A.26}
$$

while the following behavior for the final epoch holds:

$$
\begin{aligned}
&\tau \to \infty,\ t = \frac{a_0^2\,|\lambda|}{4q\left|p_{\bar 1}\right|(1+2p_{\bar 1})} e^{-q(1+2p_{\bar 1})\tau} \to 0, \\
a^2 &= \frac{16 b_0^2 c_0^2 p_{\bar 1}^2}{\lambda^2} e^{2qp_{\bar 1}\tau} = a_0^2 \left(\frac{16q^2 p_{\bar 1}^2}{\lambda^2 a_0^4}\right)^{\frac{1+p_{\bar 1}}{1+2p_{\bar 1}}} (1+2p_{\bar 1})^{-\frac{2p_{\bar 1}}{1+2p_{\bar 1}}} t^{-\frac{2p_{\bar 1}}{1+2p_{\bar 1}}}, \\
b^2 &= \frac{a_0^2\lambda^2}{16 c_0^2 p_{\bar 1}^2} e^{-2q(p_{\bar 2}+2p_{\bar 1})\tau} = b_0^2 \left(\frac{16q^2 p_{\bar 1}^2}{\lambda^2 a_0^4}\right)^{\frac{p_{\bar 2}-1}{1+2p_{\bar 1}}} (1+2p_{\bar 1})^{\frac{2(p_{\bar 2}+2p_{\bar 1})}{1+2p_{\bar 1}}} t^{\frac{2(p_{\bar 2}+2p_{\bar 1})}{1+2p_{\bar 1}}}, \\
c^2 &= \frac{a_0^2\lambda^2}{16 b_0^2 p_{\bar 1}^2} e^{-2q(p_{\bar 3}+2p_{\bar 1})\tau} = c_0^2 \left(\frac{16q^2 p_{\bar 1}^2}{\lambda^2 a_0^4}\right)^{\frac{p_{\bar 3}-1}{1+2p_{\bar 1}}} (1+2p_{\bar 1})^{\frac{2(p_{\bar 3}+2p_{\bar 1})}{1+2p_{\bar 1}}} t^{\frac{2(p_{\bar 3}+2p_{\bar 1})}{1+2p_{\bar 1}}}.
\end{aligned}
\tag{A.27}
$$

A.4 The Derivation of the Rotation Effect of the Kasner Axes

The appearance of the rotation effect of the Kasner axes can be understood from the following analysis. As was already explained, in the transition region, the diagonal frame metric (with respect to the first epoch frame $l_{\alpha}^{\bar{\alpha}}$) $\eta_{\bar{\alpha}\bar{\beta}} = diag(a^2, b^2, c^2)$ is a good approximation because in this region the non-diagonal projections of the Ricci tensor $P_{\bar{\alpha}\bar{\beta}}$ can be neglected. Nevertheless, they produce some small non-diagonal corrections $\eta_{\bar{1}\bar{2}}, \eta_{\bar{1}\bar{3}}, \eta_{\bar{2}\bar{3}}$ which, while being negligibly small around $t \sim t_c$, start to grow during the subsequent stages of evolution at $t \ll t_c$, sooner or later making the approximation (1.67) invalid. This happens during the second epoch when b^2 becomes essentially larger than a^2.

The way to calculate these non-diagonal corrections follows the same general pattern described in Appendix A.1. We consider the metric

$$g_{\alpha\beta} = \eta_{\bar{\alpha}\bar{\beta}} l_{\alpha}^{\bar{\alpha}} l_{\beta}^{\bar{\beta}} \ , \ \eta_{\bar{\alpha}\bar{\beta}} = diag(a^2, b^2, c^2) \tag{A.28}$$

with an exact solution for the scale factors a^2, b^2, c^2 covering both epochs (see Appendix A.3) and with a non-rotating frame $l_{\alpha}^{\bar{\alpha}}$ for both epochs as the leading approximation $g_{\alpha\beta}^{(0)}$. We write the corrected metric (foreseeing generation of corrections due to the rotation of the axes) as:

$$g_{\alpha\beta} = \eta_{\bar{\alpha}\bar{\beta}} \acute{l}_{\alpha}^{\bar{\alpha}} \acute{l}_{\beta}^{\bar{\beta}} \ , \ \eta_{\bar{\alpha}\bar{\beta}} = diag(a^2, b^2, c^2) \tag{A.29}$$

with some new frame $\acute{l}_{\alpha}^{\bar{\alpha}}$ but with the same scale factors a^2, b^2, c^2, which are valid for both epochs. The new vectors $\acute{l}_{\alpha}^{\bar{\alpha}}$ can be represented as linear combinations of the original vectors $l_{\alpha}^{\bar{\alpha}}$. If we project all tensors in both epochs onto the original vectors $l_{\alpha}^{\bar{\alpha}}$, it is clear that the rotation of the Kasner axes can be described as the appearance during the second epoch of non-diagonal projections $\eta_{\bar{\alpha}\bar{\beta}}$, which behave in time like linear combinations of the functions a^2, b^2, c^2. We write the corrected metric (A.29) as $g_{\alpha\beta} = g_{\alpha\beta}^{(0)} + h_{\alpha\beta}$ and this formula, once projected into the frame $l_{\alpha}^{\bar{\alpha}}$, can be written as $\eta_{\bar{\alpha}\bar{\beta}} = \eta_{\bar{\alpha}\bar{\beta}}^{(0)} + h_{\bar{\alpha}\bar{\beta}}$, where $\eta_{\bar{\alpha}\bar{\beta}}^{(0)} = diag(a^2, b^2, c^2)$.

To get the equations for $h_{\bar{\alpha}\bar{\beta}}$ ($\bar{\alpha} \neq \bar{\beta}$) we have to project the gravitational equations (1.28) into the triad $l_{\alpha}^{\bar{\alpha}}$ and take only the non-diagonal components of these projections, retaining in their right-hand sides the non-diagonal components of the Ricci tensor $P_{\bar{\alpha}\bar{\beta}}^{(0)}$ calculated from the leading approximation to the metric tensor (A.28). These equations are:

$$\ddot{h}_{\bar{1}\bar{2}} + \left(\frac{\dot{c}}{c} - \frac{\dot{a}}{a} - \frac{\dot{b}}{b}\right)\dot{h}_{\bar{1}\bar{2}} + \frac{4\dot{a}\dot{b}}{ab} h_{\bar{1}\bar{2}} = -2P_{\bar{1}\bar{2}}^{(0)} \ , \tag{A.30}$$

$$\ddot{h}_{\bar{1}\bar{3}} + \left(\frac{\dot{b}}{b} - \frac{\dot{c}}{c} - \frac{\dot{a}}{a}\right)\dot{h}_{\bar{1}\bar{3}} + \frac{4\dot{a}\dot{c}}{ac} h_{\bar{1}\bar{3}} = -2P_{\bar{1}\bar{3}}^{(0)} \ ,$$

$$\ddot{h}_{\bar{2}\bar{3}} + \left(\frac{\dot{a}}{a} - \frac{\dot{b}}{b} - \frac{\dot{c}}{c}\right)\dot{h}_{\bar{2}\bar{3}} + \frac{4\dot{b}\dot{c}}{bc} h_{\bar{2}\bar{3}} = -2P_{\bar{2}\bar{3}}^{(0)} \ .$$

The right-hand side in these formulas can be found in a way identical to the one we followed before for the calculation of the diagonal components of the Ricci tensor. We should retain in $P^{(0)}_{\bar{\alpha}\bar{\beta}}$ $(\bar{\alpha} \neq \bar{\beta})$ in (A.30) only the largest terms in the transition region. For the identification of these terms, we can use the fact that during the first epoch, the scale factors followed the laws $a^2, b^2, c^2 \sim t^{2p_{\bar{1}}}, t^{2p_{\bar{2}}}, t^{2p_{\bar{3}}}$ under the conditions that $p_{\bar{1}}$ is negative and $p_{\bar{2}}$, $p_{\bar{3}}$ are positive and around the critical time $t \sim t_c$ the orders $a^2(t_c) \gg b^2(t_c)$ and $a^2(t_c) \gg c^2(t_c)$ have settled. These largest terms are:

$$\begin{aligned}
P^{(0)}_{\bar{1}\bar{2}} &= -ab\sigma^{\bar{1}}_{\bar{2}\bar{3}}\sigma^{\bar{1}}_{\bar{1}\bar{3}} - \frac{ab}{2c}\left(\sigma^{\bar{1}}_{\bar{2}\bar{3}}\right)_{,\nu} l^{\nu}_{\bar{3}}\,, \\
P^{(0)}_{\bar{1}\bar{3}} &= ac\sigma^{\bar{1}}_{\bar{2}\bar{3}}\sigma^{\bar{1}}_{\bar{1}\bar{2}} - \frac{ac}{2b}\left(\sigma^{\bar{1}}_{\bar{2}\bar{3}}\right)_{,\nu} l^{\nu}_{\bar{2}}\,, \\
P^{(0)}_{\bar{2}\bar{3}} &= -\frac{bc}{2}\left[\sigma^{\bar{1}}_{\bar{2}\bar{3}}\left(\sigma^{\bar{2}}_{\bar{1}\bar{2}} - \sigma^{\bar{3}}_{\bar{1}\bar{3}}\right) + 2\sigma^{\bar{1}}_{\bar{1}\bar{2}}\sigma^{\bar{1}}_{\bar{1}\bar{3}} + \sigma^{\bar{1}}_{\bar{1}\bar{2}}\sigma^{\bar{2}}_{\bar{3}\bar{2}} + \sigma^{\bar{1}}_{\bar{1}\bar{3}}\sigma^{\bar{3}}_{\bar{2}\bar{3}}\right] \\
&\quad - \frac{b}{2}\left(\sigma^{\bar{1}}_{\bar{1}\bar{2}}\right)_{,\nu} l^{\nu}_{\bar{3}} - \frac{c}{2}\left(\sigma^{\bar{1}}_{\bar{1}\bar{3}}\right)_{,\nu} l^{\nu}_{\bar{2}}\,.
\end{aligned} \tag{A.31}$$

All omitted terms in these formulas are small in the sense that their ratios to the leading terms, which we have explicitly written out, have the orders of magnitude $b^2(t_c)/a^2(t_c)$ or $c^2(t_c)/a^2(t_c)$ so that such terms can indeed be discarded in the transition region around $t \sim t_c$. The quantities $\sigma^{\bar{\alpha}}_{\bar{\beta}\bar{\gamma}}$ in (A.31) are:

$$\sigma^{\bar{\alpha}}_{\bar{\beta}\bar{\gamma}} = \left(L^{\bar{\alpha}}_{\mu},_{\nu} - L^{\bar{\alpha}}_{\nu},_{\mu}\right) L^{\mu}_{\bar{\beta}} L^{\nu}_{\bar{\gamma}}\,, \tag{A.32}$$

where $L^{\bar{1}}_{\alpha} = al^{\bar{1}}_{\alpha}$, $L^{\bar{2}}_{\alpha} = bl^{\bar{2}}_{\alpha}$, $L^{\bar{3}}_{\alpha} = cl^{\bar{3}}_{\alpha}$ and $L^{\alpha}_{\bar{1}} = a^{-1}l^{\alpha}_{\bar{1}}$, $L^{\alpha}_{\bar{2}} = b^{-1}l^{\alpha}_{\bar{2}}$, $L^{\alpha}_{\bar{3}} = c^{-1}l^{\alpha}_{\bar{3}}$.

In equations (A.30)–(A.32) we should use the exact solution for the functions a^2, b^2, c^2 derived in Appendix A.3. The integration procedure is a little bit sophisticated. This procedure have been described in [20] and [21] and consists of the following steps. First we should find the qualitative character of the change of axes, that is, we need to find the general structure of the linear transformation between $\acute{l}^{\bar{\alpha}}_{\alpha}$ and $l^{\bar{\alpha}}_{\alpha}$. Then we have to calculate exactly the coefficients of this transformation. Estimating the orders of magnitude of the right-hand sides of the expressions (A.31), we find that

$$P^{(0)}_{\bar{1}\bar{2}} \sim k^2a^2/c^2,\; P^{(0)}_{\bar{1}\bar{3}} \sim k^2a^2/b^2,\; P^{(0)}_{\bar{2}\bar{3}} \sim k^2. \tag{A.33}$$

We recall that k^{-1} denotes, as in estimation (1.45), the order of magnitude of spatial distances over which the metric changes substantially. Then we solve the equations (A.30) for the initial and final epochs and we next match the solutions at $t \sim t_c$. Consider the first of equations (A.30). During the initial epoch, with a, b, c taken from (1.54), we have

$$h_{\bar{1}\bar{2}} = C_{\bar{1}\bar{2}}a^2 + D_{\bar{1}\bar{2}}b^2 + \frac{b^2}{p_{\bar{1}} - p_{\bar{2}}}\int P^{(0)}_{\bar{1}\bar{2}}\frac{tdt}{b^2} - \frac{a^2}{p_{\bar{1}} - p_{\bar{2}}}\int P^{(0)}_{\bar{1}\bar{2}}\frac{tdt}{a^2}, \tag{A.34}$$

where $C_{\bar{1}\bar{2}}$ and $D_{\bar{1}\bar{2}}$ are arbitrary three-dimensional parameters. By assumption, in this epoch, the Kasner axes coincide with the directions $l^{\bar{\alpha}}_{\alpha}$. This means that

$h_{\bar{1}\bar{2}}$ cannot contain terms proportional to a^2 or to b^2, i.e., we must have $C_{\bar{1}\bar{2}} = D_{\bar{1}\bar{2}} = 0$. With $P^{(0)}_{\bar{1}\bar{2}}$ taken from (A.33) we then find the order of magnitude:

$$h_{\bar{1}\bar{2}} \sim b^2 a^4 k^2 / \left(a_0 b_0 c_0\right)^2 \sim b^2 a^4 / a^4_{\max}, \tag{A.35}$$

where $a_{\max}$ is taken from (1.58). (We note that owing to (A.35) and the condition $a_{\max} = a\left(t_c\right) >> b\left(t_c\right)$ we have $h_{\bar{1}\bar{2}} << ab$ in the transition region, as it should be.)

For the final epoch, the solution has the same form (A.34) but with new constants $\acute{C}_{\bar{1}\bar{2}}, \acute{D}_{\bar{1}\bar{2}}$. The latter should be determined from the condition of matching to the solution (A.35) for $t \sim t_c$. This implies $\acute{C}_{\bar{1}\bar{2}} \sim b^2\left(t_c\right)/a^2\left(t_c\right)$ and $\acute{D}_{\bar{1}\bar{2}} \sim 1$. Thus, for the final epoch we have

$$h_{\bar{1}\bar{2}} \sim a^2 \frac{b^2\left(t_c\right)}{a^2\left(t_c\right)} + b^2 + b^2 a^4 \frac{1}{a^4\left(t_c\right)} \tag{A.36}$$

(the sum here should, of course, be understood only as a symbolic way of enumerating the types of terms that occur in $h_{\bar{1}\bar{2}}$). Since, however, for $t < t_c$ the function a^2 decreases with the decrease of t, the last term will become small compared to b^2 (after leaving the intermediate period $t \sim t_c$ between the two epochs), so that we remain only with

$$h_{\bar{1}\bar{2}}(t) \sim a^2(t) \frac{b^2\left(t_c\right)}{a^2\left(t_c\right)} + b^2(t). \tag{A.37}$$

The same procedure being applied to the second and third equations (A.30) gives:

$$h_{\bar{1}\bar{3}}(t) \sim a^2(t) \frac{c^2\left(t_c\right)}{a^2\left(t_c\right)} + c^2(t)\,, \tag{A.38}$$

$$h_{\bar{2}\bar{3}}(t) \sim b^2(t) \frac{c^2\left(t_c\right)}{a^2\left(t_c\right)} + c^2(t) \frac{b^2\left(t_c\right)}{a^2\left(t_c\right)}\,. \tag{A.39}$$

The expression (A.37) signifies a rotation in the $(l^{\bar{1}}_{\alpha}, l^{\bar{2}}_{\alpha})$ "plane" of the second Kasner axis $l^{\bar{2}}_{\alpha}$ by a large "angle" (~ 1), and a rotation of the first Kasner axis $l^{\bar{1}}_{\alpha}$ by a small "angle" ($\sim b^2(t_c)/a^2\left(t_c\right) << 1$). The small rotations are, however, outside the approximation considered. Taking only the big rotations into account, we find from all the expressions (A.37)–(A.39), that the new Kasner axes indeed are related to the old ones by the formulas (1.69) where the coefficients $\sigma_{\bar{2}}, \sigma_{\bar{3}}$ are of the order of unity.

We stress the fact that the use of the linearized equations (A.30), which led to the expressions (A.37)–(A.39), is legitimate only as long as all $h_{\bar{1}\bar{2}}, h_{\bar{1}\bar{3}}, h_{\bar{2}\bar{3}}$ are small (in the sense $h_{\bar{\alpha}\bar{\beta}} \ll \sqrt{\eta_{\bar{\alpha}\bar{\alpha}}\eta_{\bar{\beta}\bar{\beta}}}, \bar{\alpha} \neq \bar{\beta}$). This condition is obviously violated when, during the evolution of the metric towards a new epoch, the function b^2 stops being small compared to a^2. But by this time all the components of the three-dimensional Ricci tensor will be damped out to such an extent that they will disappear from the Einstein equations, and after that these equations will be satisfied by the generalized Kasner solution with arbitrary orientation of the axes.*

For an exact quantitative determination of the coefficients $\sigma_{\bar{2}}, \sigma_{\bar{3}}$ in (1.69), the evaluation of the $h_{\bar{\alpha}\bar{\beta}}$ through the order of magnitude used above is not sufficient. We need an exact solution of the equations (A.30). Fortunately, one can get away from this task, making use of the existence of the following exact first integral of the equations (1.15) and (1.17) in vacuum:

$$\frac{1}{2}(\kappa^{\beta}_{\alpha;\beta} - \kappa_{,\alpha}) = C_{\alpha}/\sqrt{g} \tag{A.40}$$

where C_{α} are arbitrary functions of the three spatial coordinates. This relation is a consequence of the Bianchi identity for the three-dimensional Ricci tensor: $P^{\beta}_{\alpha;\beta} = P_{,\alpha}/2$, and is obtained by substituting into it the expressions of P^{β}_{α} and P in terms of κ^{β}_{α} according to (1.17), followed by some simple transformations which take into account (1.15).

The expression in the left-hand side of (A.40) is nothing other than $R_{0\alpha}$. If the matter energy–momentum tensor were exactly zero, the Einstein equations $R_{0\alpha} = T_{0\alpha}$ would imply that one has to set the functions C_{α} equal to zero. The nontrivial circumstance for our approximation expressed by the relation (A.40) in the case when matter is present is related to the fact that the absence of the energy-momentum tensor in the right-hand sides of the equations (1.17) and (1.15) is only approximate, and admissible only in the approximation under discussion. Owing to the equations $R_{0\alpha} = T_{0\alpha}$ the relation (A.40) shows that in this approximation, throughout the whole evolution of the metric, including the transition period between two the epochs, the quantities $\sqrt{g}T_{0\alpha}$ remain constant in time. We stress, however, that the values of the constants C_{α} (zero or not) have no consequence for the calculation of the coefficients $\sigma_{\bar{2}}, \sigma_{\bar{3}}$. For this calculation, the only thing that matters is that the left-hand side of the expression (A.40) does not change during the transition from one epoch to another.

Computing the expressions $\sqrt{g}(\kappa^{\beta}_{\alpha;\beta} - \kappa_{,\alpha})$ for two successive epochs, with the metrics (1.67) and (1.68) and equating the results we obtain:

$$\sum_{\bar{\mu}} l^{\bar{\mu}}_{\alpha}\left[p_{\bar{\mu},\beta} l^{\beta}_{\bar{\mu}} + \sum_{\bar{\nu}} (p_{\bar{\nu}} - p_{\bar{\mu}})\,\lambda^{\bar{\nu}}_{\bar{\mu}\bar{\nu}} - \sum_{\bar{\nu}} (p_{\bar{\nu}} - p_{\bar{\mu}})\left(\ln F^{\bar{\nu}}\right)_{,\beta} l^{\beta}_{\bar{\mu}}\right] \tag{A.41}$$
$$= (1 + 2p_{\bar{1}})\left\{\sum_{\bar{\mu}} \acute{l}^{\bar{\mu}}_{\alpha}\left[\acute{p}_{\bar{\mu},\beta} \acute{l}^{\beta}_{\bar{\mu}} + \sum_{\bar{\nu}} (\acute{p}_{\bar{\nu}} - \acute{p}_{\bar{\mu}})\,\acute{\lambda}^{\bar{\nu}}_{\bar{\mu}\bar{\nu}} - \sum_{\bar{\nu}} (\acute{p}_{\bar{\nu}} - \acute{p}_{\bar{\mu}})\left(\ln \acute{F}^{\bar{\nu}}\right)_{,\beta} \acute{l}^{\beta}_{\bar{\mu}}\right]\right\}$$

where the usual summation over repeated vectorial indices is assumed but summation over frame indices is shown explicitly. The quantities $\lambda^{\bar{\alpha}}_{\bar{\beta}\bar{\gamma}}$ in this formula

* In this reasoning it is understood that the oscillation amplitudes (i.e., the ratios $a^2(t_c)/b^2(t_c)$ and $a^2(t_c)/c^2(t_c)$) are sufficiently large so that there is room for all required conditions to be true. We recall that in the asymptotic region of arbitrarily close approach to the singularity, these amplitudes increase without bounds.

are those defined by (1.71) and correspondingly $\acute{\lambda}^{\bar{\alpha}}_{\bar{\beta}\bar{\gamma}} = \left(\acute{l}^{\bar{\alpha}}_{\mu,\nu} - \acute{l}^{\bar{\alpha}}_{\nu,\mu}\right)\acute{l}^{\mu}_{\bar{\beta}}\acute{l}^{\nu}_{\bar{\gamma}}$. The factor $1 + 2p_{\bar{1}}$ in the right-hand side of this equation appears because during the first epoch, as it follows from (1.67) and (1.25), one has $\sqrt{g} = ta_0 b_0 c_0 \varepsilon^{\alpha\beta\gamma} l^{\bar{1}}_{\alpha} l^{\bar{2}}_{\beta} l^{\bar{3}}_{\gamma}$, while during the second epoch, as it can be seen from (1.68) and (1.57), one has

$$\sqrt{g} = t\acute{a}_0\acute{b}_0\acute{c}_0\varepsilon^{\alpha\beta\gamma}\acute{l}^{\bar{1}}_{\alpha}\acute{l}^{\bar{2}}_{\beta}\acute{l}^{\bar{3}}_{\gamma} = ta_0 b_0 c_0 \left(1 + 2p_{\bar{1}}\right)\varepsilon^{\alpha\beta\gamma} l^{\bar{1}}_{\alpha} l^{\bar{2}}_{\beta} l^{\bar{3}}_{\gamma} \tag{A.42}$$

(taking into account that $\varepsilon^{\alpha\beta\gamma}\acute{l}^{\bar{1}}_{\alpha}\acute{l}^{\bar{2}}_{\beta}\acute{l}^{\bar{3}}_{\gamma} = \varepsilon^{\alpha\beta\gamma} l^{\bar{1}}_{\alpha} l^{\bar{2}}_{\beta} l^{\bar{3}}_{\gamma}$, as follows from the transformation law (1.69)). The quantities $F^{\bar{\nu}}$ and $\acute{F}^{\bar{\nu}}$ are:

$$F^{\bar{\nu}} = (a_0, b_0, c_0)\,, \quad \acute{F}^{\bar{\nu}} = \left(\acute{a}_0, \acute{b}_0, \acute{c}_0\right)\,. \tag{A.43}$$

It is worth remarking that all terms in the expressions $\sqrt{g}(\kappa^{\beta}_{\alpha;\beta} - \kappa_{,\alpha})$ containing the factor $\ln t$ for both epochs exactly cancel each other and all the other terms are pure three-dimensional functions. If in the right-hand side of (A.41) we express $\acute{l}^{\bar{\alpha}}_{\alpha}, \acute{p}_{\bar{\alpha}}, \acute{a}_0, \acute{b}_0, \acute{c}_0$ in terms of $l^{\bar{\alpha}}_{\alpha}, p_{\bar{\alpha}}, a_0, b_0, c_0$ in accordance with (1.69), (1.56), (1.57) and project, after this is done, the equations (A.41) into the triad $l^{\bar{\alpha}}_{\alpha}$, we obtain three relations, two of which give exactly the result (1.70) while the third one then becomes an identity.

Appendix B

Homogeneous Spaces and Bianchi Classification

B.1 Homogeneous Three-Dimensional Spaces

Homogeneous three-dimensional spaces are generalizations of the three-dimensional Euclidean space with metric:

$$dl^2 = dx^2 + dy^2 + dz^2. \tag{B.1}$$

Homogeneity means equivalence of all geometrical and physical properties of the space at all points. Mathematically, this means the presence of a transitive group of motion of the space. For the flat space (B.1), this group is realized by the translations

$$\acute{x} = x + C_1, \ \acute{y} = y + C_2, \ \acute{z} = z + C_3 \tag{B.2}$$

with three arbitrary constants C_1, C_2, C_3. Homogeneity manifests itself through the fact that the new expression for the metric in the new coordinates $\acute{x}, \acute{y}, \acute{z}$

$$dl^2 = d\acute{x}^2 + d\acute{y}^2 + d\acute{z}^2 \tag{B.3}$$

has the same functional form as in the old coordinates x, y, z. For the flat case this is trivially evident because $g_{\alpha\beta}^{(old)} = diag(1,1,1)$ and also $g_{\alpha\beta}^{(new)} = diag(1,1,1)$. By analogy, for the general three-dimensional homogeneous spaces with metric $dl^2 = g_{\alpha_\beta}(x^\mu)dx^\alpha dx^\beta$, there should exist a transformation $\acute{x}^\alpha = \acute{x}^\alpha(x^\mu, C_1, C_2, C_3)$ depending on three arbitrary constants C_1, C_2, C_3 under which the interval acquires the *same functional form*, also with respect to the new coordinates; that is, one must have $dl^2 = g_{\alpha_\beta}(\acute{x}^\mu)d\acute{x}^\alpha d\acute{x}^\beta$. This is the *definition* of a three-dimensional homogeneous space.

If we use the frame $l_\alpha^{\bar{\alpha}}(x^\mu)$ and write the three-dimensional interval as

$$dl^2 = \eta_{\bar{\alpha}\bar{\beta}} l_\alpha^{\bar{\alpha}}(x^\mu) l_\beta^{\bar{\beta}}(x^\mu) dx^\alpha dx^\beta, \tag{B.4}$$

where $\eta_{\bar{\alpha}\bar{\beta}} = const$ is an arbitrary symmetric matrix, the homogeneity requirement implies that

$$l_\alpha^{\bar{\alpha}}(x^\mu)dx^\alpha = l_\alpha^{\bar{\alpha}}(\acute{x}^\mu)d\acute{x}^\alpha. \tag{B.5}$$

This relation puts severe restrictions on the possible forms of the functions $l_\alpha^{\bar{\alpha}}(x^\mu)$. Equations (B.5) can be written as the following differential equations for the transformation functions $\acute{x}^\alpha = \acute{x}^\alpha(x^\mu)$:

$$\frac{\partial \acute{x}^{\alpha}}{\partial x^{\beta}} = l^{\alpha}_{\bar{\gamma}}(\acute{x}^{\mu}) l^{\bar{\gamma}}_{\beta}(x^{\mu}). \tag{B.6}$$

For infinitesimal transformations $\acute{x}^{\alpha} = x^{\alpha} + \xi^{\alpha}(x^{\mu})$, the last equation is equivalent to the well-known Killing equations $l^{\bar{\alpha}}_{\alpha}\xi^{\alpha}{}_{,\beta} + l^{\bar{\alpha}}_{\beta,\alpha}\xi^{\alpha} = 0$. For finite transformations, the integrability conditions $\frac{\partial \acute{x}^{\alpha}}{\partial x^{\beta}\partial x^{\gamma}} = \frac{\partial \acute{x}^{\alpha}}{\partial x^{\gamma}\partial x^{\beta}}$ for the functions $\acute{x}^{\alpha}(x^{\mu})$, following from (B.6), can be reduced to the form:

$$\left[\frac{\partial l^{\bar{\gamma}}_{\lambda}(\acute{x})}{\partial \acute{x}^{\mu}} - \frac{\partial l^{\bar{\gamma}}_{\mu}(\acute{x})}{\partial \acute{x}^{\lambda}}\right] l^{\lambda}_{\bar{\alpha}}(\acute{x}) l^{\mu}_{\bar{\beta}}(\acute{x}) = \left[\frac{\partial l^{\bar{\gamma}}_{\lambda}(x)}{\partial x^{\mu}} - \frac{\partial l^{\bar{\gamma}}_{\mu}(x)}{\partial x^{\lambda}}\right] l^{\lambda}_{\bar{\alpha}}(x) l^{\mu}_{\bar{\beta}}(x). \tag{B.7}$$

Transitivity means that any two points x^{μ} and $\acute{x}^{\mu}$ can be connected by a transformation $\acute{x}^{\alpha} = \acute{x}^{\alpha}(x^{\mu})$ satisfying the equations (B.6). But equations (B.7) can be satisfied for any arbitrary pair $\acute{x}^{\mu}$ and x^{μ} if and only if their left and right parts are equal to the same set of constants; that is, one should have

$$\left[\frac{\partial l^{\bar{\gamma}}_{\lambda}(x)}{\partial x^{\mu}} - \frac{\partial l^{\bar{\gamma}}_{\mu}(x)}{\partial x^{\lambda}}\right] l^{\lambda}_{\bar{\alpha}}(x) l^{\mu}_{\bar{\beta}}(x) = C^{\bar{\gamma}}_{\bar{\alpha}\bar{\beta}}, \quad (C^{\bar{\gamma}}_{\bar{\alpha}\bar{\beta}} = const.). \tag{B.8}$$

We stress that these last equations automatically contain the Jacobi identity for the constants $C^{\bar{\gamma}}_{\bar{\alpha}\bar{\beta}}$:

$$C^{\bar{\lambda}}_{\bar{\alpha}\bar{\beta}} C^{\bar{\mu}}_{\bar{\lambda}\bar{\gamma}} + C^{\bar{\lambda}}_{\bar{\gamma}\bar{\alpha}} C^{\bar{\mu}}_{\bar{\lambda}\bar{\beta}} + C^{\bar{\lambda}}_{\bar{\beta}\bar{\gamma}} C^{\bar{\mu}}_{\bar{\lambda}\bar{\alpha}} = 0. \tag{B.9}$$

This can be seen by introducing the linear differential operators $X_{\bar{\alpha}} = l^{\alpha}_{\bar{\alpha}}\partial_{\alpha}$ (using the inverse triad $l^{\alpha}_{\bar{\alpha}}$) and by observing that the equation (B.8) is equivalent to the commutation rule $[X_{\bar{\alpha}} X_{\bar{\beta}}] = X_{\bar{\alpha}} X_{\bar{\beta}} - X_{\bar{\beta}} X_{\bar{\alpha}} = C^{\bar{\gamma}}_{\bar{\alpha}\bar{\beta}} X_{\bar{\alpha}}$. Inserting this into the operator Jacobi identity $[[X_{\bar{\alpha}} X_{\bar{\beta}}]\, X_{\bar{\gamma}}] + [[X_{\bar{\gamma}} X_{\bar{\alpha}}]\, X_{\bar{\beta}}] + [[X_{\bar{\beta}} X_{\bar{\gamma}}]\, X_{\bar{\alpha}}] \equiv 0$ we come to the relation (B.9). The constants $C^{\bar{\gamma}}_{\bar{\alpha}\bar{\beta}}$ are the structure constants of the group of motion of the three-dimensional homogeneous space under consideration.

The set of constants $C^{\bar{\gamma}}_{\bar{\alpha}\bar{\beta}}$ cannot be unique. Any linear transformation of the triad vectors with arbitrary constant coefficients is permissible. Such a transformation replaces the constant frame metric tensor $\eta_{\bar{\alpha}\bar{\beta}}$ in (B.4) by another one of the same type. Let the linear transformation of the triad vectors be $l^{\bar{\alpha}}_{\alpha} = A^{\bar{\alpha}}_{\bar{\beta}} \acute{l}^{\bar{\beta}}_{\alpha}$ where $A^{\bar{\alpha}}_{\bar{\beta}}$ are arbitrary constants. Then it is easy to derive the corresponding transformation law for the structure constants $C^{\bar{\gamma}}_{\bar{\alpha}\bar{\beta}}$:

$$C^{\bar{\gamma}}_{\bar{\alpha}\bar{\beta}} = A^{\bar{\gamma}}_{\bar{\mu}} D^{\bar{\lambda}}_{\bar{\alpha}} D^{\bar{\nu}}_{\bar{\beta}} \acute{C}^{\bar{\mu}}_{\bar{\lambda}\bar{\nu}}, \tag{B.10}$$

where $\acute{C}^{\bar{\gamma}}_{\bar{\alpha}\bar{\beta}} = \left[\frac{\partial \acute{l}^{\bar{\gamma}}_{\lambda}(x)}{\partial x^{\mu}} - \frac{\partial \acute{l}^{\bar{\gamma}}_{\mu}(x)}{\partial x^{\lambda}}\right] \acute{l}^{\lambda}_{\bar{\alpha}}(x) \acute{l}^{\mu}_{\bar{\beta}}(x)$ and the matrix $\mathbf{D}$ is inverse to the matrix $\mathbf{A}$; that is, $D^{\bar{\alpha}}_{\bar{\gamma}} A^{\bar{\gamma}}_{\bar{\beta}} = \delta^{\bar{\alpha}}_{\bar{\beta}}$. One can use this freedom to choose the triad vectors in some special way in order to reduce the structure constants to some canonical form. But there are invariant properties of the structure constants that cannot be changed by the aforementioned transformation of the triad vectors. The Bianchi classification for homogeneous spaces is precisely based on the existence of such invariants.

B.2 Bianchi Classification

We can describe the standard Bianchi classification in the following six steps:

1. Due to the antisymmetry $C^{\bar{\gamma}}_{\bar{\alpha}\bar{\beta}} = -C^{\bar{\gamma}}_{\bar{\beta}\bar{\alpha}}$, there are nine independent constants $C^{\bar{\gamma}}_{\bar{\alpha}\bar{\beta}}$. These independent constants $C^{\bar{\gamma}}_{\bar{\alpha}\bar{\beta}}$ can be equivalently represented by the nine components of an arbitrary constant matrix $C^{\bar{\alpha}\bar{\beta}}$:
$$C^{\bar{\gamma}}_{\bar{\alpha}\bar{\beta}} = \varepsilon_{\bar{\alpha}\bar{\beta}\bar{\lambda}} C^{\bar{\lambda}\bar{\gamma}}, \tag{B.11}$$
where $\varepsilon_{\bar{\alpha}\bar{\beta}\bar{\lambda}}$ is again the totally antisymmetric three-dimensional Levi-Civita symbol ($\varepsilon_{\bar{1}\bar{2}\bar{3}} = 1$). If we then substitute this expression for $C^{\bar{\gamma}}_{\bar{\alpha}\bar{\beta}}$ into the Jacobi identity (B.9), the result will be:
$$\varepsilon_{\bar{\alpha}\bar{\beta}\bar{\lambda}} C^{\bar{\beta}\bar{\lambda}} C^{\bar{\alpha}\bar{\gamma}} = 0. \tag{B.12}$$
2. The transformations law (B.10) for the structure constants correspond to the following transformations for the constants $C^{\bar{\alpha}\bar{\beta}}$:
$$C^{\bar{\alpha}\bar{\beta}} = (\det \mathbf{A})^{-1} A^{\bar{\alpha}}_{\bar{\mu}} A^{\bar{\beta}}_{\bar{\nu}} \acute{C}^{\bar{\mu}\bar{\nu}}. \tag{B.13}$$
Appearance of $\det \mathbf{A}$ in this formula is due to the fact that the symbol $\varepsilon_{\bar{\alpha}\bar{\beta}\bar{\lambda}}$ transforms as tensor density: $\varepsilon_{\bar{\alpha}\bar{\beta}\bar{\gamma}} = (\det \mathbf{A}) D^{\bar{\mu}}_{\bar{\alpha}} D^{\bar{\nu}}_{\bar{\beta}} D^{\bar{\lambda}}_{\bar{\gamma}} \acute{\varepsilon}_{\bar{\mu}\bar{\nu}\bar{\lambda}}$, where $\acute{\varepsilon}_{\bar{\mu}\bar{\nu}\bar{\lambda}} \equiv \varepsilon_{\bar{\mu}\bar{\nu}\bar{\lambda}}$. By this transformation it is always possible to reduce the matrix $C^{\bar{\alpha}\bar{\beta}}$ to the form:
$$C^{\bar{\alpha}\bar{\beta}} = \begin{pmatrix} n_{\bar{1}} & 0 & 0 \\ 0 & C^{\bar{2}\bar{2}} & C^{\bar{2}\bar{3}} \\ o & C^{\bar{3}\bar{2}} & C^{\bar{3}\bar{3}} \end{pmatrix}. \tag{B.14}$$
After such a choice, we still have the freedom of making triad transformations but with the restrictions $A^{\bar{1}}_{\bar{2}} = A^{\bar{1}}_{\bar{3}} = 0$ and $A^{\bar{2}}_{\bar{1}} = A^{\bar{3}}_{\bar{1}} = 0$.
3. Now, the Jacobi identities (B.12) give only one constraint:
$$\left(C^{\bar{2}\bar{3}} - C^{\bar{3}\bar{2}}\right) n_{\bar{1}} = 0\,. \tag{B.15}$$
4. If $n_{\bar{1}} \neq 0$ we have $C^{\bar{2}\bar{3}} - C^{\bar{3}\bar{2}} = 0$ and by the remaining transformations with $A^{\bar{a}}_{\bar{b}} \neq 0$, $\bar{a}, \bar{b} = \bar{2}, \bar{3}$, the 2×2 matrix $C^{\bar{a}\bar{b}}$ in (B.14) can be made diagonal. Then
$$C^{\bar{\alpha}\bar{\beta}} = \begin{pmatrix} n_{\bar{1}} & 0 & 0 \\ 0 & n_{\bar{2}} & 0 \\ o & 0 & n_{\bar{3}} \end{pmatrix}. \tag{B.16}$$
The diagonality condition for $C^{\bar{\alpha}\bar{\beta}}$ is preserved under the transformations with diagonal $A^{\bar{\alpha}}_{\bar{\beta}}$. Under these transformations, the three parameters $n_{\bar{1}}, n_{\bar{2}}, n_{\bar{3}}$ change in the following way:
$$n_{\bar{\alpha}} = \left(A^{\bar{1}}_{\bar{1}} A^{\bar{2}}_{\bar{2}} A^{\bar{3}}_{\bar{3}}\right)^{-1} \left(A^{\bar{\alpha}}_{\bar{\alpha}}\right)^2 \acute{n}_{\bar{\alpha}}\ , \quad \textit{no summation over } \bar{\alpha}. \tag{B.17}$$

By these diagonal transformations, the modulus of any $n_{\bar{\alpha}}$ (if it is not zero) can be made equal to unity. Taking into account that the simultaneous change of sign of all $n_{\bar{\alpha}}$ produce nothing new, we come to the following invariantly different sets for the numbers $n_{\bar{1}}, n_{\bar{2}}, n_{\bar{3}}$ (invariantly different in the sense that there is no way to pass from one to another by some transformation of the triad $l^{\bar{\alpha}}_{\alpha} = A^{\bar{\alpha}}_{\bar{\beta}} \acute{l}^{\bar{\beta}}_{\alpha}$), that is to the following different types of homogeneous spaces with diagonal matrix $C^{\bar{\alpha}\bar{\beta}}$ from (B.16):

$$\begin{aligned} Bianchi\ IX: &\ (n_{\bar{1}}, n_{\bar{2}}, n_{\bar{3}}) = (1,1,1), \\ Bianchi\ VIII: &\ (n_{\bar{1}}, n_{\bar{2}}, n_{\bar{3}}) = (1,1,-1), \\ Bianchi\ VII_0: &\ (n_{\bar{1}}, n_{\bar{2}}, n_{\bar{3}}) = (1,1,0), \\ Bianchi\ VI_0: &\ (n_{\bar{1}}, n_{\bar{2}}, n_{\bar{3}}) = (1,-1,0), \\ Bianchi\ II: &\ (n_{\bar{1}}, n_{\bar{2}}, n_{\bar{3}}) = (1,0,0). \end{aligned} \tag{B.18}$$

5. Consider now the case $n_{\bar{1}} = 0$. It can also happen in that case that $C^{\bar{2}\bar{3}} - C^{\bar{3}\bar{2}} = 0$. We return then to the situation already analyzed in the previous section but with the additional condition $n_{\bar{1}} = 0$. Now, all essentially different types for the sets $n_{\bar{1}}, n_{\bar{2}}, n_{\bar{3}}$ are $(0,1,1)$, $(0,1-1)$, $(0,0,1)$ and $(0,0,0)$. The first three repeat the types $B\ VII_0$, $B\ VI_0$, $B\ II$. Consequently, only one new type arises:

$$Bianchi\ I: \ (n_{\bar{1}}, n_{\bar{2}}, n_{\bar{3}}) = (0,0,0). \tag{B.19}$$

6. We are left with the case $n_{\bar{1}} = 0$ and $C^{\bar{2}\bar{3}} - C^{\bar{3}\bar{2}} \neq 0$. Now the 2×2 matrix $C^{\bar{a}\bar{b}}$ $(\bar{a}, \bar{b} = \bar{2}, \bar{3})$ is non-symmetric and it cannot be made diagonal by a transformations using $A^{\bar{a}}_{\bar{b}} \neq 0$. However, its symmetric part can be diagonalized, that is the 3×3 matrix $C^{\bar{\alpha}\bar{\beta}}$ can be reduced to the form:

$$C^{\bar{\alpha}\bar{\beta}} = \begin{pmatrix} 0 & 0 & 0 \\ 0 & n_{\bar{2}} & a \\ o & -a & n_{\bar{3}} \end{pmatrix}, \tag{B.20}$$

where a is an arbitrary number. After this is done, there still remains the possibility to perform transformations with diagonal $A^{\bar{a}}_{\bar{b}}$, under which the quantities $n_{\bar{2}}, n_{\bar{3}}$ and a change as follows:

$$n_{\bar{2}} = \left(A^{\bar{1}}_{\bar{1}} A^{\bar{2}}_{\bar{2}} A^{\bar{3}}_{\bar{3}}\right)^{-1} (A^{\bar{2}}_{\bar{2}})^2 \acute{n}_{\bar{2}}, \ n_{\bar{3}} = \left(A^{\bar{1}}_{\bar{1}} A^{\bar{2}}_{\bar{2}} A^{\bar{3}}_{\bar{3}}\right)^{-1} (A^{\bar{3}}_{\bar{3}})^2 \acute{n}_{\bar{3}}, a = (A^{\bar{1}}_{\bar{1}})^{-1} \acute{a}. \tag{B.21}$$

These formulas shows that for nonzero $n_{\bar{2}}, n_{\bar{3}}, a$, the combination $a^2 (n_{\bar{2}} n_{\bar{3}})^{-1}$ is an invariant quantity. By a choice of $A^{\bar{1}}_{\bar{1}}$, we can impose the condition $a > 0$ and after this is done, the choice of the sign of $A^{\bar{3}}_{\bar{3}} (A^{\bar{2}}_{\bar{2}})^{-1}$ permits one to change both signs of $n_{\bar{2}}$ and $n_{\bar{3}}$ simultaneously, that is the set $(n_{\bar{2}}, n_{\bar{3}})$ is equivalent to the set $(-n_{\bar{2}}, -n_{\bar{3}})$. It follows that we have the following four different possibilities:

$$(a, n_{\bar{2}}, n_{\bar{3}}) = (a,0,0), (a,0,1), (a,1,1), (a,1,-1). \tag{B.22}$$

For the first two, the number a can be transformed to unity by a choice of the parameters $A_{\bar{1}}^{\bar{1}}$ and $A_{\bar{3}}^{\bar{3}}(A_{\bar{2}}^{\bar{2}})^{-1}$. For the second two possibilities, both of these parameters are already fixed and a remains an invariant and arbitrary positive number. Historically these four types of homogeneous spaces with matrix $C^{\bar{\alpha}\bar{\beta}}$ from (B.20) have been classified as:

$$\begin{aligned} &Bianchi\ V: \ n_{\bar{1}} = 0,\ (a, n_{\bar{2}}, n_{\bar{3}}) = (1,0,0), \\ &Bianchi\ IV: \ n_{\bar{1}} = 0,\ (a, n_{\bar{2}}, n_{\bar{3}}) = (1,0,1), \\ &Bianchi\ VII: \ n_{\bar{1}} = 0,\ (a, n_{\bar{2}}, n_{\bar{3}}) = (a,1,1), \\ &Bianchi\ III: \ n_{\bar{1}} = 0,\ (a, n_{\bar{2}}, n_{\bar{3}}) = (1,1,-1), \\ &Bianchi\ VI: \ n_{\bar{1}} = 0,\ (a, n_{\bar{2}}, n_{\bar{3}}) = (a,1,-1). \end{aligned} \tag{B.23}$$

We see that type $B\ III$ is just a particular case of type $B\ VI$ corresponding to $a = 1$. The types $B\ VII$ and $B\ VI$ in fact contain an infinity of invariantly different types of algebras corresponding to the arbitrariness of the continuous parameter a. The type $B\ VII_0$ from (B.18) is a particular case of $B\ VII$ corresponding to $a = 0$ while the type $B\ VI_0$ from (B.18) is a particular case of $B\ VI$ corresponding also to $a = 0$.

The triad vectors $l_{\alpha}^{\bar{\alpha}}$ can be found from the differential equations (B.8) for each given set of structure constants. Then (B.4) gives the metric for the homogeneous space under interest.

The three-dimensional Ricci tensor $P_{\alpha}^{\beta} = g^{\beta\gamma}P_{\alpha\gamma}$ for the homogeneous metric (B.4), together with its components $P_{\bar{\alpha}}^{\bar{\beta}} = \eta^{\bar{\beta}\bar{\gamma}}P_{\bar{\alpha}\bar{\gamma}}$ into the triad $l_{\alpha}^{\bar{\alpha}}(x^{\mu})$, follows from the formulas of Appendix A.2 applied to the case $N = 3$. One gets:

$$\begin{aligned} 2\eta P_{\bar{\alpha}}^{\bar{\beta}} =& 2C^{\bar{\beta}\bar{\gamma}}\eta_{\bar{\lambda}\bar{\alpha}}\eta_{\bar{\mu}\bar{\gamma}}C^{\bar{\lambda}\bar{\mu}} + C^{\bar{\gamma}\bar{\beta}}\eta_{\bar{\lambda}\bar{\alpha}}\eta_{\bar{\mu}\bar{\gamma}}C^{\bar{\lambda}\bar{\mu}} \\ &+ C^{\bar{\beta}\bar{\gamma}}\eta_{\bar{\lambda}\bar{\gamma}}\eta_{\bar{\mu}\bar{\alpha}}C^{\bar{\lambda}\bar{\mu}} - \eta_{\bar{\lambda}\bar{\mu}}C^{\bar{\lambda}\bar{\mu}}\left(\eta_{\bar{\alpha}\bar{\gamma}}C^{\bar{\beta}\bar{\gamma}} + \eta_{\bar{\alpha}\bar{\gamma}}C^{\bar{\gamma}\bar{\beta}}\right) \\ &+ \delta_{\bar{\alpha}}^{\bar{\beta}}\left[\left(\eta_{\bar{\lambda}\bar{\mu}}C^{\bar{\lambda}\bar{\mu}}\right)^2 - 2C^{\bar{\lambda}\bar{\mu}}\eta_{\bar{\lambda}\bar{\nu}}\eta_{\bar{\mu}\bar{\gamma}}C^{\bar{\nu}\bar{\gamma}}\right], \end{aligned} \tag{B.24}$$

where $\eta = \det(\eta_{\bar{\alpha}\bar{\beta}})$ and where we showed the product $2\eta P_{\bar{\alpha}}^{\bar{\beta}}$ instead of $P_{\bar{\alpha}}^{\bar{\beta}}$ to avoid too much brackets in the right-hand side of this formula.

The three-dimensional Bianchi identities $P_{\bar{\alpha};\bar{\beta}}^{\bar{\beta}} - \frac{1}{2}P_{\bar{\beta};\bar{\alpha}}^{\bar{\beta}} \equiv 0$ take the form:

$$P_{\bar{\gamma}}^{\bar{\beta}}C_{\bar{\beta}\bar{\alpha}}^{\bar{\gamma}} + P_{\bar{\alpha}}^{\bar{\gamma}}C_{\bar{\gamma}\bar{\beta}}^{\bar{\beta}} \equiv 0 \tag{B.25}$$

for homogeneous spaces.

B.3 Frame Vectors

We are interested in the most general models, that is in types $B\ IX$ and $B\ VIII$. For these two types, all structure constants vanish except the following three: $C^{\bar{1}}_{\bar{2}\bar{3}}, C^{\bar{2}}_{\bar{3}\bar{1}}, C^{\bar{3}}_{\bar{1}\bar{2}}$. The corresponding frame vectors $l^{\bar{\alpha}}_{\alpha}$ can be found from the differential equations (B.8). One of the forms of the solution has been presented in [14]. It is:

$$\begin{aligned} l^{\bar{1}}_{\alpha} &= \left(l^{\bar{1}}_{1}, l^{\bar{1}}_{2}, l^{\bar{1}}_{3}\right) = \left(F\cos\omega z\ ,\ F^{-1}\omega^{-1}\lambda\sin\omega z\ ,\ 0\right), \\ l^{\bar{2}}_{\alpha} &= \left(l^{\bar{2}}_{1}, l^{\bar{2}}_{2}, l^{\bar{2}}_{3}\right) = \left(-F\omega^{-1}\mu\sin\omega z\ ,\ F^{-1}\cos\omega z\ ,\ 0\right), \\ l^{\bar{3}}_{\alpha} &= \left(l^{\bar{3}}_{1}, l^{\bar{3}}_{2}, l^{\bar{3}}_{3}\right) = \left(\nu y\ ,\ 0\ ,\ 1\right), \end{aligned} \tag{B.26}$$

where the quantities λ, μ, ν are arbitrary constants and where we used the following notations:

$$\left(x^1, x^2, x^3\right) = (x, y, z);\ F = \sqrt{1 - \lambda\nu y^2};\ \omega = \sqrt{\lambda\mu}. \tag{B.27}$$

The vectors $l^{\alpha}_{\bar{\alpha}}$ which are inverse to the $l^{\bar{\alpha}}_{\alpha}$ are:

$$\begin{aligned} l^{\alpha}_{\bar{1}} &= \left(l^{1}_{\bar{1}}, l^{2}_{\bar{1}}, l^{3}_{\bar{1}}\right) = \left(F^{-1}\cos\omega z\ ,\ F\omega^{-1}\mu\sin\omega z\ ,\ -\nu y F^{-1}\cos\omega z\right), \\ l^{\alpha}_{\bar{2}} &= \left(l^{1}_{\bar{2}}, l^{2}_{\bar{2}}, l^{3}_{\bar{2}}\right) = \left(-F^{-1}\omega^{-1}\lambda\sin\omega z\ ,\ F\cos\omega z\ ,\ \nu y F^{-1}\omega^{-1}\lambda\sin\omega z\right), \\ l^{\alpha}_{\bar{3}} &= \left(l^{1}_{\bar{3}}, l^{2}_{\bar{3}}, l^{3}_{\bar{3}}\right) = \left(0\ ,\ 0\ ,\ 1\right). \end{aligned} \tag{B.28}$$

It is easy to check that $l^{\bar{\alpha}}_{\alpha} l^{\beta}_{\bar{\alpha}} = \delta^{\beta}_{\alpha}$, $l^{\bar{\alpha}}_{\alpha} l^{\alpha}_{\bar{\beta}} = \delta^{\bar{\beta}}_{\bar{\alpha}}$ as it should be. It is also easy to check that the nonzero structure constants, which can be calculated from (B.8), are:

$$C^{\bar{1}}_{\bar{2}\bar{3}} = \lambda\ ,\ C^{\bar{2}}_{\bar{3}\bar{1}} = \mu\ ,\ C^{\bar{3}}_{\bar{1}\bar{2}} = \nu \tag{B.29}$$

and for the matrix $C^{\bar{\alpha}\bar{\beta}}$ we have $C^{\bar{\alpha}\bar{\beta}} = diag(n_{\bar{1}}, n_{\bar{2}}, n_{\bar{3}}) = (\lambda, \mu, \nu)$. The absolute values of the numbers λ, μ, ν are of no importance and we can put them equal to one. However, it is useful to keep these parameters to be arbitrary for computational convenience and also for including those limiting cases when some of them tend to zero (it is worth mentioning that the frame vectors (B.26) have well defined limits in such cases). With this generalization the set of structure constants (B.29) covers not only types IX and VIII but all six Bianchi types (B.18)–(B.19).

The case $\lambda = \mu = \nu = 0$ corresponds to the type $B\ I$ and from (B.26) one gets:

$$l^{\bar{1}}_{\alpha} = (1, 0, 0)\ ;\ l^{\bar{2}}_{\alpha} = (0, 1, 0)\ ;\ l^{\bar{3}}_{\alpha} = (0, 0, 1)\,. \tag{B.30}$$

If $\lambda = \mu = 0$, $\nu \neq 0$, we have type $B\ II$ with frame:

$$l^{\bar{1}}_{\alpha} = (1, 0, 0)\ ;\ l^{\bar{2}}_{\alpha} = (0, 1, 0)\ ;\ l^{\bar{3}}_{\alpha} = (\nu y, 0, 1)\,. \tag{B.31}$$

In the case $\lambda \neq 0, \mu \neq 0$, $\nu = 0$, we obtain from (B.26) the frame vectors:

$$\begin{aligned} l^{\bar{1}}_{\alpha} &= \left(\cos\omega z,\ \omega^{-1}\lambda\sin\omega z,\ 0\right), \\ l^{\bar{2}}_{\alpha} &= \left(-\omega^{-1}\mu\sin\omega z,\ \cos\omega z,\ 0\right), \\ l^{\bar{3}}_{\alpha} &= \left(0,\ 0,\ 1\right). \end{aligned} \tag{B.32}$$

These last expressions give type $B\ VI_0$ if $\lambda\mu < 0$ and type $B\ VII_0$ if $\lambda\mu > 0$.

For the type $B\ IX$, we should take all λ, μ, ν positive and for the type $B\ VIII$, two of the λ, μ, ν positive and one negative.

B.4 On the Freezing Effect in Bianchi IX Model

In this sub-Appendix we will confirm that the fact that the angular velocities $\frac{\partial \psi}{\partial \tau}, \frac{\partial \theta}{\partial \tau}, \frac{\partial \varphi}{\partial \tau}$ go to zero does indeed lead to the existence of the limits given in (2.51) for the angles ψ, θ, φ. Let us consider, for example, the angle ψ, writing it, as in equation (2.67), in the form $\psi = \psi_0 + \delta\psi$. From equation (2.70) it follows that the total contribution to ψ of the correction $\delta\psi$, from some initial moment τ_0, to the singularity $\tau = \infty$, is

$$\int_{\tau_0}^{\infty} \left(\frac{B^2}{A^2} \frac{1}{\cos^2 \theta_0} - \frac{C^2}{B^2} \frac{1}{\sin^2 \theta_0} \right) \frac{\cos \theta_0}{C^2} d\tau. \tag{B.33}$$

It is sufficient to analyze the contribution due to the first term of the integrand which is proportional to

$$S = \int_{\tau_0}^{\infty} \frac{B^2}{A^2} d\tau \tag{B.34}$$

because the contribution due to the second term can be considered analogously, as can the contributions of the corrections $\delta\theta$ and $\delta\varphi$. Our statements will be proved if we can show that S goes to zero as $\tau_0 \to \infty$.

With this goal in mind we make the following estimate. First of all, we note that the function B^2/A^2 is always positive, bounded from above, and goes to zero as $\tau \to \infty$, making an infinite number of oscillations. The major contribution to S is made, therefore, near the maxima of the function B^2/A^2, where the function is given exactly by the solution (2.57). Numbering the regions near the maxima by $n = 0, 1, 2, \ldots$ (beginning from the region lying nearest the initial moment $\tau = \tau_0$) we find

$$S = \sum_{n=0}^{\infty} S_n\,, \tag{B.35}$$

$$\begin{aligned} S_n &= \int_{\tau_0}^{\infty} \frac{Q_n^2 \left[p_A^2(u_n) - p_B^2(u_n)\right] d\tau}{\cosh^2 \left\{Q_n \left[p_A(u_n) - p_B(u_n)\right] \tau + const\right\}} \\ &= 2Q_n \left[p_A(u_n) - p_B(u_n)\right]. \end{aligned}$$

Here Q_n and u_n, are the values of Q and u in the n-th region. Each term S_n of the series is obviously a positive number (because $p_A > p_B$) satisfying the inequality $S_n \leqslant (4/\sqrt{3})\, Q_n$ (because the modulus of the difference between any two Kasner exponents is, as we have already noted, less than $2/\sqrt{3}$). From this we have

$$S \leqslant \frac{4}{\sqrt{3}} \sum_{n=0}^{\infty} Q_n\,. \tag{B.36}$$

The rule by which one goes from Q_n, to Q_{n+1} is based on the law (1.57). If, in the initial epoch (near $\tau = \tau_0$), $Q = Q_0$, then the next terms will be

$$Q_1 = Q_0 \left[1 + 2p_{\bar{1}}(u_0)\right], \quad Q_2 = Q_1 \left[1 + 2p_{\bar{1}}(u_0 - 1)\right], \quad \ldots \tag{B.37}$$

Finally we will arrive at a term in which the parameter u is less than 1. Making the substitution $\acute{u} = 1/u$ in that term (which we can do because of the invariance of the function $p_{\bar{1}}(u)$ under such a substitution), we are able to continue to follow Q until $\acute{u} < 1$, whereupon we let $u^{(new)} = 1/\acute{u}$ and so on. (Note that we do not investigate the case where one value of Q corresponds to more than one maximum of the function B/A. It can be shown that such maxima will only occur in pairs, which can only double the upper bound in our estimate. This does affect the result.) The ratio of any term to the one just before it, Q_{n+1}/Q_n, is

$$Q_{n+1}/Q_n = 1 + 2p_{\bar{1}}(u_n) = \frac{1 - u_n + u_n^2}{1 + u_n + u_n^2}\,. \tag{B.38}$$

This ratio is less than 1 for $0 < u_n < \infty$, and equal to 1 for $u_n = 0, \infty$. We first assume that in the process of evolution we never encounter a value of u greater than some (large) positive number N. Because of the invariance of our sequence under the transformation $u \to 1/u$, the possible values of u lie in the interval

$$1/N \leqslant u \leqslant N\,. \tag{B.39}$$

With this condition the "ratio test" for convergence indicates that the series in the right-hand side of (B.36) converges, and it is easy to obtain the limit

$$\sum_{n=0}^{\infty} Q_n < \frac{Q_0}{2\left|p_{\bar{1}}(N)\right|}\,. \tag{B.40}$$

Now from (B.36) and (B.40) we see that if we let $\tau_0 \to \infty$ in the integral (B.34) then the limiting value of S will be zero because $Q_0 \to 0$ due to the systematic decrease of the parameter Q in the course of approaching the singularity.

We must now consider the condition (B.39). It is impossible to fulfill such a condition strictly, but we can show that the probability of its being violated is extremely small. In Chapter 3, it was shown that we approach a stationary statistical distribution for the parameter u very rapidly. It can be shown that the probability of a regime for which the random variable u is greater than N is of order $(1/N) \ll 1$ (we have taken $N \gg 1$). If, in the process of evolution, the condition (B.39) fails a finite number of times, then we need only choose a new value of N, the largest one. We are forced, however, to consider the case in which $u > N$ an infinite number of times. The probability of this is obviously strictly equal to zero. These considerations verify our initial statements.

We close this appendix by pointing out that the Hamiltonian formulation of the Bianchi models has been extensively studied in [104, 105].

Appendix C
Spinor Field

We have shown that in the presence of a scalar field, the general solution of the gravitational equations in the vicinity of a cosmological singularity acquires a monotonous power-law asymptotical behavior instead of the oscillation regime. In the present section we shall perform a similar task for the Dirac spinor field, which we shall regard as classical (as a wave function of a single particle). It will be shown in what follows that the spinor field changes the asymptotics of the general cosmological solution in a way analogous to the role played by a scalar field: instead of the oscillation regime we obtain simple power-law asymptotics, characterized by three positive exponents $p_{\bar{\alpha}}$ which satisfy the same relations (4.4) (with the appropriate interpretation of the value p_φ) [17]. However, in the case of a spinor field, the metric has a more complicated form than (1.42)–(1.43): the spatial metric tensor $g_{\alpha\beta}$ is now a linear combination (with time-independent coefficients) of the six powers: $t^{2p_{\bar{1}}}, t^{2p_{\bar{2}}}, t^{2p_{\bar{3}}}, t^{p_{\bar{1}}+p_{\bar{2}}}, t^{p_{\bar{1}}+p_{\bar{3}}}, t^{p_{\bar{2}}+p_{\bar{3}}}$.

This phenomenon is due to the following circumstance. In the presence of the spinor field one can, as before, indicate three time-independent directions $l^{\bar{\alpha}}_{\alpha}$ along which the asymptotic scales vary in accordance with three different power laws $t^{p_{\bar{1}}}, t^{p_{\bar{2}}}, t^{p_{\bar{3}}}$. Now, however, these directions cannot form the directions of an orthogonal triad in the space with the metric $g_{\alpha\beta}$. If we denote the orthogonal triad vectors by $U^{\bar{\alpha}}_{\alpha}$, so that $g_{\alpha\beta}$ has the form

$$g_{\alpha\beta} = U^{\bar{1}}_{\alpha}U^{\bar{1}}_{\beta} + U^{\bar{2}}_{\alpha}U^{\bar{2}}_{\beta} + U^{\bar{3}}_{\alpha}U^{\bar{3}}_{\beta} \,, \tag{C.1}$$

then in the vicinity of the singularity, as it will be shown, the vectors $U^{\bar{\alpha}}_{\alpha}$ have the structure

$$U^{\bar{\alpha}}_{\alpha} = \sum_{\bar{\beta}} A^{\bar{\alpha}}_{\bar{\beta}} t^{p_{\bar{\beta}}} l^{\bar{\beta}}_{\alpha} \,, \tag{C.2}$$

where the values of $A^{\bar{\alpha}}_{\bar{\beta}}$ (as well as the vectors $l^{\bar{\beta}}_{\alpha}$) do not depend on time. In the case of the scalar field, the matrix $A^{\bar{\alpha}}_{\bar{\beta}}$ can be reduced to diagonal form by a time-independent orthogonal rotation of the triad $U^{\bar{\alpha}}_{\alpha}$, which yields the metric (1.42)–(1.43). In the case of a spinor field, such a reduction is not possible and, as a result, the metric tensor $g_{\alpha\beta}$ will involve besides the three powers $t^{2p_{\bar{1}}}, t^{2p_{\bar{2}}}, t^{2p_{\bar{3}}}$ also the cross terms $t^{p_{\bar{1}}+p_{\bar{2}}}, t^{p_{\bar{1}}+p_{\bar{3}}}, t^{p_{\bar{2}}+p_{\bar{3}}}$.

C.1 Equations of the Gravitational and Spinor Fields

The set of equations describing the coupled gravitational and spinor fields evidently consists of the Dirac equation in curved space and of the Einstein equations with the right-hand side given by the energy–momentum tensor of the spinor field. The rules for writing these equations are well known [73], [31]. Let $L^{\bar{\alpha}}_{\alpha}$ be a certain orthogonal triad in the synchronous reference system, such that the metric has the form:

$$-ds^2 = -dt^2 + \delta_{\bar{\alpha}\bar{\beta}} L^{\bar{\alpha}}_{\alpha} L^{\bar{\beta}}_{\beta} dx^{\alpha} dx^{\beta}, \tag{C.3}$$

where $\delta_{\bar{\alpha}\bar{\beta}}$ ($\delta_{\bar{\alpha}\bar{\beta}} \equiv \delta^{\bar{\alpha}\bar{\beta}}$) is the Kroneker symbol. Below we will use the inverse vectors $L^{\alpha}_{\bar{\alpha}}$ which, as usual, are defined by the relations $L^{\bar{\alpha}}_{\alpha} L^{\alpha}_{\bar{\beta}} = \delta^{\bar{\alpha}}_{\bar{\beta}}$, $L^{\alpha}_{\bar{\alpha}} L^{\bar{\alpha}}_{\beta} = \delta^{\alpha}_{\beta}$ and also the same triad vectors but written in the forms $L_{\bar{\alpha}\alpha}$ and $L^{\alpha\bar{\alpha}}$, where $L_{\bar{\alpha}\alpha} = \delta_{\bar{\alpha}\bar{\beta}} L^{\bar{\beta}}_{\alpha}$ and $L^{\alpha\bar{\alpha}} = \delta^{\bar{\alpha}\bar{\beta}} L^{\alpha}_{\bar{\beta}}$. Then let $\gamma^0, \gamma^{\bar{\alpha}}$ be the ordinary Dirac matrices of the special relativity theory, which satisfy the equations

$$\gamma^{\bar{\alpha}}\gamma^{\bar{\beta}} + \gamma^{\bar{\beta}}\gamma^{\bar{\alpha}} = 2\delta^{\bar{\alpha}\bar{\beta}} \ , \ \gamma^0\gamma^{\bar{\alpha}} + \gamma^{\bar{\alpha}}\gamma^0 = 0 \ , \ (\gamma^0)^2 = -1 \ , \tag{C.4}$$

$$(\gamma^0)^{\dagger} = -\gamma^0 \ , \ \left(\gamma^{\bar{\alpha}}\right)^{\dagger} = \gamma^{\bar{\alpha}} \ , \tag{C.5}$$

where the dagger symbol stands for Hermitian conjugation. Then the Dirac equation in the space with the metric (C.3) takes the form:

$$\gamma^0\dot{\Psi} + \gamma^{\bar{\alpha}}\Psi_{,\alpha} L^{\alpha}_{\bar{\alpha}} - (\gamma^0\Gamma_0 + \gamma^{\bar{\alpha}}\Gamma_{\bar{\alpha}})\Psi + m\Psi = 0 \ . \tag{C.6}$$

The four matrices $\Gamma_0, \Gamma_{\bar{\alpha}}$ are the well-known analogues of the Christoffel symbols for the spinor fields. A simple calculation shows that in the given case they acquire the following form:

$$\Gamma_{\bar{\alpha}} = \frac{1}{4}\kappa_{\bar{\alpha}\bar{\beta}}\gamma^0\gamma^{\bar{\beta}} + \frac{1}{8}\left(\sigma_{\bar{\gamma}\bar{\beta}\bar{\alpha}} - \sigma_{\bar{\beta}\bar{\gamma}\bar{\alpha}} + \sigma_{\bar{\alpha}\bar{\beta}\bar{\gamma}}\right) s^{\bar{\gamma}\bar{\beta}} \ , \tag{C.7}$$

$$\Gamma_0 = \frac{1}{8}\left(\dot{L}_{\bar{\alpha}\alpha} L^{\alpha}_{\bar{\beta}} - \dot{L}_{\bar{\beta}\alpha} L^{\alpha}_{\bar{\alpha}}\right) s^{\bar{\alpha}\bar{\beta}} \ . \tag{C.8}$$

The projections of the second fundamental form are defined as usual:

$$\kappa_{\bar{\alpha}\bar{\beta}} = \kappa_{\alpha\beta} L^{\alpha}_{\bar{\alpha}} L^{\beta}_{\bar{\beta}} = \dot{g}_{\alpha\beta} L^{\alpha}_{\bar{\alpha}} L^{\beta}_{\bar{\beta}} \ , \ \kappa^{\bar{\beta}}_{\bar{\alpha}} = \delta^{\bar{\beta}\bar{\mu}} \kappa_{\bar{\alpha}\bar{\mu}} \ . \tag{C.9}$$

The definitions of the quantities $s^{\bar{\alpha}\bar{\beta}}$ and $\sigma_{\bar{\alpha}\bar{\beta}\bar{\gamma}}$ are:

$$s^{\bar{\alpha}\bar{\beta}} = \frac{1}{2}\left(\gamma^{\bar{\alpha}}\gamma^{\bar{\beta}} - \gamma^{\bar{\beta}}\gamma^{\bar{\alpha}}\right) \ , \tag{C.10}$$

$$\sigma^{\bar{\alpha}}_{\bar{\beta}\bar{\gamma}} = \left(L^{\bar{\alpha}}_{\alpha,\beta} - L^{\bar{\alpha}}_{\beta,\alpha}\right) L^{\alpha}_{\bar{\beta}} L^{\beta}_{\bar{\gamma}} \ , \ \sigma_{\bar{\alpha}\bar{\beta}\bar{\gamma}} = \delta_{\bar{\alpha}\bar{\mu}} \sigma^{\bar{\mu}}_{\bar{\beta}\bar{\gamma}} \ . \tag{C.11}$$

It is easy to see that $\kappa_{\bar{\alpha}\bar{\beta}} = \dot{L}_{\bar{\alpha}\alpha} L^{\alpha}_{\bar{\beta}} + \dot{L}_{\bar{\beta}\alpha} L^{\alpha}_{\bar{\alpha}}$. In order to simplify the subsequent analysis, we use the local rotation freedom in the choice of the orthogonal triad $L^{\bar{\alpha}}_{\alpha}$, i.e., we impose the restriction:

$$\dot{L}_{\bar{\alpha}\alpha} L^{\alpha}_{\bar{\beta}} = \dot{L}_{\bar{\beta}\alpha} L^{\alpha}_{\bar{\alpha}} \ . \tag{C.12}$$

In this case

$$\kappa_{\bar{\alpha}\bar{\beta}} = 2\dot{L}_{\bar{\alpha}\alpha}L^{\alpha}_{\bar{\beta}} , \tag{C.13}$$

and the matrix Γ_0 is found to be zero (this choice of orthogonal frames is analogous to the choice of the synchronous reference system in which those Christoffel symbols containing the index zero are essentially simple; see (1.9)). In the gauge (C.12), the Dirac equation (C.6) acquires a simpler form:

$$\gamma^0\dot{\Psi} + \gamma^{\bar{\alpha}}\Psi_{,\alpha}L^{\alpha}_{\bar{\alpha}} - \gamma^{\bar{\alpha}}\Gamma_{\bar{\alpha}}\Psi + m\Psi = 0. \tag{C.14}$$

The condition (C.12) does not fix the triad completely, since after it has been imposed, certain orthogonal rotations are still available, namely, those that are independent of time and that contain therefore three arbitrary local parameters (three functions of the spatial coordinates). It is necessary to keep in mind this feature while evaluating the number of physically arbitrary three-dimensional functions in the solution.

The energy–momentum tensor of the field Ψ is given in [31]. Writing its components $T_{00}, T_{0\alpha}, T_{\alpha\beta}$ and projecting them on the directions of the triad ($T_{0\bar{\alpha}} = T_{0\alpha}L^{\alpha}_{\bar{\alpha}}$, $T_{\bar{\alpha}\bar{\beta}} = T_{\alpha\beta}L^{\alpha}_{\bar{\alpha}}L^{\beta}_{\bar{\beta}}$) with the condition (C.12) taken into account, we obtain:

$$T_{00} = \frac{i}{2}\left(\Psi^{\dagger}\dot{\Psi} - \dot{\Psi}^{\dagger}\Psi\right) , \tag{C.15}$$

$$\begin{aligned} T_{0\bar{\alpha}} &= \frac{i}{4}\left(\Psi^{\dagger}\Psi_{,\alpha} - \Psi^{\dagger}_{,\alpha}\Psi\right)L^{\alpha}_{\bar{\alpha}} \\ &+ \frac{i}{4}\left(\Psi^{\dagger}\gamma^0\gamma_{\bar{\alpha}}\dot{\Psi} - \dot{\Psi}^{\dagger}\gamma^0\gamma_{\bar{\alpha}}\Psi + \Psi^{\dagger}\gamma^0\Gamma_{\bar{\alpha}}\gamma^0\Psi - \Psi^{\dagger}\Gamma_{\bar{\alpha}}\Psi\right) , \end{aligned} \tag{C.16}$$

$$\begin{aligned} T_{\bar{\alpha}\bar{\beta}} &= \frac{i}{4}\left(\Psi^{\dagger}\gamma^0\gamma_{\bar{\alpha}}\Psi_{,\mu}L^{\mu}_{\bar{\beta}} + \Psi^{\dagger}\gamma^0\gamma_{\bar{\beta}}\Psi_{,\mu}L^{\mu}_{\bar{\alpha}} - \Psi^{\dagger}_{,\mu}\gamma^0\gamma_{\bar{\alpha}}\Psi L^{\mu}_{\bar{\beta}} - \Psi^{\dagger}_{,\mu}\gamma^0\gamma_{\bar{\beta}}\Psi L^{\mu}_{\bar{\alpha}}\right) \\ &- \frac{i}{4}\Psi^{\dagger}\gamma^0\left(\gamma_{\bar{\alpha}}\Gamma_{\bar{\beta}} + \gamma_{\bar{\beta}}\Gamma_{\bar{\alpha}} + \Gamma_{\bar{\beta}}\gamma_{\bar{\alpha}} + \Gamma_{\bar{\alpha}}\gamma_{\bar{\beta}}\right)\Psi , \end{aligned} \tag{C.17}$$

(here $\gamma_{\bar{\alpha}} = \delta_{\bar{\alpha}\bar{\beta}}\gamma^{\bar{\beta}}$). The trace of the written tensor with the Dirac equation taken into account reduces to the form:

$$T = T^0_0 + T^{\bar{\alpha}}_{\bar{\alpha}} = -im\Psi^{\dagger}\gamma^0\Psi . \tag{C.18}$$

To write down the Einstein equation we have to calculate the projections $R_{0\bar{\alpha}} = R_{0\alpha}L^{\alpha}_{\bar{\alpha}}$, $R_{\bar{\alpha}\bar{\beta}} = R_{\alpha\beta}L^{\alpha}_{\bar{\alpha}}L^{\beta}_{\bar{\beta}}$ of the Ricci tensor and the component R_{00}. A simple calculation based on the formulae (1.11)–(1.13) gives:

$$R_{00} = -\frac{1}{2}(\kappa^{\bar{\alpha}}_{\bar{\alpha}})^{\cdot} - \frac{1}{4}\kappa^{\bar{\beta}}_{\bar{\alpha}}\kappa^{\bar{\alpha}}_{\bar{\beta}} , \tag{C.19}$$

$$R_{0\alpha} = \frac{1}{2}(\kappa^{\bar{\beta}}_{\bar{\alpha},\mu}L^{\mu}_{\bar{\beta}} - \kappa^{\bar{\beta}}_{\bar{\beta},\mu}L^{\mu}_{\bar{\alpha}} + \kappa^{\bar{\mu}}_{\bar{\beta}}\sigma^{\bar{\beta}}_{\bar{\alpha}\bar{\mu}} - \kappa^{\bar{\mu}}_{\bar{\alpha}}\sigma^{\bar{\beta}}_{\bar{\mu}\bar{\beta}}) , \tag{C.20}$$

$$R_{\bar{\alpha}\bar{\beta}} = \frac{1}{2}\left(\kappa_{\bar{\alpha}\bar{\beta}}\right)^{\cdot} + \frac{1}{4}\kappa^{\bar{\mu}}_{\bar{\mu}}\kappa_{\bar{\alpha}\bar{\beta}} + P_{\bar{\alpha}\bar{\beta}} . \tag{C.21}$$

Note that in the derivation of the expression (C.21) from (1.13), it is essential to use the condition (C.12) in virtue of which we have:

$$L^{\alpha}_{\bar{\alpha}} L^{\bar{\beta}}_{\beta} \left(\kappa^{\beta}_{\alpha}\right)^{\cdot} = \left(L^{\alpha}_{\bar{\alpha}} L^{\bar{\beta}}_{\beta} \kappa^{\beta}_{\alpha}\right)^{\cdot} = \left(\kappa^{\bar{\beta}}_{\bar{\alpha}}\right)^{\cdot} . \tag{C.22}$$

The values $P_{\bar{\alpha}\bar{\beta}}$ are the triad projections of the three-dimensional Ricci tensor and are expressed in terms of the coefficients $\sigma^{\bar{\alpha}}_{\bar{\beta}\bar{\gamma}}$ by the following formula (this formula results from (1.20), (1.37) and (1.39) taking there formally $p_{\bar{\alpha}} = 0$ or follows directly from the formulae (A.18)–(A.19)):

$$\begin{aligned} P_{\bar{\alpha}\bar{\beta}} = -\frac{1}{2}[&(\sigma_{\bar{\beta}\bar{\alpha}\bar{\mu},\nu} + \sigma_{\bar{\alpha}\bar{\beta}\bar{\mu},\nu}) L^{\nu}_{\bar{\rho}} \delta^{\bar{\rho}\bar{\mu}} \\ &+ \sigma^{\bar{\mu}}_{\bar{\mu}\bar{\alpha},\nu} L^{\nu}_{\bar{\beta}} + \sigma^{\bar{\mu}}_{\bar{\mu}\bar{\beta},\nu} L^{\nu}_{\bar{\alpha}} + (\sigma^{\bar{\nu}}_{\bar{\rho}\bar{\alpha}} \delta^{\bar{\rho}\bar{\mu}} + \sigma^{\bar{\mu}}_{\bar{\rho}\bar{\alpha}} \delta^{\bar{\rho}\bar{\nu}}) \sigma_{\bar{\mu}\bar{\nu}\bar{\beta}} \\ &+ (\sigma_{\bar{\beta}\bar{\alpha}\bar{\rho}} \delta^{\bar{\rho}\bar{\nu}} + \sigma_{\bar{\alpha}\bar{\beta}\bar{\rho}} \delta^{\bar{\rho}\bar{\nu}}) \sigma^{\bar{\mu}}_{\bar{\mu}\bar{\nu}} - \frac{1}{2} \sigma_{\bar{\alpha}\bar{\mu}\bar{\nu}} \sigma_{\bar{\beta}\bar{\rho}\bar{\lambda}} \delta^{\bar{\mu}\bar{\rho}} \delta^{\bar{\nu}\bar{\lambda}}] . \end{aligned} \tag{C.23}$$

Thus we have all what is necessary to write down the Einstein equations $R_{00} = T_{00} + \frac{1}{2}T$, $R_{0\bar{\alpha}} = T_{0\bar{\alpha}}$, $R_{\bar{\alpha}\bar{\beta}} = T_{\bar{\alpha}\bar{\beta}} - \frac{1}{2}\delta_{\bar{\alpha}\bar{\beta}} T$.

C.2 An Exact Homogeneous Solution for the Massless Case

To find the asymptotics of the solution of the Einstein–Dirac equations in the vicinity of the singularity, we should start with the construction of the exact solution of these equations for the case when ψ and $g_{\alpha\beta}$ (as well as the triad $L^{\bar{\alpha}}_{\alpha}$) depend only on time. It appears that such an exact solution is easy to find and moreover, as in case of a scalar field, this solution already describes the asymptotic character of the general solution near the singularity as well. To get the asymptotics for the general case, it is sufficient to replace in the homogeneous solution all the arbitrary constants by arbitrary three-dimensional functions.

Let $L^{\bar{\alpha}}_{\alpha}$ and Ψ be dependent only on time and let the mass be equal to zero: $m = 0$ (near the singularity we can neglect the mass due to the same reason that was explained in Section 4.5 where we treated the scalar field). In this case all the values $\sigma^{\bar{\alpha}}_{\bar{\beta}\bar{\gamma}}$ are equal to zero and

$$\Gamma_{\bar{\alpha}} = \frac{1}{4} \kappa_{\bar{\alpha}\bar{\beta}} \gamma^0 \gamma^{\bar{\beta}} \ , \quad \gamma^{\bar{\alpha}} \Gamma_{\bar{\alpha}} = -\frac{1}{4} \kappa^{\bar{\alpha}}_{\bar{\alpha}} \gamma^0 . \tag{C.24}$$

The Dirac equation (C.14) gives:

$$\dot{\Psi} + \frac{1}{4} \kappa^{\bar{\alpha}}_{\bar{\alpha}} \Psi = 0 . \tag{C.25}$$

The trace $\kappa^{\bar{\alpha}}_{\bar{\alpha}}$ is expressed, as is known, in terms of the determinant $g = \det g_{\alpha\beta}$ by the formula $\kappa^{\bar{\alpha}}_{\bar{\alpha}} = \dot{g}/g$. Equation (C.25) is then easy to solve:

$$\Psi = W g^{-1/4} , \tag{C.26}$$

where W is an arbitrary constant spinor.

The Einstein equation $R_{0\bar{\alpha}} = T_{0\bar{\alpha}}$ is satisfied identically since its left-hand side (as is evident from (C.20)) is identically equal to zero. The right-hand side also identically turns out to be zero, which is not difficult to check, using relations (C.16), (C.24) and (C.26). In the equation $R_{00} = T_{00} + \frac{1}{2}T$, the right-hand side also turns out to be zero, as it follows from (C.15), (C.26) and from the fact that the trace T for zero mass is equal to zero. Thus, one gets from this equation:

$$(\kappa^{\bar{\alpha}}_{\bar{\alpha}})^{\cdot} + \frac{1}{2}\kappa^{\bar{\beta}}_{\bar{\alpha}}\kappa^{\bar{\alpha}}_{\bar{\beta}} = 0. \tag{C.27}$$

The equation $R_{\bar{\alpha}\bar{\beta}} = T_{\bar{\alpha}\bar{\beta}} - \frac{1}{2}\delta_{\bar{\alpha}\bar{\beta}}T$ now takes the form:

$$\left(\kappa_{\bar{\alpha}\bar{\beta}}\right)^{\cdot} + \frac{1}{2}\kappa^{\bar{\mu}}_{\bar{\mu}}\kappa_{\bar{\alpha}\bar{\beta}} = \frac{i}{4\sqrt{g}}W^{\dagger}\left(\kappa^{\bar{\mu}}_{\bar{\alpha}}s_{\bar{\mu}\bar{\beta}} + \kappa^{\bar{\mu}}_{\bar{\beta}}s_{\bar{\mu}\bar{\alpha}}\right)W, \tag{C.28}$$

(here $s_{\bar{\alpha}\bar{\beta}} = \delta_{\bar{\alpha}\bar{\mu}}\delta_{\bar{\beta}\bar{\nu}}s^{\bar{\mu}\bar{\nu}}$). After taking the trace of the last equation, we get $\left(\sqrt{g}\right)^{\cdot\cdot} = 0$. It then follows:

$$\sqrt{g} = \Lambda t, \tag{C.29}$$

where we have fixed the integration constant so that the singularity takes place at $t = 0$. The value Λ is an arbitrary constant.

If we now introduce the constants $\omega_{\bar{\alpha}\bar{\beta}}$ according to the definition

$$\omega_{\bar{\alpha}\bar{\beta}} = -\frac{i}{4\Lambda}W^{\dagger}s_{\bar{\alpha}\bar{\beta}}W, \tag{C.30}$$

then we can rewrite equation (C.28) in the form:

$$\dot{\kappa}_{\bar{\alpha}\bar{\beta}} + \frac{1}{t}\kappa_{\bar{\alpha}\bar{\beta}} + \frac{1}{t}\left(\kappa^{\bar{\mu}}_{\bar{\alpha}}\omega_{\bar{\mu}\bar{\beta}} + \kappa^{\bar{\mu}}_{\bar{\beta}}\omega_{\bar{\mu}\bar{\alpha}}\right) = 0. \tag{C.31}$$

Two conditions follow from the previous results (C.29) and (C.27), in addition to this equation. These are:

$$\kappa^{\bar{\alpha}}_{\bar{\alpha}} = \frac{2}{t}, \quad \kappa^{\bar{\beta}}_{\bar{\alpha}}\kappa^{\bar{\alpha}}_{\bar{\beta}} = \frac{4}{t^2}. \tag{C.32}$$

In what follows it will be convenient to use a matrix notation. Let κ be the symmetric matrix with components $\kappa_{\bar{\alpha}\bar{\beta}}$, ω be the antisymmetric matrix with components $\omega_{\bar{\alpha}\bar{\beta}}$ and let us denote by $\mathbf{L}$ the matrix with components $L^{\bar{\alpha}}_{\alpha}$ (or $L_{\bar{\alpha}\alpha}$) and $\mathbf{L}^{-1}$ the inverse to $\mathbf{L}$ matrix with components $L^{\alpha}_{\bar{\alpha}}$ (or $L^{\alpha\bar{\alpha}}$). Let us again consider the first index (or the upper index in case of mixed indices) as numbering the lines and the second (or lower) index as numbering the columns (this is the same rule as the one we have adopted earlier). Then equation (C.31) can be rewritten as

$$\dot{\kappa} + \frac{1}{t}\kappa + \frac{1}{t}\left(\kappa\omega - \omega\kappa\right) = 0. \tag{C.33}$$

From (C.32) it follows that

$$Sp\,\kappa = \frac{2}{t}, \quad Sp\left(\kappa^2\right) = \frac{4}{t^2}. \tag{C.34}$$

Here Sp denotes again the trace. The solution of the matrix equation (C.33) should be sought in the form:

$$\kappa = \frac{2}{t}\mathbf{O}\mathbf{C}_0\mathbf{O}^{-1}, \tag{C.35}$$

where $\mathbf{C}_0$ is a symmetric constant matrix and $\mathbf{O}$ is an orthogonal matrix dependent of time. The conditions (C.34) are immediately fulfilled, if we require

$$Sp\mathbf{C}_0 = 1, \quad Sp\left(\mathbf{C}_0^2\right) = 1. \tag{C.36}$$

Substituting (C.35) into equation (C.33), we find that the latter is fulfilled for an arbitrary matrix $\mathbf{C}_0$, if the matrix $\mathbf{O}$ satisfies the equation:

$$\dot{\mathbf{O}}\mathbf{O}^{-1} = \frac{1}{t}\omega\ . \tag{C.37}$$

It is not difficult to construct the solution of this equation. If the orthogonal matrix $\mathbf{O}$ is written down in terms of the Euler angles ψ, θ, φ in the standard form (2.20), then (C.37) reduces to the following three equations for the angles:

$$\begin{aligned} \dot{\varphi}\sin\psi\sin\theta + \dot{\theta}\cos\psi &= \frac{1}{t}\omega_{\bar{2}\bar{3}}\ , \\ \dot{\varphi}\cos\psi\sin\theta - \dot{\theta}\sin\psi &= \frac{1}{t}\omega_{\bar{3}\bar{1}}\ , \\ \dot{\varphi}\cos\theta + \dot{\psi} &= \frac{1}{t}\omega_{\bar{1}\bar{2}}\ . \end{aligned} \tag{C.38}$$

It is sufficient for us to take any particular solution of the equations (C.38).* We shall chose the following:

$$\psi = \psi_0\ , \quad \theta = \theta_0\ , \quad \varphi = \varphi_0 \ln t\ . \tag{C.39}$$

Here $\psi_0, \theta_0\ , \varphi_0$ are constants determined by the three constants $\omega_{\bar{1}\bar{2}}, \omega_{\bar{3}\bar{1}}, \omega_{\bar{2}\bar{3}}$ through the relations

$$\varphi_0 \sin\psi_0 \sin\theta_0 = \omega_{\bar{2}\bar{3}}, \ \ \varphi_0 \cos\psi_0 \sin\theta_0 = \omega_{\bar{3}\bar{1}}, \ \ \varphi_0 \cos\theta_0 = \omega_{\bar{1}\bar{2}}\ . \tag{C.40}$$

Substituting the obtained solution for the matrix $\mathbf{O}$ into (C.35), we get a solution for the matrix κ involving four arbitrary constants (6 elements of the symmetric matrix $\mathbf{C}_0$, subject to the two conditions (C.36)), i.e., as many as the general solution of equations (C.33)–(C.34) must contain.

The next step for getting the complete solution is to find the matrix $\mathbf{L}$. The equation for $\mathbf{L}$ results from (C.13) and (C.35):

$$\dot{\mathbf{L}}\mathbf{L}^{-1} = \frac{1}{t}\mathbf{O}\mathbf{C}_0\mathbf{O}^{-1}. \tag{C.41}$$

* The general solution for $\mathbf{O}$ must involve three arbitrary constants. If a certain particular solution $\mathbf{O}_0$ is found, then the general solution is constructed as a product $\mathbf{O}_0\mathbf{S}$ where $\mathbf{S}$ is an orthogonal and time-independent arbitrary matrix. However without restricting the generality, $\mathbf{S}$ can be assumed to be the unit matrix since the substitution into (C.35) of the matrix $\mathbf{S}$ does not give any other arbitrary parameter above those already present in the matrix $\mathbf{C}_0$.

We search for its solution in the form:

$$\mathbf{L} = \mathbf{O}\mathbf{U}\,. \tag{C.42}$$

Then for the matrix $\mathbf{U}$ we get

$$\dot{\mathbf{U}}\mathbf{U}^{-1} = \frac{1}{t}\mathbf{C}_0 - \mathbf{O}^{-1}\dot{\mathbf{O}}\,. \tag{C.43}$$

From (2.20) and (C.39) we obtain

$$\mathbf{O}^{-1}\dot{\mathbf{O}} = \frac{\varphi_0}{t}\mathbf{E}\,, \tag{C.44}$$

where

$$\mathbf{E} = \begin{pmatrix} 0 & 1 & 0 \\ -1 & 0 & 0 \\ 0 & 0 & 0 \end{pmatrix}\,. \tag{C.45}$$

From (C.40) it follows that

$$\varphi_0 = \sqrt{(\omega_{\bar{1}\bar{2}})^2 + (\omega_{\bar{3}\bar{1}})^2 + (\omega_{\bar{2}\bar{3}})^2}, \tag{C.46}$$

and finally we can write down the equation (C.43) in the form:

$$\dot{\mathbf{U}}\mathbf{U}^{-1} = \frac{1}{t}\left(\mathbf{C}_0 - \varphi_0\mathbf{E}\right). \tag{C.47}$$

If thc three eigenvalues of the matrix $\mathbf{C}_0 - \varphi_0\mathbf{E}$ are denoted by $p_{\bar{1}}, p_{\bar{2}}, p_{\bar{3}}$ we can write:

$$\mathbf{C}_0 - \varphi_0\mathbf{E} = \mathbf{A}\mathbf{P}\mathbf{A}^{-1}, \tag{C.48}$$

where

$$\mathbf{P} = \begin{pmatrix} p_{\bar{1}} & 0 & 0 \\ 0 & p_{\bar{2}} & 0 \\ 0 & 0 & p_{\bar{3}} \end{pmatrix} \tag{C.49}$$

and the matrix $\mathbf{A}$ (with components $A_{\bar{\alpha}}^{\bar{\beta}}$) is *defined* by the relation (C.48). To make this definition unique, let us require that the diagonal elements of the matrix $\mathbf{A}$ are equal to unity (we have this freedom due to the invariance of the equation (C.48) with respect to the transformation $\mathbf{A} \to \acute{\mathbf{A}}\mathbf{D}$ with an arbitrary diagonal matrix $\mathbf{D}$):

$$A_{\bar{1}}^{\bar{1}} = A_{\bar{2}}^{\bar{2}} = A_{\bar{3}}^{\bar{3}} = 1. \tag{C.50}$$

Thus, the matrix $\mathbf{A}$ does not involve any new arbitrary element and is determined by the same four parameters of $\mathbf{C}_0 - \varphi_0\mathbf{E}$. It is now essential to note that $\mathbf{A}$ *cannot be orthogonal* since by means of this matrix we transform the non-symmetric matrix $\mathbf{C}_0 - \varphi_0\mathbf{E}$ to the diagonal form (the latter is symmetric only at $\varphi_0 = 0$, i.e., when the matrix ω is zero, which means the absence of the spinor field). This peculiarity leads to the difference in the algebraic structure of the metric $g_{\alpha\beta}$ for the case of a spinor field.

It is easy to verify that the eigenvalues $p_{\bar{1}}, p_{\bar{2}}, p_{\bar{3}}$ of the matrix $\mathbf{C}_0 - \varphi_0 \mathbf{E}$ satisfy two relations:

$$p_{\bar{1}} + p_{\bar{2}} + p_{\bar{3}} = 1\ , \quad p_{\bar{1}}^2 + p_{\bar{2}}^2 + p_{\bar{3}}^2 = 1 - 2\varphi_0^2\ . \tag{C.51}$$

The exponents of the power asymptotics of the solution in the case of the scalar field satisfy the same relations and, consequently, all their properties can be transferred unchanged to the case of the spinor field. In particular, near the singularity all three $p_{\bar{\alpha}}$ become positive.

It is easy to see that the solution of Eq. (C.47) is:

$$\mathbf{U} = \mathbf{A}\tau\mathbf{l}\ , \tag{C.52}$$

where

$$\tau = \begin{pmatrix} t^{p_{\bar{1}}} & 0 & 0 \\ 0 & t^{p_{\bar{2}}} & 0 \\ 0 & 0 & t^{p_{\bar{3}}} \end{pmatrix} \tag{C.53}$$

and the matrix $\mathbf{l}$ has components $l_{\alpha}^{\bar{\alpha}}$, which represent nine arbitrary constants. As follows from (C.3) and (C.42), the metric matrix $\mathbf{g}$ (with components $g_{\alpha\beta}$) is $\mathbf{g} = \tilde{\mathbf{L}}\mathbf{L}$ (the tilde indicates transposition) so that (C.42) gives $\mathbf{g} = \tilde{\mathbf{U}}\mathbf{U}$. Substituting here $\mathbf{U}$ from (C.52), we obtain:

$$\mathbf{g} = \tilde{\mathbf{l}}\tau\tilde{\mathbf{A}}\mathbf{A}\tau\mathbf{l}\,. \tag{C.54}$$

This result clearly shows that due to the non-orthogonality of the matrix $\mathbf{A}$ (i.e., because $\tilde{\mathbf{A}}\mathbf{A}$ is not unity), the metric tensor is a linear combination of the all six powers $t^{p_{\bar{\alpha}}+p_{\bar{\beta}}}$. In the component representation, the formula (C.52) is nothing other than the relation (C.2) and the metric tensor from (C.1) is equivalent to its matrix form (C.54), which can be written as

$$g_{\alpha\beta} = \eta_{\bar{\alpha}\bar{\beta}} l_{\alpha}^{\bar{\alpha}} l_{\beta}^{\bar{\beta}}\ , \tag{C.55}$$

where

$$\eta_{\bar{\alpha}\bar{\beta}} = t^{p_{\bar{\alpha}}+p_{\bar{\beta}}} \sum_{\bar{\mu}} A_{\bar{\alpha}}^{\bar{\mu}} A_{\bar{\beta}}^{\bar{\mu}}\ , \quad \textit{no summation over } \bar{\alpha}, \bar{\beta}. \tag{C.56}$$

C.3 The General Solution in the Vicinity of the Singularity

As has already been noted, the relations (C.51) admit the possibility that all the three indices $p_{\bar{\alpha}}$ take positive values. As in the case of the scalar field, this fact determines the simple power asymptotics also of the general solution of the Einstein–Dirac equations in the vicinity of the singularity. It can be shown that for $p_{\bar{\alpha}} > 0$, the asymptotics of the general solution is obtained by the replacement of all integration constants in the exact solution described above, by three-dimensional functions of the spatial coordinates. To confirm this, let us first note that this replacement results in the appearance of non-vanishing values $\sigma_{\bar{\alpha}\bar{\beta}\bar{\gamma}}$, given by expression (C.11). It is easy to see that all orders in time (at $t \to 0$) which might appear in these expressions, can be estimated to behave as $t^{p_{\bar{\alpha}}+p_{\bar{\beta}}-1} \ln t$. Thus, if all $p_{\bar{\alpha}}$ are positive, we have $\sigma_{\bar{\alpha}\bar{\beta}\bar{\gamma}} \ll t^{-1}$. The values $\kappa_{\bar{\alpha}\bar{\beta}}$ have the order t^{-1} and it follows from (C.7) that in the matrices $\Gamma_{\bar{\alpha}}$ the first term is the leading one. That is, in the first approximation, the equations (C.24) are correct. It is evident from (C.14) that the right-hand side in the first of the equations (C.24) is of the order of Ψ and may be omitted in comparison with the left-hand side, which has the order $t^{-1}\Psi$. Consequently, the Dirac equation in the leading approximation has the form (C.25) and the solution (C.26) is correct.

Let us now consider the 00 and $\bar{\alpha}\bar{\beta}$ components of the Einstein equations. Since $\sqrt{g} = \Lambda t$ and $\Psi = W(\Lambda t)^{-1/2}$ it is evident from (C.15) that the component T_{00} vanishes at order t^{-2}. The trace of the energy–momentum tensor has the order only t^{-1}, as is seen from (C.18). Thus the right-hand side of the equation $R_{00} = T_{00} + \frac{1}{2}T$ is small in comparison with its left-hand side (which has the order t^{-2}) and may be neglected. Consequently, in the first approximation this equation has the form (C.27). It is also evident that the asymptotic form of the equation $R_{\bar{\alpha}\bar{\beta}} = T_{\bar{\alpha}\bar{\beta}} - \frac{1}{2}\delta_{\bar{\alpha}\bar{\beta}}T$ will be represented by the equation (C.28). Actually, in the right-hand side of the equation $R_{\bar{\alpha}\bar{\beta}} = T_{\bar{\alpha}\bar{\beta}} - \frac{1}{2}\delta_{\bar{\alpha}\bar{\beta}}T$ we can omit the term $\delta_{\bar{\alpha}\bar{\beta}}T \sim t^{-1}$ and in the components $T_{\bar{\alpha}\bar{\beta}}$ (see expression (C.17)) we can omit the first parentheses having the order t^{-1} in comparison with the second parentheses (of order t^{-2}). In the left-hand side, we can omit the three-dimensional Ricci tensor, as it involves only the terms linear and quadratic in $\sigma_{\bar{\alpha}\bar{\beta}\bar{\gamma}}$. Both of them are much smaller than t^{-2}. The remaining terms (of order t^{-2}) give equation (C.28).

As for the equation $R_{0\bar{\alpha}} = T_{0\bar{\alpha}}$, its role reduces to imposing three additional conditions upon the arbitrary three-dimensional functions appearing in the solutions. We shall not write these conditions in explicit form.

Now let us count the number of physically arbitrary three-dimensional functions in the solution. The metric is described by the triad $U^{\bar{\alpha}}_{\alpha}$ (C.2), containing thirteen arbitrary three-dimensional functions (four functions determine the exponents $p_{\bar{\alpha}}$ and the matrix $\mathbf{A}$ and nine functions introduce the components $l^{\bar{\alpha}}_{\alpha}$). Apart from this, the spinor W involving four functions remains arbitrary. Consequently, we have seventeen functions in total. However, it is still permissible to perform arbitrary three-dimensional orthogonal rotations of the triad, which means that there are three nonphysical three-dimensional functions in the solution. Besides, there are still left admissible three-dimensional transformations of the coordinates and the equation $R_{0\bar{\alpha}} = T_{0\bar{\alpha}}$ imposes three more constraints on the arbitrary functions. Thus, the physical arbitrariness of the solution is determined by eight arbitrary three-dimensional functions. This indeed is the number of functions that must be contained in the general solution of the Einstein–Dirac equations. To complete the analysis, it should be shown that if at a certain instant the general solution is close to the one described in the preceding subsection (with three-dimensional functions instead of constants) but one of the indices $p_{\bar{\alpha}}$ is negative, then the series of the finite number of oscillations finally results in the asymptotic regime with all positive indices. In the case of the scalar field this phenomenon actually takes place, as it follows from [16], but we shall not dwell upon a similar analysis in case of the spinor field.

To close this appendix, we note that the coupled gravitational-Dirac system has been investigated in the case of homogeneous cosmological models of general Bianchi type in [86, 87]. We also note that (quantum) spinor fields have been discussed in the billiard context in the work [53, 56]. How to incorporate fermions in the hidden symmetry suggested by the billiards has been investigated in [58, 59, 54, 55].

Appendix D

Lorentzian Kac–Moody Algebras

The purpose of this appendix is to explain some basic material on Kac–Moody algebras. Our goal is not to give a comprehensive survey of this vast and fascinating branch of mathematics. It is much more modest: we aim just to develop the concepts necessary to understand the connection between Coxeter groups and Kac–Moody algebras, so that the reader can appreciate why the billiard dynamics and the remarkable properties of the billiard tables uncovered by the BKL limit points to the existence of a much deeper symmetry structure underlying gravitational theories, which remains, however, to be discovered (and, in particular, to be made precise). This is standard material which has already been reviewed in many places. We closely follow the presentation given in the review [90] written in collaboration with Daniel Persson and Philippe Spindel. More information on Kac–Moody algebras can be found in the classic monographs [111, 135].

D.1 Definitions

Cartan matrices were defined in Chapter 7. Given a Cartan matrix A (with $\det A \neq 0$), one defines the corresponding Kac–Moody algebra $\mathfrak{g} = \mathfrak{g}(A)$ as the algebra generated by $3n$ generators h_i, e_i, f_i subject to the following "Chevalley–Serre" relations (in addition to the Jacobi identity and antisymmetry of the Lie bracket),

$$\begin{aligned} [h_i, h_j] &= 0, && \\ [h_i, e_j] &= A_{ij} e_j && \text{(no summation on } j\text{)}, \\ [h_i, f_j] &= -A_{ij} f_j && \text{(no summation on } j\text{)}, \\ [e_i, f_j] &= \delta_{ij} h_j && \text{(no summation on } j\text{)}, \end{aligned} \tag{D.1}$$

and

$$\mathrm{ad}_{e_i}^{1-A_{ij}}(e_j) = 0, \qquad \mathrm{ad}_{f_i}^{1-A_{ij}}(f_j) = 0, \qquad i \neq j. \tag{D.2}$$

The operator ad_x is a derivation defined through $\mathrm{ad}_x y = [x, y]$. So, the relations (D.2), called Serre relations, read explicitly

$$\underbrace{[e_i, [e_i, [e_i, \cdots, [e_i, e_j]] \cdots]}_{1-A_{ij}\ \text{commutators}} = 0 \tag{D.3}$$

(and likewise for the f_ks).

When $\det A = 0$, the definition of the Kac–Moody algebra is slightly different [111]. This case is discussed in Section D.5 below.

Any multicommutator can be reduced, using the Jacobi identity and the above relations, to a multicommutator involving only the e_is, or only the f_is. Hence, the Kac–Moody algebra splits as a direct sum ("triangular decomposition")

$$\mathfrak{g} = \mathfrak{n}_- \oplus \mathfrak{h} \oplus \mathfrak{n}_+, \tag{D.4}$$

where $\mathfrak{n}_-$ is the subalgebra involving the multicommutators

$$[f_{i_1}, [f_{i_2}, \cdots, [f_{i_{k-1}}, f_{i_k}] \cdots],$$

$\mathfrak{n}_+$ is the subalgebra involving the multicommutators

$$[e_{i_1}, [e_{i_2}, \cdots, [e_{i_{k-1}}, e_{i_k}] \cdots]$$

and $\mathfrak{h}$ is the abelian subalgebra containing the h_is. This is called the *Cartan subalgebra* and its dimension n is the *rank* of the Kac–Moody algebra $\mathfrak{g}$. It should be stressed that the direct sum Equation (D.4) is a direct sum of $\mathfrak{n}_-$, $\mathfrak{h}$ and $\mathfrak{n}_+$ as vector spaces, not as subalgebras (since these subalgebras do not commute).

A priori, the numbers of the multicommutators

$$[f_{i_1}, [f_{i_2}, \cdots, [f_{i_{k-1}}, f_{i_k}] \cdots] \qquad \text{and} \qquad [e_{i_1}, [e_{i_2}, \cdots, [e_{i_{k-1}}, e_{i_k}] \cdots]$$

are infinite, even after one has taken into account the Jacobi identity. However, the Serre relations impose nontrivial relations among them, which, in some cases, make the Kac–Moody algebra finite-dimensional. Three worked examples are given in Section D.4 to illustrate the use of the Serre relations.

We shall assume that the Cartan matrix is symmetrizable (see Chapter 7). One can show [111] that the Kac–Moody algebra is finite-dimensional if and only if the symmetrization S of A is positive definite. In that case, the algebra is one of the finite-dimensional simple Lie algebras given by the Cartan classification.

When the Cartan matrix A is of Lorentzian signature the Kac–Moody algebra $\mathfrak{g}(A)$, constructed from A using the Chevalley–Serre relations, is called a *Lorentzian Kac–Moody algebra*. This is the case of main interest for the gravitational billiards.

D.2 Roots

The adjoint action of the Cartan subalgebra on $\mathfrak{n}_+$ and $\mathfrak{n}_-$ is diagonal. Explicitly,

$$[h, e_i] = \alpha_i(h) e_i \qquad \text{(no summation on } i) \tag{D.5}$$

for any element $h \in \mathfrak{h}$, where α_i is the linear form on $\mathfrak{h}$ (i.e., the element of the dual $\mathfrak{h}^*$) defined by $\alpha_i(h_j) = A_{ji}$. The α_is are called the simple roots. Similarly,

$$\begin{aligned}&[h, [e_{i_1}, [e_{i_2}, \cdots, [e_{i_{k-1}}, e_{i_k}] \cdots]] \\ &\qquad = (\alpha_{i_1} + \alpha_{i_2} + \cdots \alpha_{i_k})(h) \, [e_{i_1}, [e_{i_2}, \cdots, [e_{i_{k-1}}, e_{i_k}] \cdots]\end{aligned} \tag{D.6}$$

and, if

$$[e_{i_1}, [e_{i_2}, \cdots, [e_{i_{k-1}}, e_{i_k}] \cdots]$$

is nonzero, one says that $\alpha_{i_1} + \alpha_{i_2} + \cdots \alpha_{i_k}$ is a positive root. On the negative side, $\mathfrak{n}_-$, one has

$$\begin{aligned}&[h, [f_{i_1}, [f_{i_2}, \cdots, [f_{i_{k-1}}, f_{i_k}] \cdots]] \\ &\qquad = -(\alpha_{i_1} + \alpha_{i_2} + \cdots \alpha_{i_k})(h) \, [f_{i_1}, [f_{i_2}, \cdots, [f_{i_{k-1}}, f_{i_k}] \cdots]\end{aligned} \tag{D.7}$$

and $-(\alpha_{i_1} + \alpha_{i_2} + \cdots \alpha_{i_k})(h)$ is called a negative root when

$$[f_{i_1}, [f_{i_2}, \cdots, [f_{i_{k-1}}, f_{i_k}]$$

is nonzero. This occurs if and only if $[e_{i_1}, [e_{i_2}, \cdots, [e_{i_{k-1}}, e_{i_k}] \cdots]$ is nonzero: $-\alpha$ is a negative root if and only if α is a positive root.

We see from the construction that the roots (linear forms α such that $[h, x] = \alpha(h)x$ has nonzero solutions x) are either positive (linear combinations of the simple roots α_i with integer nonnegative coefficients) or negative (linear combinations of the simple roots with integer non-positive coefficients). The set of positive roots is denoted by Δ_+; that of negative roots by Δ_-. The set of all roots is Δ, so we have $\Delta = \Delta_+ \cup \Delta_-$. The simple roots are positive and form a basis of $\mathfrak{h}^*$. One sometimes denotes the h_i by $\alpha_i^\vee$ (and thus, $[\alpha_i^\vee, e_j] = A_{ij} e_j$ etc). Similarly, one also uses the notation $\langle \cdot, \cdot \rangle$ for the standard pairing between $\mathfrak{h}$ and its dual $\mathfrak{h}^*$, i.e., $\langle \alpha, h \rangle = \alpha(h)$. In this notation the entries of the Cartan matrix can be written as

$$A_{ij} = \alpha_j(\alpha_i^\vee) = \langle \alpha_j, \alpha_i^\vee \rangle \, . \tag{D.8}$$

Finally, the root lattice Q is the set of linear combinations with integer coefficients of the simple roots,

$$Q = \sum_i \mathbb{Z} \alpha_i . \tag{D.9}$$

All roots belong to the root lattice, of course, but the converse is not true. There are elements of Q that are not roots.

D.3 The Chevalley Involution

The symmetry between the positive and negative subalgebras $\mathfrak{n}_+$ and $\mathfrak{n}_-$ of the Kac–Moody algebra can be precisely reformulated as the invariance of the Kac–Moody algebra under the Chevalley involution τ, which is defined on the generators as

$$\tau(h_i) = -h_i, \qquad \tau(e_i) = -f_i, \qquad \tau(f_i) = -e_i. \tag{D.10}$$

The Chevalley involution is in fact an algebra automorphism that exchanges the positive and negative sides of the algebra.

Finally, we quote the following useful theorem.

Theorem 1 *The Kac–Moody algebra $\mathfrak{g}$ defined by the relations (D.1), (D.2) is simple.*

Proof The proof may be found in [111], page 12.

We note that invertibility and indecomposability of the Cartan matrix A are central ingredients in the proof. In particular, the theorem does not hold in the affine case, for which the Cartan matrix is degenerate and has nontrivial ideals* (see [111] and Section D.5).

D.4 Three Examples

The implications of the Serre relations are quite subtle. This can be illustrated by means of three examples of Kac–Moody algebras of rank 2.

- A_2: We start with A_2, the Cartan matrix of which is

$$A[A_2] = \begin{pmatrix} 2 & -1 \\ -1 & 2 \end{pmatrix}. \tag{D.11}$$

From this matrix, we read the following defining relations:

$$\begin{aligned}
&[h_1, h_2] = 0, \\
&[h_1, e_1] = 2e_1, \quad [h_1, e_2] = -e_2, \\
&[h_1, f_1] = -2f_1, \quad [h_1, f_2] = f_2, \\
&[h_2, e_1] = -e_1, \quad [h_2, e_2] = 2e_2, \\
&[h_2, f_1] = f_1, \quad [h_2, f_2] = -2f_2, \\
&[e_1, [e_1, e_2]] = 0, \quad [e_2, [e_2, e_1]] = 0, \\
&[f_1, [f_1, f_2]] = 0, \quad [f_2, [f_2, f_1]] = 0, \\
&[e_i, f_j] = \delta_{ij} h_j.
\end{aligned}$$

* We recall that an ideal $\mathfrak{i}$ is a subalgebra such that $[\mathfrak{i}, \mathfrak{g}] \subset \mathfrak{i}$. A simple algebra has no nontrivial ideals.

The commutator $[e_1, e_2]$ is not killed by the defining relations and hence is not equal to zero (the defining relations are *all* the relations). All the commutators with three (or more) es are, however, zero. A similar phenomenon occurs on the negative side. Hence, the algebra A_2 is eight-dimensional and one may take as basis $\{h_1, h_2, e_1, e_2, [e_1, e_2], f_1, f_2, [f_1, f_2]\}$. The vector $[e_1, e_2]$ corresponds to the positive root $\alpha_1 + \alpha_2$. This is $sl(2, \mathbb{R})$.

- B_2: The algebra B_2, the Cartan matrix of which is

$$A[B_2] = \begin{pmatrix} 2 & -2 \\ -1 & 2 \end{pmatrix}, \tag{D.12}$$

is defined by the same set of generators, but the Serre relations are now $[e_1, [e_1, [e_1, e_2]]] = 0$ and $[e_2, [e_2, e_1]] = 0$ (and similar relations for the fs). The algebra is still finite-dimensional and contains, besides the generators, the commutators $[e_1, e_2]$, $[e_1, [e_1, e_2]]$, their negative counterparts $[f_1, f_2]$ and $[f_1, [f_1, f_2]]$, and nothing else. The triple commutator $[e_1, [e_1, [e_1, e_2]]]$ vanishes by the Serre relations. The other triple commutator $[e_2, [e_1, [e_1, e_2]]]$ vanishes also by the Jacobi identity and the Serre relations,

$$[e_2, [e_1, [e_1, e_2]]] = [[e_2, e_1], [e_1, e_2]] + [e_1, [e_2, [e_1, e_2]]] = 0.$$

(Each term on the right-hand side is zero: the first by antisymmetry of the bracket and the second because $[e_2, [e_1, e_2]] = -[e_2, [e_2, e_1]] = 0$.) The algebra is ten-dimensional and is isomorphic to $so(3, 2)$.

- A_1^+: We now turn to A_1^+, the Cartan matrix of which is

$$A[A_1^+] = \begin{pmatrix} 2 & -2 \\ -2 & 2 \end{pmatrix}. \tag{D.13}$$

This algebra is defined by the same set of generators as A_2, but with Serre relations given by

$$\begin{aligned} [e_1, [e_1, [e_1, e_2]]] &= 0, \\ [e_2, [e_2, [e_2, e_1]]] &= 0 \end{aligned} \tag{D.14}$$

(and similar relations for the fs). This innocent-looking change in the Serre relations has dramatic consequences because the corresponding algebra is infinite-dimensional. (We analyze here the algebra generated by the hs, es and fs, which is in fact the derived Kac–Moody algebra – see Section D.5 on affine Kac–Moody algebras. The derived algebra is already infinite-dimensional.) To see this, consider the $sl(2, \mathbb{R})$ current algebra, defined by

$$[J^a_m, J^b_n] = f^{ab}{}_c J^c_{m+n} + m k^{ab} c \delta_{m+n,0}, \tag{D.15}$$

where the index a takes the values $a = 3, +, -$, where the $f^{ab}{}_c$s are the structure constants of $sl(2, \mathbb{R})$ and where k^{ab} is the invariant metric on $sl(2, \mathbb{R})$

which we normalize here so that $k^{-+} = 1$. The subalgebra with $n = 0$ is isomorphic to $sl(2, \mathbb{R})$,

$$[J_0^3, J_0^+] = 2J_0^+, \qquad [J_0^3, J_0^-] = -2J_0^-, \qquad [J_0^+, J_0^-] = J_0^3.$$

The current algebra (D.15) is generated by J_0^a, c, J_1^- and J_{-1}^+ since any element can be written as a multi-commutator involving them. The map

$$\begin{array}{lllllll} h_1 & \to & J_0^3, & h_2 & \to & -J_0^3 + c, \\ e_1 & \to & J_0^+, & e_2 & \to & J_1^-, \\ f_1 & \to & J_0^-, & f_2 & \to & J_{-1}^+ \end{array} \tag{D.16}$$

preserves the defining relations of the Kac–Moody algebra and defines an isomorphism of the (derived) Kac–Moody algebra with the current algebra. The Kac–Moody algebra is therefore infinite-dimensional. One can construct non-vanishing infinite multi-commutators, in which e_1 and e_2 alternate:

$$\begin{array}{lll} [e_1, [e_2, [e_1, \cdots, [e_1, e_2] \cdots]]] & \sim \; J_n^3 & (n \; e_1\text{s and } n \; e_2\text{s}), \\ [e_1, [e_2, [e_1, \cdots, [e_2, e_1] \cdots]]] & \sim \; J_n^+ & (n+1 \; e_1\text{s and } n \; e_2\text{s}), \\ [e_2, [e_1, [e_2, \cdots, [e_1, e_2] \cdots]]] & \sim \; J_{n+1}^- & (n \; e_1\text{s and } n+1 \; e_2\text{s}). \end{array} \tag{D.17}$$

The Serre relations do not cut the chains of multi-commutators to a finite number.

We see from these examples that the exact consequences of the Serre relations might be intricate to derive explicitly. This is one of the difficulties of the theory.

D.5 The Affine Case

The affine case, of which we just saw an example, is characterized by the conditions that the Cartan matrix has vanishing determinant, is symmetrizable and is such that its symmetrization S is positive semi-definite, $v^t S v \geq 0$. The matrix S has rank $n - 1$, that is, the solutions of $v^t S v = 0$ form a one-dimensional subspace. As before, we also take the Cartan matrix to be indecomposable. The radical of S is easily shown to be one-dimensional and the ranks of S and A are equal to $n - 1$.

When A is non-invertible, the definition of the corresponding Kac–Moody algebra must be slightly amended [111]. One now defines the corresponding Kac–Moody algebras in terms of $3n + 1$ generators (the $+1$ comes from the fact that the rank is $n - 1$), which are the same generators h_i, e_i, f_i subject to the same conditions (D.1, D.2) as above, plus one extra generator η which can be taken to fulfill

$$[\eta, h_i] = 0, \qquad [\eta, e_i] = \delta_{1i} e_1, \qquad [\eta, f_i] = -\delta_{1i} f_1. \tag{D.18}$$

The algebra admits the same triangular decomposition as above,

$$\mathfrak{g} = \mathfrak{n}_- \oplus \mathfrak{h} \oplus \mathfrak{n}_+, \tag{D.19}$$

but now the Cartan subalgebra $\mathfrak{h}$ has dimension $n+1$ (it contains the extra generator η).

Because the matrix A_{ij} has vanishing determinant, one can find a_i such that $\sum_i a_i A_{ij} = 0$. The element $c = \sum_i a_i h_i$ is in the center of the algebra. In fact, the center of the Kac–Moody algebra is one-dimensional and coincides with $\mathbb{C}c$ [111]. The derived algebra $\mathfrak{g}' = [\mathfrak{g}, \mathfrak{g}]$ is the subalgebra generated by h_i, e_i, f_i and has codimension one (it does not contain η). One has

$$\mathfrak{g} = \mathfrak{g}' \oplus \mathbb{C}\eta \tag{D.20}$$

(direct sum of vector spaces, not as algebras). The only proper ideals of the affine Kac–Moody algebra $\mathfrak{g}$ are $\mathfrak{g}'$ and $\mathbb{C}c$.

Affine Kac–Moody algebras appear in the BKL context as subalgebras of the relevant Lorentzian Kac–Moody algebras. Their complete list is known and is given, e.g., in [111]. The sublist of the so-called "untwisted" ones is given in Table D.3.

D.6 The Invariant Bilinear Form

D.6.1 Definition

To proceed, we come back to the assumption that the Cartan matrix is invertible and symmetrizable since these are the only cases encountered in the billiards. Under these assumptions, an invertible, invariant bilinear form is easily defined on the algebra. We denote by ϵ_i the diagonal elements of D,

$$A = DS, \qquad D = \mathrm{diag}(\epsilon_1, \epsilon_2 \cdots, \epsilon_n). \tag{D.21}$$

First, one defines an invertible bilinear form in the dual $\mathfrak{h}^*$ of the Cartan subalgebra. This is done by simply setting

$$(\alpha_i|\alpha_j) = S_{ij} \tag{D.22}$$

for the simple roots. It follows from $A_{ii} = 2$ that

$$\epsilon_i = \frac{2}{(\alpha_i|\alpha_i)} \tag{D.23}$$

and thus the Cartan matrix can be written as

$$A_{ij} = 2\frac{(\alpha_i|\alpha_j)}{(\alpha_i|\alpha_i)}. \tag{D.24}$$

It is customary to fix the normalization of S so that the longest roots have $(\alpha_i|\alpha_i) = 2$. As we shall now see, the definition (D.22) leads to an invariant bilinear form on the Kac–Moody algebra.

Since the bilinear form $(\cdot|\cdot)$ is non-degenerate on $\mathfrak{h}^*$, one has an isomorphism $\mu : \mathfrak{h}^* \to \mathfrak{h}$ defined by

$$\langle \alpha, \mu(\gamma) \rangle = (\alpha|\gamma). \tag{D.25}$$

This isomorphism induces a bilinear form on the Cartan subalgebra, also denoted by $(\cdot|\cdot)$. The inverse isomorphism is denoted by ν and is such that

$$\langle \nu(h), h' \rangle = (h|h'), \qquad h, h' \in \mathfrak{h}. \tag{D.26}$$

Since the Cartan elements $h_i \equiv \alpha_i^\vee$ obey

$$\langle \alpha_i, \alpha_j^\vee \rangle = A_{ji}, \tag{D.27}$$

it is clear from the definitions that

$$\nu(h_i) \equiv \nu(\alpha_i^\vee) = \epsilon_i \alpha_i \quad \Leftrightarrow \quad h_i \equiv \alpha_i^\vee = \frac{2\mu(\alpha_i)}{(\alpha_i|\alpha_i)}, \tag{D.28}$$

and thus also

$$(h_i|h_j) = \epsilon_i \epsilon_j S_{ij}. \tag{D.29}$$

The bilinear form $(\cdot|\cdot)$ can be uniquely extended from the Cartan subalgebra to the entire algebra by requiring that it is *invariant*, i.e., that it fulfills

$$([x, y]|z) = (x|[y, z]) \quad \forall\, x, y, z \in \mathfrak{g}. \tag{D.30}$$

For instance, for the e_is and f_is one finds

$$(h_i|e_j)A_{kj} = (h_i|[h_k, e_j]) = ([h_i, h_k]|e_j) = 0 \quad \Rightarrow \quad (h_i|e_j) = 0, \tag{D.31}$$

and similarly

$$(h_i|f_j) = 0. \tag{D.32}$$

In the same way we have

$$A_{ij}(e_j|f_k) = ([h_i, e_j]|f_k) = (h_i|[e_j, f_k]) = (h_i|h_j)\delta_{jk} = A_{ij}\epsilon_j \delta_{jk}, \tag{D.33}$$

and thus

$$(e_i|f_j) = \epsilon_i \delta_{ij}. \tag{D.34}$$

Quite generally, if e_α and e_γ are root vectors corresponding respectively to the roots α and γ,

$$[h, e_\alpha] = \alpha(h) e_\alpha, \qquad [h, e_\gamma] = \gamma(h) e_\gamma,$$

then $(e_\alpha|e_\gamma) = 0$ unless $\gamma = -\alpha$. Indeed, one has

$$\alpha(h)(e_\alpha|e_\gamma) = ([h, e_\alpha]|e_\gamma) = -(e_\alpha|[h, e_\gamma]) = -\gamma(h)(e_\alpha|e_\gamma),$$

and thus

$$(e_\alpha|e_\gamma) = 0 \qquad \text{if } \alpha + \gamma \neq 0. \tag{D.35}$$

It is proven in [111] that the invariance condition on the bilinear form defines it indeed consistently and that it is non-degenerate. Furthermore, one finds the relations

$$[h, x] = \alpha(h) x, \qquad [h, y] = -\alpha(h) y \quad \Rightarrow \quad [x, y] = (x|y)\mu(\alpha). \tag{D.36}$$

D.6.2 Real and Imaginary Roots

Consider the restriction $(\cdot|\cdot)_{\mathbb{R}}$ of the bilinear form to the real vector space $\mathfrak{h}^\star_{\mathbb{R}}$ obtained by taking the real span of the simple roots,

$$\mathfrak{h}^\star_{\mathbb{R}} = \sum_i \mathbb{R}\alpha_i. \tag{D.37}$$

This defines a scalar product with a definite signature. As we have mentioned, the signature is Euclidean if and only if the algebra is finite-dimensional [111]. In that case, all roots – and not just the simple ones – are spacelike, i.e., such that $(\alpha|\alpha) > 0$.

When the algebra is infinite-dimensional, the invariant scalar product does not have Euclidean signature. The spacelike roots are called "real roots," the non-spacelike ones are called "imaginary roots" [111]. While the real roots are non-degenerate (i.e., the corresponding eigenspaces, called "root spaces," are one-dimensional), this is not so for imaginary roots. In fact, it is a challenge to understand the degeneracy of imaginary roots for general indefinite Kac–Moody algebras, and, in particular, for Lorentzian Kac–Moody algebras.

Another characteristic feature of real roots, familiar from standard finite-dimensional Lie algebra theory, is that if α is a (real) root, no multiple of α is a root except $\pm\alpha$. This is not so for imaginary roots, where 2α (or other nontrivial multiples of α) can be a root even if α is. We shall provide explicit examples below.

Finally, while there are at most two different root lengths in the finite-dimensional case, this is no longer true even for real roots in the case of infinite-dimensional Kac–Moody algebras.* When all the real roots have the same length, one says that the algebra is "simply-laced." Note that the imaginary roots (if any) do not have the same length, except in the affine case, where they all have length squared equal to zero.

D.6.3 Fundamental Weights and the Weyl Vector

The fundamental weights $\{\Lambda_i\}$ of the Kac–Moody algebra are vectors in the dual space $\mathfrak{h}^*$ of the Cartan subalgebra defined by

$$\langle \Lambda_i, \alpha_j^\vee \rangle = \delta_{ij}. \tag{D.38}$$

This equation implies

$$(\Lambda_i|\alpha_j) = \frac{\delta_{ij}}{\epsilon_j}. \tag{D.39}$$

The Weyl vector $\rho \in \mathfrak{h}^*$ is defined by

$$(\rho|\alpha_j) = \frac{1}{\epsilon_j} \tag{D.40}$$

* Imaginary roots may have arbitrarily negative length squared in general.

and is thus equal to

$$\rho = \sum_i \Lambda_i. \tag{D.41}$$

D.7 The Weyl Group

The Weyl group $W[\mathfrak{g}]$ of a Kac–Moody algebra $\mathfrak{g}$ is a discrete group of transformations acting on $\mathfrak{h}^*$. It is the Weyl group that makes the connection between Kac–Moody algebras and cosmological billiards, because the Coxeter groups that emerge in the BKL limit can be identified with Weyl groups of Kac–Moody algebras. Thus, the Weyl groups are central to the billiard discussion.

The Weyl group $W[\mathfrak{g}]$ is defined as follows. One associates a "fundamental Weyl reflection" $r_i \in W[\mathfrak{g}]$ to each simple root through the formula

$$r_i(\lambda) = \lambda - 2\frac{(\lambda|\alpha_i)}{(\alpha_i|\alpha_i)}\alpha_i. \tag{D.42}$$

The Weyl group is just the group generated by the fundamental Weyl reflections. In particular,

$$r_i(\alpha_j) = \alpha_j - A_{ij}\alpha_i \qquad \text{(no summation on } i\text{)}. \tag{D.43}$$

The Weyl group enjoys a number of interesting properties [111]:

- It preserves the scalar product on $\mathfrak{h}^*$.
- It preserves the root lattice and hence is crystallographic.
- Two roots that are in the same orbit have identical multiplicities.
- Any real root has in its orbit (at least) one simple root and hence is non-degenerate.
- The Weyl group is a Coxeter group. The connection between the Coxeter exponents and the Cartan integers A_{ij} $(i \neq j)$ was given in Subsection 7.5.4 and is recalled in Table D.1.

This close relationship between Coxeter groups and Kac–Moody algebras is the reason for denoting both with the same notation (for instance, A_n denotes at the same time the Coxeter group with Coxeter graph of type A_n and the Kac–Moody algebra with Dynkin diagram A_n).

In the case of Lorentzian Kac–Moody algebras, the Weyl group is a subgroup of the orthochronous Lorentz group $O^+(n,1)$. It maps therefore roots within the future light cone on roots within the future light cone.

We stress again – as observed in Subsection 7.5.4 – that different Kac–Moody algebras may have the same Weyl group, because these might preserve different lattices. This is in fact already true for finite-dimensional Lie algebras, where dual algebras (obtained by reversing the arrows in the Dynkin diagram) have the same Weyl group. This property can be seen from the fact that the Coxeter exponents are related to the duality-invariant product $A_{ij}A_{ji}$.

Table D.1: *Cartan integers and Coxeter exponents.*

$A_{ij}A_{ji}$	m_{ij}
0	2
1	3
2	4
3	6
≥ 4	∞

But the ambiguity is in fact bigger. On top of this "finite-dimensional phenomenon," one sees that whenever the product $A_{ij}A_{ji}$ is greater than or equal to four, which occurs only in the infinite-dimensional case, the Coxeter exponent m_{ij} is equal to infinity, independently of the exact value of $A_{ij}A_{ji}$. Information is thus clearly lost. For example, the Cartan matrices

$$\begin{pmatrix} 2 & -2 & -2 \\ -2 & 2 & -2 \\ -2 & -2 & 2 \end{pmatrix}, \qquad \begin{pmatrix} 2 & -9 & -8 \\ -4 & 2 & -5 \\ -3 & -7 & 2 \end{pmatrix} \tag{D.44}$$

lead to the same Weyl group, even though the corresponding Kac–Moody algebras are not isomorphic or even dual to each other.

Because the Weyl groups are (crystallographic) Coxeter groups, we can use the theory of Coxeter groups to analyze them. We also note that by standard vector space duality, one can define the action of the Weyl group in the Cartan subalgebra $\mathfrak{h}$, such that

$$\langle \gamma, r_i^\vee(h) \rangle = \langle r_i(\gamma), h \rangle \qquad \text{for } \gamma \in \mathfrak{h}^\star \text{ and } h \in \mathfrak{h}. \tag{D.45}$$

One has using Equations (D.23, D.25, D.26, D.28),

$$r_i^\vee(h) = h - \langle \alpha_i, h \rangle h_i = h - 2\frac{(h|h_i)}{(h_i|h_i)} h_i. \tag{D.46}$$

It should be pointed out that the imaginary roots of the Kac–Moody algebras do not have immediate analogs on the Coxeter side.

Example

We now give the explicit construction of the Weyl groups associated with interesting gravitational models.

Consider the Cartan matrices

$$A' \equiv A[A_1^{++}] = \begin{pmatrix} 2 & -2 & 0 \\ -2 & 2 & -1 \\ 0 & -1 & 2 \end{pmatrix}, \qquad A'' \equiv A[A_2^{(2)+}] = \begin{pmatrix} 2 & -4 & 0 \\ -1 & 2 & -1 \\ 0 & -1 & 2 \end{pmatrix}$$

As we know, the first Cartan matrix defines the Lie algebra A_1^{++} met in the main text for the billiard of pure gravity. Hence the notation $A[A_1^{++}]$. Note, however, that for practical purposes, we are using a different labeling of the roots here. Similarly, the matrix A'' defines the Kac–Moody algebra $A_2^{(2)+}$, which can be shown to be relevant to the Einstein–Maxwell theory in four space-time dimensions [88]).

We denote the associated sets of simple roots by $\{\alpha_1', \alpha_2', \alpha_3'\}$ and $\{\alpha_1'', \alpha_2'', \alpha_3''\}$, respectively. In both cases, the Coxeter exponents are $m_{12} = \infty$, $m_{13} = 2$, $m_{23} = 3$.

Choosing the longest Kac–Moody roots to have squared length equal to two yields the scalar products

$$S' = \begin{pmatrix} 2 & -2 & 0 \\ -2 & 2 & -1 \\ 0 & -1 & 2 \end{pmatrix}, \qquad S'' = \begin{pmatrix} \frac{1}{2} & -1 & 0 \\ -1 & 2 & -1 \\ 0 & -1 & 2 \end{pmatrix}.$$

According to Equation (D.43), the Weyl group $W[A_1^{++}]$ acts as follows on the roots α_i'

$$\begin{array}{llll} r_1': & \alpha_1' \to -\alpha_1', & \alpha_2' \to \alpha_2' + 2\alpha_1', & \alpha_3' \to \alpha_3', \\ r_2': & \alpha_1' \to \alpha_1' + 2\alpha_2', & \alpha_2' \to -\alpha_2', & \alpha_3' \to \alpha_3' + \alpha_2', \\ r_3': & \alpha_1' \to \alpha_1', & \alpha_2' \to \alpha_2' + \alpha_3', & \alpha_3' \to -\alpha_3', \end{array}$$

while the Weyl group $W[A_2^{(2)+}]$ acts as

$$\begin{array}{llll} r_1'': & \alpha_1'' \to -\alpha_1'', & \alpha_2'' \to \alpha_2'' + 4\alpha_1'', & \alpha_3'' \to \alpha_3'', \\ r_2'': & \alpha_1'' \to \alpha_1'' + \alpha_2'', & \alpha_2'' \to -\alpha_2'', & \alpha_3'' \to \alpha_3'' + \alpha_2'', \\ r_3'': & \alpha_1'' \to \alpha_1'', & \alpha_2'' \to \alpha_2'' + \alpha_3'', & \alpha_3'' \to -\alpha_3''. \end{array}$$

We see that the reflections coincide, $r_1' = r_1''$, $r_2' = r_2''$, $r_3' = r_3''$, as well as the scalar products, provided that we set $2\alpha_1'' = \alpha_1'$, $\alpha_2'' = \alpha_2'$, $\alpha_3' = \alpha_3$.

It follows that the Weyl groups of the Kac–Moody algebras A_1^{++} and $A_2^{(2)+}$ are the same,

$$W[A_1^{++}] = W[A_2^{(2)+}]. \tag{D.47}$$

We have seen that in Chapter 5 that this group is isomorphic to $PGL(2, \mathbb{Z})$. We have here an action of $PGL(2, \mathbb{Z})$ in the three-dimensional Minkowski space of the roots. The Coxeter group $PGL(2, \mathbb{Z})$ preserves the two different lattices in three-dimensional Minkowski space.

$$Q^\pi = \sum_i \mathbb{Z}\alpha_i' \qquad \text{and} \qquad Q'' = \sum_i \mathbb{Z}\alpha_i''. \tag{D.48}$$

D.8 Hyperbolic Kac–Moody Algebras

Hyperbolic Kac–Moody algebras are by definition Lorentzian Kac–Moody algebras with the property that removing any node from their Dynkin diagram leaves one with a Dynkin diagram of affine or finite type.

This is the translation in terms of Dynkin diagrams of the criterion for hyperbolicity of Coxeter groups given in Subsection 7.5.3 in terms of Coxeter diagrams. It follows therefore that the Weyl group of hyperbolic Kac–Moody algebras is a crystallographic hyperbolic Coxeter group as defined in Section 7.5, and conversely, as we have also seen in that same section, that any crystallographic hyperbolic Coxeter group is the Weyl group of at least one hyperbolic Kac–Moody algebra.

For these reasons, the concepts developed for hyperbolic Coxeter groups (in particular, that of fundamental domain) apply directly to Weyl groups of hyperbolic Kac–Moody algebras. The hyperbolic Kac–Moody algebras have been classified in [146] and exist only up to rank 10 (see also [60]). In rank 10, there are four possibilities, known as $E_{10} \equiv E_8^{++}$, $BE_{10} \equiv B_8^{++}$, $DE_{10} \equiv D_8^{++}$ and $CE_{10} \equiv A_{15}^{(2)+}$, corresponding to the three hyperbolic Coxeter groups and the four corresponding Cartan matrices mentioned in Chapter 7. The groups BE_{10} and CE_{10} are dual to each other and possess the same Weyl group. The notation $++$ will be explained below when we discuss overextensions.

The Fundamental Weyl Chamber $\mathcal{F}$ of Hyperbolic Kac–Moody Algebras

For a hyperbolic Kac–Moody algebra, the region $\mathcal{F}$ defined by

$$\{v \in \mathcal{F} \Leftrightarrow (v|\alpha_i) \leq 0\}$$

is a polywedge that lies entirely within the (say) future light cone [95] (boundary included). The fundamental weights Λ_i are timelike or null and lie within the past light cone.

The positive imaginary roots α_K of the algebra fulfill $(\alpha_K|\Lambda_i) \geq 0$ (with, for any K, strict inequality for at least one i) and hence, since they are non-spacelike, must lie in the *future* light cone if the fundamental weights are in the past light cone. Recall indeed that the scalar product of two non-spacelike vectors with the same time orientation is non-positive.

For this reason, it is of interest to consider the action of the Weyl group inside the future light cone. In the Kac–Moody context, a fundamental domain for that action is called a "fundamental Weyl chamber." The region $\mathcal{F}$ defined above turns out to be a fundamental region for the action of the Weyl group in the future light cone and is thus a fundamental Weyl chamber. Any positive imaginary root is Weyl-conjugated to one that lies in $\mathcal{F}$.

All these properties are the exact Kac–Moody translations of the equivalent properties derived in our discussion of hyperbolic Coxeter groups. The minus sign in the definition of $\mathcal{F}$ as opposed to the plus sign appearing in the Coxeter definitions come from the fact that the α_is here are pointing outwards from the fundamental region – the orientation conventions are opposite.

One can thus reformulate the gravitational billiards in terms of motions within the fundamental Weyl chamber of a Kac–Moody algebra. The billiard ball (logarithmic scale factors and dilatons if any) is a time-dependent Cartan element that hits the dominant walls, which are defined by the simple roots.

Roots and the Root Lattice

Not all points belonging to the root lattice Q of a Kac–Moody algebras are actually roots. For instance, linear combinations of simple roots with integer coefficients of both signs are certainly not roots.

For hyperbolic algebras, there exists a simple criterion which enables one to determine whether a point on the root lattice is a root or not. We give it first in the case where all simple roots have equal length squared (assumed equal to two).

Theorem 2 *Consider a hyperbolic Kac–Moody algebra such that $(\alpha_i|\alpha_i) = 2$ for all simple roots α_i. Then, any point α on the root lattice Q with $(\alpha|\alpha) \leq 2$ is a root (note that $(\alpha|\alpha)$ is even). In particular, the set of real roots is the set of points on the root lattice with $(\alpha|\alpha) = 2$, while the set of imaginary roots is the set of points on the root lattice (minus the origin) with $(\alpha|\alpha) \leq 0$.*

Proof For a proof, see [111], Chapter 5.

The version of this theorem applicable to Kac–Moody algebras with different simple root lengths is the following.

Theorem 3 *Consider a hyperbolic algebra with root lattice Q. Let a be the smallest length squared of the simple roots, $a = \min_i(\alpha_i|\alpha_i)$. Then we have:*

- *The set of all short real roots is $\{\alpha \in Q \,|\, (\alpha|\alpha) = a\}$.*
- *The set of all real roots is*

$$\left\{\alpha = \sum_i k_i \alpha_i \in Q \,|\, (\alpha|\alpha) > 0 \text{ and } k_i \frac{(\alpha_i|\alpha_i)}{(\alpha|\alpha)} \in \mathbb{Z}\ \forall i\right\}.$$

- *The set of all imaginary roots is the set of points on the root lattice (minus the origin) with $(\alpha|\alpha) \leq 0$.*

Proof For a proof, see again [111], Chapter 5.

Note that it follows in particular from the theorems that if α is an imaginary root, all its integer multiples are also imaginary roots.

Examples

We shall briefly illustrate these theorems on the simply-laced algebra A_1^{++}, and on the non-simply-laced algebra $A_2^{(2)+}$, the Weyl group of which is $PGL(2,\mathbb{Z})$.

The Kac–Moody Algebra A_1^{++}

We consider again first the algebra A_1^{++} –also denoted AE_3, or H_3 – associated with vacuum four-dimensional Einstein gravity and the BKL billiard. Its Cartan matrix is (see Section D.7)

$$\begin{pmatrix} 2 & -2 & 0 \\ -2 & 2 & -1 \\ 0 & -1 & 2 \end{pmatrix}. \tag{D.49}$$

As it follows from our analysis of Chapter 5, the simple roots may be identified with the following linear forms $\alpha_i(\beta)$ in the three-dimensional space of the β^is,

$$\alpha_1(\beta) = 2\beta^1, \qquad \alpha_2(\beta) = \beta^2 - \beta^1, \qquad \alpha_3(\beta) = \beta^3 - \beta^2 \tag{D.50}$$

with scalar product

$$(F|G) = \sum_i F_i G_i - \frac{1}{2}\left(\sum_i F_i\right)\left(\sum_i G_i\right) \tag{D.51}$$

for two linear forms $F = F_i\beta^i$ and $G = G_i\beta^i$. It is sometimes convenient to analyze the root system in terms of an "affine" level ℓ that counts the number of times the root α_3 occurs: the root $k\alpha_1 + m\alpha_2 + \ell\alpha_3$ has by definition level ℓ. We shall consider here only positive roots for which $k, m, \ell \geq 0$.

Applying the first theorem, one can easily verify that the only positive roots at level zero are the roots $k\alpha_1 + m\alpha_2$, $|k-m| \leq 1$ $(k, m \geq 0)$ of the affine subalgebra A_1^+. When $k = m$, the root is imaginary and has length squared equal to zero. When $|k-m| = 1$, the root is real and has length squared equal to two.

Similarly, the only roots at level one are $(m+a)\alpha_1 + m\alpha_2 + \alpha_3$ with $a^2 \leq m$, i.e., $-[\sqrt{m}] \leq a \leq [\sqrt{m}]$. Whenever $\sqrt{m}$ is an integer, the roots $(m \pm \sqrt{m})\alpha_1 + m\alpha_2 + \alpha_3$ have squared length equal to two and are real. The roots $(m+a)\alpha_1 + m\alpha_2 + \alpha_3$ with $a^2 < m$ are imaginary and have squared length equal to $2(a^2 + 1 - m) \leq 0$. In particular, the root $m(\alpha_1 + \alpha_2) + \alpha_3$ has length squared equal to $2(1-m)$.

Of all the roots at level one with $m > 1$, these are the only ones that are in the fundamental domain $\mathcal{F}$ (i.e., that fulfill $(\beta|\alpha_i) \leq 0$). When $m = 1$, none of the level-one roots is in $\mathcal{F}$ and is either in the Weyl orbit of $\alpha_1 + \alpha_2$, or in the Weyl orbit of α_3.

We leave it to the reader to verify that the roots at level two that are in the fundamental domain $\mathcal{F}$ take the form $(m-1)\alpha_1+m\alpha_2+2\alpha_3$ and $m(\alpha_1+\alpha_2)+2\alpha_3$ with $m \geq 4$.

Further information on the roots of A_1^{++} may be found in [111], Chapter 11, page 215.

The Kac–Moody Algebra $A_2^{(2)+}$

This is the algebra associated with the Einstein–Maxwell theory with same Weyl group as A_1^{++} (see Section D.7). The Cartan matrix is now

$$\begin{pmatrix} 2 & -4 & 0 \\ -1 & 2 & -1 \\ 0 & -1 & 2 \end{pmatrix}, \tag{D.52}$$

and there are now two lengths for the simple roots. The scalar products are

$$(\alpha_1|\alpha_1) = \frac{1}{2}, \qquad (\alpha_1|\alpha_2) = -1 = (\alpha_2|\alpha_1), \qquad (\alpha_2|\alpha_2) = 2. \tag{D.53}$$

One may realize the simple roots as the linear forms

$$\alpha_1(\beta) = \beta^1, \qquad \alpha_2(\beta) = \beta^2 - \beta^1, \qquad \alpha_3(\beta) = \beta^3 - \beta^2 \tag{D.54}$$

in the three-dimensional space of the β^is with scalar product Equation (D.51).

The real roots, which are Weyl conjugate to one of the simple roots α_1 or α_2 (α_3 is in the same Weyl orbit as α_2), divide into long and short real roots. The long real roots are the vectors on the root lattice with squared length equal to two that fulfill the extra condition in the theorem. This condition expresses here that the coefficient of α_1 should be a multiple of 4. The short real roots are the vectors on the root lattice with length squared equal to one-half. The imaginary roots are all the vectors on the root lattice with length squared ≤ 0.

We define again the level ℓ as counting the number of times the root α_3 occurs. The positive roots at level zero are the positive roots of the twisted affine algebra $A_2^{(2)}$, namely, α_1 and $(2m+a)\alpha_1+m\alpha_2$, $m = 1, 2, 3, \cdots$, with $a = -2, -1, 0, 1, 2$ for m odd and $a = -1, 0, 1$ for m odd. Although belonging to the root lattice and of length squared equal to two, the vectors $(2m \pm 2)\alpha_1 + m\alpha_2$ are not long real roots when m is even because they fail to satisfy the condition that the coefficient $(2m \pm 2)$ of α_1 is a multiple of 4. The roots at level zero are all real, except when $a = 0$, in which case the roots $m(2\alpha_1 + \alpha_2)$ have zero norm.

To get the long real roots at level one, we first determine the vectors $\alpha = \alpha_3 + k\alpha_1 + m\alpha_2$ of squared length equal to 2. The condition $(\alpha|\alpha) = 2$ easily leads to $m = p^2$ for some integer $p \geq 0$ and $k = 2p^2 \pm 2p = 2p(p \pm 1)$. Since k is automatically a multiple of 4 for all $p = 0, 1, 2, 3, \cdots$, the corresponding vectors are all long real roots. Similarly, the short real roots at level one are found to be $(2p^2+1)\alpha_1 + (p^2+p+1)\alpha_2 + \alpha_3$ and $(2p^2+4p+3)\alpha_1 + (p^2+p+1)\alpha_2 + \alpha_3$ for p a nonnegative integer.

Table D.2: *Finite-dimensional, simple Lie algebras.*

Name	Dynkin diagram
A_n	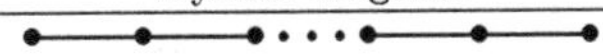
B_n	
C_n	
D_n	
G_2	
F_4	
E_6	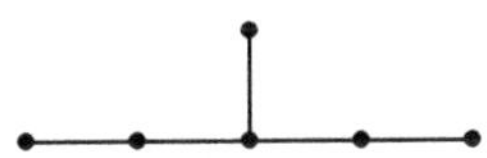
E_7	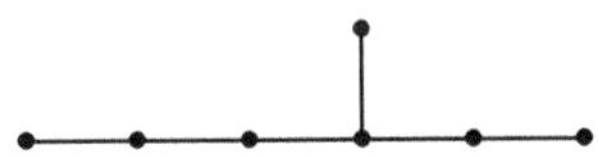
E_8	

Finally, the imaginary roots at level one in the fundamental domain $\mathcal{F}$ read $(2m-1)\alpha_1 + m\alpha_2 + \alpha_3$ and $2m\alpha_1 + m\alpha_2 + \alpha_3$ where m is an integer greater than or equal to 2. The first roots have length squared equal to $-2m + \frac{5}{2}$, the second roots have length squared equal to $-2m + 2$.

D.9 Overextensions of Finite-Dimensional Lie Algebras

An interesting class of Lorentzian Kac–Moody algebras can be constructed by adding simple roots to finite-dimensional simple Lie algebras in a particular way which will be described below. These are called "overextensions."

Let $\mathfrak{g}$ be a complex, finite-dimensional, simple Lie algebra of rank r, with simple roots $\alpha_1, \cdots, \alpha_r$. As stated above, we normalize the roots so that the long roots have length squared equal to 2 (the short roots, if any, have then length squared equal to 1 or 2/3 for G_2). The roots of simply-laced algebras are regarded as long roots, i.e., have length squared equal to 2.

For completeness, the list of all finite-dimensional, simple Lie algebras is given in Table D.2. The alert reader will notice the remarkable similarities between Tables 7.1 and D.2. This is of course not an accident but a mere reflection of the fact that crystallographic finite Coxeter groups are Weyl groups of finite-dimensional algebras.

Let $\alpha = \sum_i n_i \alpha_i$, $n_i \geq 0$ be a positive root. One defines the *height* of α as

$$\mathrm{ht}(\alpha) = \sum_i n_i. \tag{D.55}$$

Among the roots of $\mathfrak{g}$, there is a unique one that has greatest height, called the highest root. We denote it by θ. It is long and it fulfills the property that $(\theta|\alpha_i) \geq 0$ for all simple roots α_i, and

$$2\frac{(\alpha_i|\theta)}{(\theta|\theta)} \in \mathbb{Z}, \qquad 2\frac{(\theta|\alpha_i)}{(\alpha_i|\alpha_i)} \in \mathbb{Z} \tag{D.56}$$

(see, e.g., [75]). We denote by V the r-dimensional Euclidean vector space spanned by α_i $(i = 1, \cdots, r)$. Let M_2 be the two-dimensional Minkowski space with basis vectors u and v so that $(u|u) = (v|v) = 0$ and $(u|v) = 1$. The metric in the space $V \oplus M_2$ has clearly Minkowskian signature $(-, +, +, \cdots, +)$ so that any Kac–Moody algebra whose simple roots span $V \oplus M_2$ is necessarily Lorentzian.

Untwisted Overextensions

The standard "untwisted" overextensions $\mathfrak{g}^{++}$ – the only ones described here – are obtained by adding to the original roots of $\mathfrak{g}$ the roots

$$\alpha_0 = u - \theta, \qquad \alpha_{-1} = -u - v.$$

The matrix $A_{ij} = 2\frac{(\alpha_i|\alpha_j)}{(\alpha_i|\alpha_i)}$ where $i, j = -1, 0, 1, \cdots, r$ is a (generalized) Cartan matrix and defines indeed a Kac–Moody algebra.

The root α_0 is called the affine root and the algebra $\mathfrak{g}^+$ ($\mathfrak{g}^{(1)}$ in Kac's notations [111]) with roots $\alpha_0, \alpha_1, \cdots, \alpha_r$ is the untwisted affine extension of $\mathfrak{g}$. Since this subalgebra of the overextension $\mathfrak{g}^{++}$ plays an important role in deciphering the root structure of $\mathfrak{g}^{++}$, we list in Table D.3 all the untwisted affine extensions of simple Lie algebras. Note that Tables 7.2 and D.3 are the translations of each other in respective Coxeter and Dynkin conventions, as expected.

The root α_{-1} is known as the overextended root. One clearly has $\mathrm{rank}(\mathfrak{g}^{++}) = \mathrm{rank}(\mathfrak{g}) + 2$. The overextended root has vanishing scalar product with all other simple roots except α_0. One has explicitly $(\alpha_{-1}|\alpha_{-1}) = 2 = (\alpha_0|\alpha_0)$ and $(\alpha_{-1}|\alpha_0) = -1$, which shows that the overextended root is attached to the affine root (and only to the affine root) with a single link.

We show in Table D.4 the list of all untwisted overextensions obtained through the above extension procedure.

Of these Lorentzian algebras, the following are hyperbolic:

- A_k^{++} $(k \leq 7)$,
- B_k^{++} $(k \leq 8)$,
- C_k^{++} $(k \leq 4)$,

Table D.3: *Untwisted affine Kac–Moody algebras. The affine root is* α_0.

Name	Coxeter graph
A_1^+	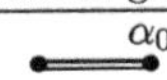
A_n^+ $(n > 1)$	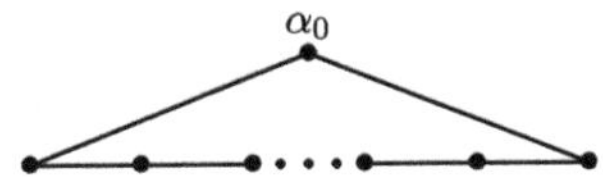
B_n^+	
C_n^+	
D_n^+	
G_2^+	
F_4^+	
E_6^+	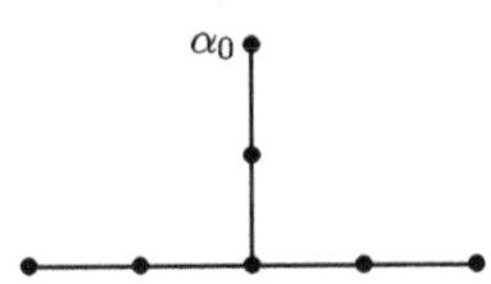
E_7^+	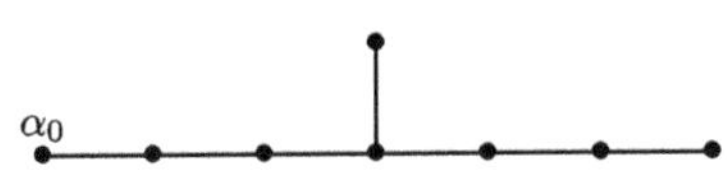
E_8^+	

- D_k^{++} $(k \leq 8)$,
- G_2^{++},
- F_4^{++},
- E_k^{++} $(k = 6, 7, 8)$.

The algebras B_8^{++}, D_8^{++} and E_8^{++} are also denoted by BE_{10}, DE_{10} and E_{10}, respectively.

A Special Property of E_{10}

Of these maximal rank hyperbolic algebras, E_{10} plays a very special role. Indeed, one can verify that the determinant of its Cartan matrix is equal to -1. It follows

Table D.4: *Untwisted overextensions. The affine root is denoted by α_0. The overextended root is α_{-1}.*

Name	Coxeter graph
A_1^{++}	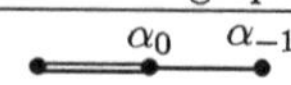
A_n^{++} $(n > 1)$	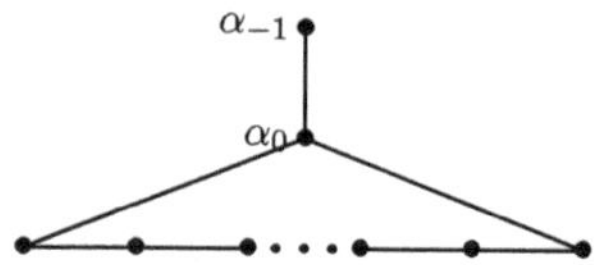
B_n^{++}	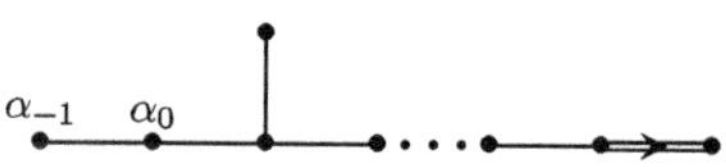
C_n^{++}	
D_n^{++}	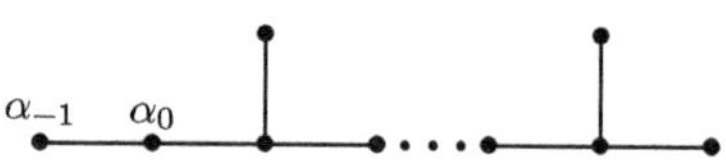
G_2^{++}	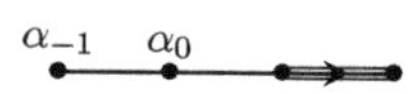
F_4^{++}	
E_6^{++}	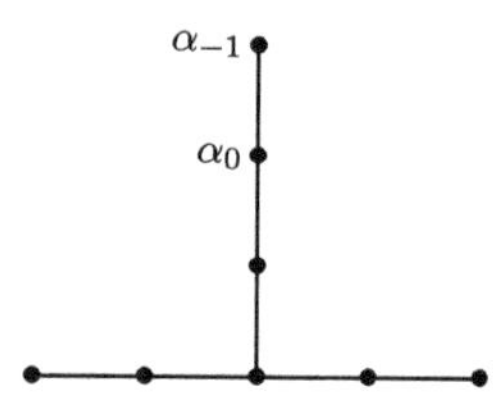
E_7^{++}	
E_8^{++}	

that the lattice of E_{10} is self-dual, i.e., that the fundamental weights belong to the root lattice of E_{10}. In view of the above theorem on roots of hyperbolic algebras and of the hyperbolicity of E_{10}, the fundamental weights of E_{10} are actually (imaginary) roots since they are non-spacelike. The root lattice of E_{10} is the only Lorentzian, even, self-dual lattice in ten dimensions (these lattices exist only in 2 mod 8 dimensions).

The Kac–Moody algebra is the algebra that controls the billiard of maximal supergravity and M-theory.

Gravitational Models with $\mathfrak{g}^{++}$*-Billiards*

The other Kac–Moody algebras also appear in other gravitational theories. Explicit Lagrangians leading to billiards with $\mathfrak{g}^{++}$ Coxeter groups have been constructed in [40, 44] for all $\mathfrak{g}$s.

Real Forms

The untwisted overextensions, although quite interesting, do not exhaust the list of all hyperbolic Kac–Moody algebras relevant to supergravities. Twisted overextensions also appear in some models [88]. These are related to "non-split real forms" of Lie algebras. We shall not review the general theory here, but shall rather refer the interested reader to the reference HPS where more information can be found (see also [89]).

References

[1] D. V. Alekseevskij, E. B. Vinberg and A. S. Solodovnikov, *Geometry of Spaces of Constant Curvature*, Volume 29 of the series *Encyclopaedia of Mathematical Sciences*, Springer 1993. *See §7.2.*

[2] E. Anderson, "Strong coupled relativity without relativity," Gen. Rel. Grav. **36**, 255 (2004) [gr-qc/0205118]. *See §6.7.*

[3] L. Andersson and A. D. Rendall, "Quiescent cosmological singularities," Commun. Math. Phys. **218**, 479 (2001) [gr-qc/0001047]. *See Introduction, §3.2, 4.7, 6.5.4.*

[4] D. V. Anosov, *Geodesic flows on Closed Riemanian Manifolds of Negative Curvature*, Trudy Mat. Inst. Steklova (ed. I. G. Petrovskii), 90 (1967). *See §3.2, 5.9, 6.6.*

[5] R. L. Arnowitt, S. Deser and C. W. Misner, "Canonical variables for general relativity," Phys. Rev. **117**, 1595 (1960). *See §5.1.*

[6] A. Ashtekar, A. Henderson and D. Sloan, "Hamiltonian general relativity and the Belinskii, Khalatnikov, Lifshitz conjecture," Class. Quant. Grav. **26**, 052001 (2009) [arXiv:0811.4160 [gr-qc]]. *See §2.5.*

[7] A. Ashtekar, A. Henderson and D. Sloan, "A Hamiltonian formulation of the BKL conjecture," Phys. Rev. D **83**, 084024 (2011) [arXiv:1102.3474 [gr-qc]]. *See §2.5.*

[8] J. D. Barrow, "Chaotic behavior in general relativity," Phys. Rept. **85**, 1 (1982). *See §3.1, 3.2.*

[9] J. D. Barrow and R. A. Matzner, "The homogeneity and isotropy of the universe," Mon. Not. R. Astr. Soc. **181**, 719 (1977). *See Preface to §4.8.*

[10] X. Bekaert, S. Cnockaert, C. Iazeolla and M. A. Vasiliev, "Nonlinear higher spin theories in various dimensions," hep-th/0503128, in *Proceedings of the first Solvay Workshop on "Higher Spin Gauge Theories,"* eds. R. Argurio, G. Barnich, G. Bonelli and M. Grigoriev, Université Libre de Bruxelles – Vrije Universiteit Brussel (2004). *See §6.1.*

[11] V. A. Belinski, "Turbulence of a gravitational field near a cosmological singularity," JETP Letters **56**, 421 (1992). *See §3.3.*

[12] V. A. Belinski, "Stabilization of the Friedmann big bang by the shear stresses," Phys. Rev. D **88**, 103521 (2013). *See §4.8.1, 4.8.2.*

[13] V. A. Belinski, E. S. Nikomarov and I. M. Khalatnikov, "Investigation of the cosmological evolution of viscoelastic matter with causal thermodynamics," Sov. Phys. JETP **50**, 213 (1979). *See §4.8.1, 4.8.2.*

[14] V. A. Belinski and I. M. Khalatnikov, "On the nature of the singularities in the general solution of the gravitational equations," Sov. Phys. JETP **29**, 911 (1969). *See §1.7, 2.1, B.3.*

[15] V. A. Belinski and I. M. Khalatnikov, "General solution of the gravitational equations with a physical singularity," Sov. Phys. JETP **30**, 1174 (1970) [Zh. Eksp.Teor. Fiz. **57**, 2163 (1969)]. *See §4.3.1.*

[16] V. A. Belinski and I. M. Khalatnikov, "Effect of scalar and vector fields on the nature of the cosmological singularity," Sov. Phys. JETP **36**, 591 (1973) [Zh. Eksp. Teor. Fiz. **63**, 1121 (1972)]. *See Introduction, §4.1, 4.3.1, 4.5, C.3.*

[17] V. A. Belinski and I. M. Khalatnikov, "On the influence of the spinor and electromagnetic field on the cosmological singularity character," Rend. Sem. Mat. Univ. Politech. Torino **35**, 159 (1977) [preprint of Landau Institute for Theoretical Physics, Chernogolovka 1976]. *See §4.1, Preface to App.C.*

[18] V. A. Belinski and I. M. Khalatnikov, "On the influence of matter and physical fields upon the nature of cosmological singularities," Soviet Science (Physics) Reviews, Harwood Acad. Publ. **A3**, 555 (1981). *See Introduction, §4.1, 4.3.1, 4.3.2.*

[19] V. A. Belinski, I. M. Khalatnikov and E. M. Lifshitz, "Oscillatory approach to a singular point in the relativistic cosmology," Adv. in Phys. **19**, 525 (1970). *See §1.7, 3.1, 4.1.*

[20] V. A. Belinski, I. M. Khalatnikov and E. M. Lifshitz, "Construction of a general cosmological solution of the Einstein equations with a time singularity," Sov. Phys. JETP **35**, 838 (1972) [Zh. Eksp.Teor. Fiz. **62**, 1606 (1972)]. *See §1.9, A.4.*

[21] V. A. Belinski, I. M. Khalatnikov and E. M. Lifshitz, "A general solution of the Einstein equations with a time singularity," Adv. in Phys. **31**, 639 (1982). *See §1.9, 4.1, A.4.*

[22] V. A. Belinski, I. M. Khalatnikov and M. P. Ryan, "The oscillatory regime near the singularity in Bianchi-type IX universes," preprint (order 469, 1971) of Landau Institute for Theoretical Physics, Moscow 1971 (unpublished); the work due to V. A. Belinski and I. M. Khalatnikov is published as sections 1 and 2 in M. P. Ryan, Ann. Phys. **70**, 301 (1971). *See §2.4.*

[23] B. K. Berger, "Numerical approaches to space-time singularities," Living Rev. Rel. **5**, 1 (2002). *See §1.4, 4.7.*

[24] B. K. Berger, "Hunting local mixmaster dynamics in spatially inhomogeneous cosmologies," Class. Quant. Grav. **21**, S81 (2004) [gr-qc/0312095]. *See §4.7.*

[25] B. K. Berger, D. Garfinkle, J. Isenberg, V. Moncrief and M. Weaver, "The singularity in generic gravitational collapse is space-like, local, and oscillatory," Mod. Phys. Lett. A **13**, 1565 (1998) [gr-qc/9805063]. *See §4.7.*

[26] B. K. Berger and V. Moncrief, "Numerical investigation of cosmological singularities," Phys. Rev. D **48**, 4676 (1993) [gr-qc/9307032]. *See §1.4.*

[27] E. Bergshoeff, M. de Roo, B. de Wit and P. van Nieuwenhuizen, "Ten-dimensional Maxwell–Einstein supergravity, its currents, and the issue of its auxiliary fields," Nucl. Phys. B **195**, 97 (1982). *See §6.1.*

[28] O. I. Bogoyavlenskii and S. P. Novikov, "Singularities of the cosmological model of the Bianchi IX type according to the qualitative theory of differential equations," Sov. Phys. JETP **37**, 747 (1973). *See §3.2.*

[29] N. Bourbaki, *Groupes et algèbres de Lie*, chapter 4, *Eléments de mathématique*, Hermann, 1968. *See §7.3, 7.4.*

[30] P. Breitenlohner and D. Maison, "On the Geroch group," Ann. Inst. Henri Poincaré **46**, 215 (1986). *See §7.7.*

[31] D. Brill and J. A. Wheeler, "Interaction of neutrinos and gravitational fields," Rev. Mod. Phys. **29**, 465 (1957). *See* §*C.1.*

[32] C. Cattaneo, "Sur une forme de l'équation de la chaleur éliminant le paradoxe d'une propagation instantanée," Comptes rendus Acad. Sci. Paris Sér. A-B **247**, 431 (1958). Based on his earlier seminar talk "Sulla conduzione del calore," Atti Semin. Mat. Fis. Univ. Modena **3**, 83 (1948). *See* §*4.8.1.*

[33] G. F. Chapline and N. S. Manton, "Unification of Yang–Mills theory and supergravity in ten dimensions," Phys. Lett. B **120**, 105 (1983). *See* §*6.1.*

[34] D. F. Chernoff and J. D. Barrow, "Chaos in the mixmaster universe," Phys. Rev. Lett. **50**, 134 (1983). *See* §*3.1, 3.2.*

[35] D. M. Chitre, Ph.D. Thesis, University of Maryland (1972). *See Introduction,* §*2.4, 3.2, 6.7.*

[36] D. Christodoulou and S. Klainerman, *The Global Nonlinear Stability of the Minkowski Space*, Princeton Mathematical Series, **41** (1993). *See* §*1.10.*

[37] N. J. Cornish and J. J. Levin, "The mixmaster universe is chaotic," Phys. Rev. Lett. **78**, 998 (1997). *See* §*3.1, 3.2.*

[38] N. J. Cornish and J. J. Levin, "Mixmaster universe: a chaotic Farey tale," Phys. Rev. D **55**, 7489 (1997). *See* §*3.1, 3.2.*

[39] E. Cremmer and B. Julia, "The SO(8) supergravity," Nucl. Phys. B **159**, 141 (1979). *See* §*7.7.*

[40] E. Cremmer, B. Julia, H. Lu and C. N. Pope, "Higher-dimensional origin of D = 3 coset symmetries," arXiv:hep-th/9909099. *See* §*D.9.*

[41] E. Cremmer, B. Julia and J. Scherk, "Supergravity theory in 11 dimensions," Phys. Lett. B **76**, 409 (1978). *See* §*6.1, 7.6.2, 7.7.*

[42] E. Czuchry, D. Garfinkle, J. R. Klauder and W. Piechocki, "Do spikes persist in a quantum treatment of space-time singularities?" Phys. Rev. D **95**, 024014 (2017) [arXiv:1605.04648 [gr-qc]]. *See* §*1.4.*

[43] E. Czuchry and W. Piechocki, "Bianchi IX model: reducing phase space," Phys. Rev. D **87**, 084021 (2013) [arXiv:1202.5448 [gr-qc]]. *See* §*2.4.*

[44] T. Damour, S. de Buyl, M. Henneaux and C. Schomblond, "Einstein billiards and overextensions of finite dimensional simple Lie algebras," JHEP **0208** (2002) 030 [hep-th/0206125]. *See* §*7.6.2, 7.6.3, D.9.*

[45] T. Damour and M. Henneaux, "Chaos in superstring cosmology," Phys. Rev. Lett. **85**, 920 (2000) [arXiv:hep-th/0003139]. [See also short version in Gen. Rel. Grav. **32**, 2339 (2000).] *See Introduction,* §*4.6.2, 6.7.*

[46] T. Damour and M. Henneaux, "Oscillatory behaviour in homogeneous string cosmology models," Phys. Lett. B **488**, 108 (2000) [arXiv:hep-th/0006171]. *See Introduction,* §*4.6.2, 6.7, 7.6.2.*

[47] T. Damour and M. Henneaux, "E_{10}, BE_{10} and arithmetical chaos in superstring cosmology," Phys. Rev. Lett. **86**, 4749 (2001) [arXiv:hep-th/0012172]. *See Introduction,* §*5.6, 6.2, 6.7, 7.7.*

[48] T. Damour, M. Henneaux, B. Julia and H. Nicolai, "Hyperbolic Kac–Moody algebras and chaos in Kaluza–Klein models," Phys. Lett. B **509**, 323 (2001) [arXiv:hep-th/0103094]. *See* §*5.10.2, 6.7, 7.6.1.*

[49] T. Damour, M. Henneaux and H. Nicolai, "E_{10} and a 'small tension' expansion of M theory," Phys. Rev. Lett. **89**, 221601 (2002) [arXiv:hep-th/0207267]. *See* §*7.7.*

[50] T. Damour, M. Henneaux and H. Nicolai, "Billiard dynamics of Einstein-matter systems near a spacelike singularity," in *Lectures on Quantum Gravity*, Proceedings of the School on Quantum Gravity, Valdivia, Chile, January 4–14, 2002,

A. Gomberoff and D. Marolf eds, *Series of the Centro de Estudios Científicos*, Springer 2005. *See* §*5.4*.

[51] T. Damour, M. Henneaux and H. Nicolai, "Cosmological billiards," Class. Quant. Grav. **20**, R145 (2003) [arXiv:hep-th/0212256]. *See Introduction,* §*3.3, 5.2, 5.5, 5.6, 5.8, 5.10.2, 6.4, 6.7, 7.7.*

[52] T. Damour, M. Henneaux, A. D. Rendall and M. Weaver, "Kasner like behavior for subcritical Einstein matter systems," Ann. Inst. Henri Poincaré **3**, 1049 (2002) [gr-qc/0202069]. *See Introduction,* §*3.2, 4.7, 6.5.4.*

[53] T. Damour and C. Hillmann, "Fermionic Kac–Moody billiards and supergravity," JHEP **0908**, 100 (2009) [arXiv:0906.3116 [hep-th]]. *See* §*C.3.*

[54] T. Damour, A. Kleinschmidt and H. Nicolai, "Hidden symmetries and the fermionic sector of eleven-dimensional supergravity," Phys. Lett. B **634**, 319 (2006) [hep-th/0512163]. *See* §*C.3.*

[55] T. Damour, A. Kleinschmidt and H. Nicolai, "K(E(10)), supergravity and fermions," JHEP **0608**, 046 (2006) [hep-th/0606105]. *See* §*C.3.*

[56] T. Damour and P. Spindel, "Quantum supersymmetric cosmology and its hidden Kac–Moody structure," Class. Quant. Grav. **30**, 162001 (2013) [arXiv:1304.6381 [gr-qc]]. *See* §*C.3.*

[57] G. Dautcourt, "On the ultrarelativistic limit of general relativity," Acta Phys. Polon. B **29**, 1047 (1998) [arXiv:gr-qc/9801093]. *See* §*6.7.*

[58] S. de Buyl, M. Henneaux and L. Paulot, "Hidden symmetries and Dirac fermions," Class. Quant. Grav. **22**, 3595 (2005) [hep-th/0506009]. *See* §*C.3.*

[59] S. de Buyl, M. Henneaux and L. Paulot, "Extended E(8) invariance of 11-dimensional supergravity," JHEP **0602**, 056 (2006) [hep-th/0512292]. *See* §*C.3.*

[60] S. de Buyl and C. Schomblond, "Hyperbolic Kac–Moody algebras and Einstein billiards," J. Math. Phys. **45**, 4464 (2004) [hep-th/0403285]. *See* §*D.8.*

[61] J. Demaret, Y. De Rop and M. Henneaux, "Chaos in nondiagonal spatially homogeneous cosmological models in space-time dimensions ≤ 10," Phys. Lett. B **211**, 37 (1988). *See* §*4.6.2.*

[62] J. Demaret, J. L. Hanquin, M. Henneaux, P. Spindel and A. Taormina, "The fate of the mixmaster behavior in vacuum inhomogeneous Kaluza–Klein cosmological models," Phys. Lett. **B 175**, 129 (1986). *See Introduction,* §*3.2, 4.1, 4.6.2, 7.6.1.*

[63] J. Demaret, M. Henneaux and P. Spindel, "Nonoscillatory behavior in vacuum Kaluza–Klein cosmologies," Phys. Lett. **B 164**, 27 (1985). *See Introduction,* §*3.2, 4.1, Preface to 4.6, 4.6.1, 4.6.2, 7.6.1.*

[64] B. S. DeWitt, "Quantum theory of gravity. 1. The canonical theory," Phys. Rev. **160**, 1113 (1967). *See* §*5.2.*

[65] P. A. M. Dirac, "The theory of gravitation in Hamiltonian form," Proc. Roy. Soc. Lond. A **246**, 333 (1958). *See* §*5.1.*

[66] D. Eardley, E. Liang and R. Sachs, "Velocity-dominated singularities in irrotational dust cosmologies," J. Math. Phys. **13**, 99 (1972). *See Introduction,* §*4.7.*

[67] C. Eckart, "The thermodynamics of irreversible processes III. Relativistic theory of the simple fluid," Phys. Rev. **58**, 919 (1940). *See* §*4.8.1.*

[68] L. P. Eisenhart, *Riemannian Geometry*, Princeton University Press, NJ, 1926. *See* §*A.2.*

[69] J. Ehlers, Dissertation Hamburg University (1957). *See* §*7.7.*

[70] Y. Elskens and M. Henneaux, "Chaos in Kaluza–Klein models," Class. Quant. Grav. **4**, L161 (1987). *See* §*4.6.2.*

[71] Y. Elskens and M. Henneaux, "Ergodic theory of the mixmaster model in higher space-time dimensions," Nucl. Phys. B **290**, 111 (1987). *See §4.6.2.*

[72] A. Eskin and C. McMullen, "Mixing, counting and equidistribution in Lie groups," Duke Math. J. **71**, 181–209 (1993). *See §3.2, 5.9, 6.6.*

[73] V. A. Fock and D. Ivanenko, "Géométrie quantique linéaire et déplacement parallèle," Compt. Rend. Acad. Sci. Paris **188**, 1470 (1929). *See §C.1.*

[74] E. S. Fradkin and M. A. Vasiliev, "On the gravitational interaction of massless higher spin fields," Phys. Lett. B **189**, 89 (1987). *See §6.1.*

[75] J. Fuchs, *Affine Lie Algebras and Quantum Groups: An Introduction, with Applications in Conformal Field Theory*, Cambridge Monographs on Mathematical Physics, Cambridge University Press, 1995. *See §D.9.*

[76] D. Garfinkle, "Numerical simulations of generic singuarities," Phys. Rev. Lett. **93**, 161101 (2004) [gr-qc/0312117]. *See §4.7.*

[77] D. Garfinkle, "Numerical relativity beyond astrophysics," Rept. Prog. Phys. **80**, no. 1, 016901 (2017) [arXiv:1606.02999 [gr-qc]]. *See §4.7.*

[78] R. P. Geroch, "A method for generating solutions of Einstein's equations," J. Math. Phys. **12**, 918 (1971). *See §7.7.*

[79] R. P. Geroch, "A method for generating new solutions of Einstein's equation. 2," J. Math. Phys. **13**, 394 (1972). *See §7.7.*

[80] G. Gibbons, K. Hashimoto and P. Yi, "Tachyon condensates, Carrollian contraction of Lorentz group, and fundamental strings," JHEP **0209**, 061 (2002) [arXiv:hep-th/0209034]. *See §6.7.*

[81] S. W. Goode, A. A. Coley and J. Wainwright "The isotropic singularity in cosmology," Class. Quant. Grav. **9**, 445 (1992). *See Preface to 4.8, §4.8.2.*

[82] A. J. S. Hamilton, "Inflation followed by BKL collapse inside accreting, rotating black holes," arXiv:1703.01921 [gr-qc]. *See §4.7.*

[83] J. M. Heinzle and C. Uggla, "Mixmaster: fact and belief," Class. Quant. Grav. **26**, 075016 (2009) [arXiv:0901.0776 [gr-qc]]. *See §2.5.*

[84] S. Helgason, *Differential Geometry, Lie Groups, and Symmetric Spaces*, Graduate Studies in Mathematics vol. 34, American Mathematical Society, Providence 2001. *See §5.2.*

[85] M. Henneaux, "Geometry of zero signature space-times," Print-79-0606 (Princeton), published in Bull. Soc. Math. Belg. **31**, 47 (1979) (note the misprints in the published version, absent in the preprint version). *See §6.7.*

[86] M. Henneaux, "Bianchi type I cosmologies and spinor fields," Phys. Rev. D **21**, 857 (1980). *See §C.3.*

[87] M. Henneaux, "Bianchi universes and spinor fields" (in French), Ann. Inst. H. Poincare Phys. Theor. **34**, 329 (1981). *See §C.3.*

[88] M. Henneaux and B. Julia, "Hyperbolic billiards of pure D = 4 supergravities," JHEP **0305**, 047 (2003) [hep-th/0304233]. *See §7.6.3, D.7, D.9.*

[89] M. Henneaux, A. Kleinschmidt and H. Nicolai, "Real forms of extended Kac–Moody symmetries and higher spin gauge theories," Gen. Rel. Grav. **44**, 1787 (2012) [arXiv:1110.4460 [hep-th]]. *See §D.9.*

[90] M. Henneaux, D. Persson and P. Spindel, "Spacelike singularities and hidden symmetries of gravity," Living Rev. Rel. **11**, 1 (2008) [arXiv:0710.1818 [hep-th]]. *See §7.3, 7.4, 7.5.3.*

[91] M. Henneaux, M. Pilati and C. Teitelboim, "Explicit solution for the zero signature (strong coupling) limit of the propagation amplitude in quantum gravity," Phys. Lett. B **110**, 123 (1982). *See §5.2.*

[92] D. Hobill, A. Burd and A. Coley, (eds) *Deterministic Chaos in General Relativity*, NATO ASI Series B, Physics Vol. **332**, Plenum Press, New York 1994. *See §3.2.*

[93] E. Hopf, "Statistik der geodätishen Linien in Mannigfaltigkeiten negativer Krümmung,"Berlin Verh. Sächs. Akad. Wiss. Leipzig **91**, 261 (1939). *See §3.2, 5.9, 6.6.*

[94] R. E. Howe and C. C. Moore, "Asymptotic properties of unitary representations," J. Functional Analysis **32**, 72–96 (1979). *See §3.2, 5.9, 6.6.*

[95] J. E. Humphreys, *Reflection Groups and Coxeter Groups, Cambridge Studies in Advanced Mathematics 29*, Cambridge University Press, 1990. *See §7.3, 7.4, 7.5.3, 7.5.4, D.8.*

[96] G. Imponente and G. Montani, "On the Covariance of the Mixmaster Chaoticity," Phys. Rev. D **63**, 103501 (2001) [arXiv:astro-ph/0102067]. *See §3.2.*

[97] C. J. Isham, "Some quantum field theory aspects of the superspace quantization of general relativity," Proc. Roy. Soc. Lond. A **351**, 209 (1976). *See §6.7.*

[98] W. Israel, "Nonstationary irreversible thermodynamics: a causal relativistic theory," Ann. Phys. **100**, 310 (1976). *See §4.8.1, 4.8.2.*

[99] W. Israel and J. M. Stewart, "Transient relativistic thermodynamics and kinetic theory," Ann. Phys. **118**, 341 (1979). *See §4.8.1.*

[100] V. D. Ivashchuk, A. A. Kirillov and V. N. Melnikov, "Stochastic properties of multidimensional cosmological models near a singular point," JETP Lett. **60**, 235 (1994) [Pisma Zh. Eksp. Teor. Fiz. **60**, 225 (1994)]. *See §6.7.*

[101] V. D. Ivashchuk and V. N. Melnikov, "Billiard representation for multidimensional cosmology with multicomponent perfect fluid near the singularity," Class. Quant. Grav. **12**, 809 (1995) [gr-qc/9407028]. *See §6.7.*

[102] V. D. Ivashchuk and V. N. Melnikov, "Billiard representation for multidimensional cosmology with intersecting p-branes near the singularity," J. Math. Phys. **41**, 6341 (2000) [arXiv:hep-th/9904077]. *See §6.7.*

[103] V. D. Ivashchuk and V. N. Melnikov, "Exact solutions in multidimensional gravity with antisymmetric forms," Class. Quant. Grav. **18**, R87 (2001) [hep-th/0110274]. *See §6.7.*

[104] R. T. Jantzen, "The dynamical degrees of freedom in spatially homogeneous cosmology," Commun. Math. Phys. 64, 211 (1979). *See §B.4.*

[105] R. T. Jantzen, "Spatially homogeneous dynamics: a unified picture," arXiv:gr-qc/0102035. *See §B.4.*

[106] B. Julia, "Group disintegrations," LPTENS 80/16, Invited paper presented at Nuffield Gravity Workshop, Cambridge, England, June 22–July 12, 1980. *See §7.7.*

[107] B. Julia, "Infinite dimensional groups acting on (super)gravity phase spaces," in *Proceedings of the Johns Hopkins Workshop on Current Problems in Particle Physics "Unified Theories and Beyond"* (Johns Hopkins University, Baltimore, 1984). *See §7.7.*

[108] B. Julia, "On infinite dimensional symmetry groups in physics," LPTENS-85/18, in *Proceedings of the Symposium on the Occasion of the Niels Bohr Centennial: Recent Developments in Quantum Field Theory*, edited by J. Ambjorn, B. J. Durhuus, J. L. Petersen, Amsterdam, North-Holland, 1985. *See §7.7.*

[109] B. L. Julia, "Dualities in the classical supergravity limits: dualizations, dualities and a detour via (4k+2)-dimensions," in *Cargese 1997, Strings, branes and dualities*, pp. 121–139 [hep-th/9805083]. *See §7.7.*

[110] B. Julia and H. Nicolai, "Conformal internal symmetry of 2-d sigma models coupled to gravity and a dilaton," Nucl. Phys. B **482**, 431 (1996) [hep-th/9608082]. *See* §*7.7.*

[111] V. G. Kac, *Infinite Dimensional Lie Algebra*, 3rd edition, Cambridge University Press, 1990. *See* §*5.10.2, 7.5.5, Preface to App.D, D.1, D.3, D.5, D.6.1, D.6.2, D.7, D.8, D.9.*

[112] E. Kasner, "Geometrical theorems on Einstein's cosmological equations," Am. J. Math. **43**, 217 (1921). *See* §*1.5.*

[113] I. M. Khalatnikov, E. M. Lifshitz, K. M. Khanin, L. N. Shchur and Ya. G. Sinai, "On the stochasticity in relativistic cosmology," Journ. Stat. Phys. **38**, 97 (1985). *See* §*3.1, 3.2.*

[114] A. Ya. Khinchin, "Continued fractions," University of Chicago Press, 1964; [Russian edition: Fizmatlit, Moscow, 1960]. *See* §*3.1.*

[115] A. A. Kirillov, "On the nature of the spatial distribution of metric inhomogeneities in the general solution of the Einstein equations near a cosmological singularity," Sov. Phys. JETP **76**, 355 (1993). *See* §*6.7.*

[116] A. A. Kirillov and A. A. Kochnev, "Cellular structure of space near a singularity in time in Einstein's equations," JETP Letters **46**, 436 (1987). *See* §*3.3.*

[117] A. A. Kirillov and V. N. Melnikov, "Dynamics of inhomogeneities of metric in the vicinity of a singularity in multidimensional cosmology," Phys. Rev. D **52**, 723 (1995) [gr-qc/9408004]. *See* §*6.7.*

[118] A. A. Kirillov and G. V. Serebryakov, "Origin of a classical space in quantum cosmologies," Grav. Cosmol. **7**, 211 (2001) [arXiv:hep-th/0012245]. *See* §*6.7.*

[119] R. O. Kuzmin, "Sur un problème de Gauss," Atti del Congresso Internazionale dei Matematici, Bologna, Vol. 6, p. 83 (1928); [Russian publication: Doklady Akademii Nauk, serie A, p.375 (1928)]. *See* §*3.1.*

[120] L. D. Landau and E. M. Lifshitz, *Classical Theory of Fields*, Pergamon Press, Oxford, 1962. *See* §*1.1, 1.4.*

[121] L. D. Landau and E. M. Lifshitz, *Fluid Mechanics*, first English edition, Reading, Mass., 1958. [The first Russian edition appeared in 1944.] *See* §*4.8.1, 4.8.2.*

[122] E. M. Lifshitz, "On the gravitational stability of the expanding universe," ZhETP **16**, 587 (1946) (in Russian); reprinted: Journ. Phys. (USSR) **10**, 116 (1946). *See Preface to 4.8,* §*4.8.2.*

[123] E. M. Lifshitz, I. M. Lifshitz and I. M. Khalatnikov, "Asymptotic analysis of oscillatory mode of approach to a singularity in homogeneous cosmological models," Sov. Phys. JETP **32**, 173, (1971) [Zh. Eksp. Teor. Fiz., **59,** 322 (1970)]. *See* §*3.1, 3.2, 4,3,1.*

[124] E. M. Lifshitz and I. M. Khalatnikov, "Investigations in relativistic cosmology," Adv. Phys. **12**, 185 (1963). *See* §*1.5, 1.6, 4.1, 4.8.2.*

[125] W. C. Lim, L. Andersson, D. Garfinkle and F. Pretorius, "Spikes in the mixmaster regime of G(2) cosmologies," Phys. Rev. D **79**, 123526 (2009) [arXiv:0904.1546 [gr-qc]]. *See* §*1.4.*

[126] U. Lindström and H. G. Svendsen, "A pedestrian approach to high energy limits of branes and other gravitational systems," Int. J. Mod. Phys. A **16**, 1347 (2001) [arXiv:hep-th/0007101]. *See* §*6.7.*

[127] V. N. Lukash, "Homogeneous cosmological models with gravitational waves and rotation," Pis'ma Zh. Eksp. Teor. Fiz. **19**, 449 (1974). *See* §*4.2.*

[128] V. N. Lukash, "Physical interpretation of homogeneous cosmological models," Nuovo Cimento **35**, 268 (1976). *See* §*4.2.*

[129] G. A. Margulis, "Applications of ergodic theory to the investigation of manifolds of negative curvature," Funct. Anal. Appl. **4**, 335 (1969). *See* §*3.2, 5.9, 6.6.*

[130] C. W. Misner, "Mixmaster universe," Phys. Rev. Lett. **22**, 1071–1074 (1969). *See Introduction,* §*2.4.*

[131] C. W. Misner, "Quantum cosmology. 1," Phys. Rev. **186**, 1319 (1969); also in "Minisuperspace," in *Magic Without Magic*, pp. 441–473, J R Klauder ed., Freeman, San Francisco 1972. *See* §*6.7.*

[132] C. W. Misner, in: D. Hobill et al. (Eds), *Deterministic Chaos in General Relativity*, Plenum, 1994, pp. 317–328 [gr-qc/9405068]. *See* §*6.7.*

[133] C. W. Misner, K. S. Thorne, J. A. Wheeler, *Gravitation*, Freeman, 1973. *See* §*5.1.*

[134] G. Montani, "On the asymptotic regime of approach to a singular point in the general cosmological solution of the Einstein equations," Tesi di Laurea, Università di Roma, Facolta di Fisica, 1992. *See* §*3.3.*

[135] R. V. Moody and A. Pianzola, *Lie Algebras with Triangular Decomposition*, Wiley, New York, 1995. *See Preface to App. D.*

[136] H. Nicolai, "A hyperbolic Lie algebra from supergravity," Phys. Lett. B **276**, 333 (1992). *See* §*7.7.*

[137] H. Nicolai, in *Recent Aspects of Quantum Fields*, Proceedings Schladming 1991, Lecture Notes in Physics, Springer Verlag, 1991. *See* §*7.7.*

[138] R. Penrose, "Gravitational collapse and space-time singularities," Phys. Rev. Lett. **14**, 57 (1965). *See* §*1.10.*

[139] R. Penrose, "Singularities and time-asymmetry," in *General Relativity: An Einstein Centenary Survey*, p. 581, Cambridge University Press (1979). *See Preface to 4.8.*

[140] A. A. Peresetskii, "Singularity of homogeneous Einstein metrics," Mat. Zametki **21**, 71 (1977). *See* §*4.2.*

[141] W. Piechocki and G. Plewa, "Structures arising in the asymptotic dynamics of the Bianchi IX model," arXiv:1611.05262 [gr-qc]. *See* §*2.4.*

[142] J. Polchinski, *String Theory*, two volumes, Cambridge University Press, Cambridge, 1998. *See* §*6.1, 7.6.3.*

[143] J. G. Ratcliffe, *Foundations of Hyperbolic Manifolds*, *Graduate Texts in Mathematics 149*, 2nd edition, Springer, 2006. *See* §*7.3, 7.4.*

[144] H. Ringström, "The Bianchi IX attractor," Ann. Inst. Henri Poincaré **2**, 405–500 (2001). *See* §*4.7.*

[145] S. E. Rugh, "Chaos in the Einstein equations – characterisation and importance?" in *Deterministic Chaos in General Relativity*, pp. 359–422, edited by D. Hobill et al., Plenum Press, 1994. *See* §*3.2.*

[146] C.Saçlioğlu, "Dynkin diagrams for hyperbolic Kac–Moody algebras," J. Phys A: Math. Gen. **22**, 3753 (1989). *See* §*5.10.2, D.8.*

[147] J. M. M. Senovilla, "Singularity theorems and their consequences," Gen. Rel. Grav. **30**, 701 (1998). *See* §*1.10.*

[148] J. M. M. Senovilla, "Singularity theorems in general relativity: achievements and open questions," Einstein Stud. **12**, 305 (2012) [physics/0605007]. *See* §*1.10.*

[149] Ya. G. Sinai, "The stochasticity of dynamical systems," Selecta Math. Soviet., 1(1), pp. 100–119 (1981). *See* §*3.2, 5.9, 6.6.*

[150] Ya. G. Sinai, "Geodesic flows on manifolds of negative curvature," in *Algoritms, Fractals and Dynamics* pp. 201–215 (Okayama/Kyoto, 1992). Plenum, New York, 1995. *See* §*3.2, 5.9, 6.6.*

[151] C. Teitelboim, "The Hamiltonian structure of space-time," PRINT-78-0682 (Princeton), in *General Relativity and Gravitation*, vol. 1, A. Held ed., Plenum Press, 1980. *See §6.7.*

[152] C. Uggla, "Recent developments concerning generic spacelike singularities," Gen. Rel. Grav. **45**, 1669 (2013) [arXiv:1304.6905 [gr-qc]]. *See §2.5.*

[153] C. Uggla, H. van Elst, J. Wainwright and G. F. R. Ellis, "The past attractor in inhomogeneous cosmology," Phys. Rev. D **68**, 103502 (2003) [gr-qc/0304002]. *See §2.5.*

[154] E. B. Vinberg, O. B. Shvartsman, "Discrete groups of motions of spaces of constant curvature," *Encyclopaedia of Mathematical Sciences*, vol. 29, Springer, 1991. *See §7.3, 7.4, 7.5.1, 7.5.2.*

[155] P. C. West, "E_{11} and M theory," Class. Quant. Grav. **18**, 4443–4460 (2001), [arXiv:hep-th/0104081]. *See §7.7.*

[156] A. Zardecki, "Modelling in chaotic relativity," Phys. Rev. D **28**, 1235 (1983). *See §3.2.*

[157] Ya. B. Zeldovich, "The equation of state at ultrahigh densities and its relativistic limitation," Sov. Phys. JETP **14**, 1143 (1962) [Zh. Eksp. Teor. Fiz. **41**, 1609 (1961)]. *See §4.1.*

[158] R. Zimmer, *Ergodic Theory and Semisimple Groups*, Birkhauser, Boston, 1984. *See §3.2, 5.9, 6.6.*

Index

For EU product safety concerns, contact us at Calle de José Abascal, 56–1°,
28003 Madrid, Spain or eugpsr@cambridge.org.

www.ingramcontent.com/pod-product-compliance
Lightning Source LLC
LaVergne TN
LVHW062208200926
848484LV00006B/1350

* 9 7 8 1 1 0 7 0 4 7 4 7 1 *